ELECTROANALYTICAL CHEMISTRY

VOLUME 16

ELECTROANALYTICAL CHEMISTRY

A SERIES OF ADVANCES

Edited by
ALLEN J. BARD

DEPARTMENT OF CHEMISTRY
UNIVERSITY OF TEXAS
AUSTIN, TEXAS

VOLUME 16

MARCEL DEKKER, INC., New York and Basel

The Library of Congress Catalogued the First
Issue of This Title as Follows:

Electroanalytical chemistry: a series of advances, v. 1-

New York, M. Dekker, 1966-
  v. 23 cm.
Editor: 1966- A. J. Bard

  1. Electrochemical analysis-Addresses, essays, lectures.
1. Bard, Allen J., ed.

QD115E499   545.3   66-11287

Library of Congress
ISBN 0-8247-7994-0 (v. 16)

COPYRIGHT © 1989 by MARCEL DEKKER, INC. ALL RIGHTS RESERVED

Neither this book nor any part may be reproduced or transmitted
in any form or by any means, electronic or mechanical, including
photocopying, microfilming, and recording, or by any information
storage and retrieval system, without permission in writing from
the publisher.

MARCEL DEKKER, INC.
270 Madison Avenue, New York, New York 10016

Current printing (last digit):
10 9 8 7 6 5 4 3 2 1

PRINTED IN THE UNITED STATES OF AMERICA

Introduction to the Series

This series is designed to provide authoritative reviews in the field of modern electroanalytical chemistry defined in its broadest sense. Coverage will be comprehensive and critical. Enough space will be devoted to each chapter of each volume so that derivations of fundamental equations, detailed descriptions of apparatus and techniques, and complete discussions of important articles can be provided, so that the chapters may be useful without repeated reference to the periodical literature. Chapters will vary in length and subject area. Some will be reviews of recent developments and applications of well-established techniques, whereas others will contain discussion of the background and problems in areas still being investigated extensively and in which many statements may still be tentative. Finally, chapters on techniques generally outside the scope of electroanalytical chemistry, but which can be applied fruitfully to electrochemical problems, will be included.

Electroanalytical chemists and others are concerned not only with the application of new and classical techniques to analytical problems but also with the fundamental theoretical principles upon which these techniques are based. Electroanalytical techniques are proving useful in such diverse fields as electro-organic synthesis, fuel cell studies, and radical ion formation, as well as such problems as the kinetics and mechanisms of electrode reactions, and the effects of electrode surface phenomena, adsorption, and the electrical double layer on electrode reactions.

It is hoped that the series will prove useful to the specialist and nonspecialist alike—that it will provide a background and a starting point for graduate students undertaking research in the areas mentioned, and that it will also prove valuable to practicing analytical chemists interested in learning about and applying electroanalytical techniques. Furthermore, electrochemists and industrial chemists with problems of electrosynthesis, electroplating, corrosion, and fuel cells, as well as other chemists wishing to apply electrochemical techniques to chemical problems, may find useful material in these volumes.

A.J.B.

Contributors to Volume 16

BERNARD FLEET Scada Systems Inc., Rexdale, Ontario, Canada

HARI GUNASINGHAM Chemistry Department, National University of Singapore, Kent Ridge, Singapore

JOSEPH WANG Department of Chemistry, New Mexico State University, Las Cruces, New Mexico

MICHAEL D. WARD E. I. du Pont de Nemours & Company, Inc., Wilmington, Delaware

Contents of Volume 16

Introduction to the Series	iii
Contributors to Volume 16	v
Contents of Other Volumes	ix

VOLTAMMETRY FOLLOWING NONELECTROLYTIC
PRECONCENTRATION   1
Joseph Wang

I.	Introduction and Scope	2
II.	Adsorptive Stripping Voltammetry	4
III.	Voltammetry Following Preconcentration at Chemically Modified Electrodes	53
IV.	Conclusion	80
	References	81

HYDRODYNAMIC VOLTAMMETRY IN CONTINUOUS-FLOW
ANALYSIS   89
Hari Gunasingham and Bernard Fleet

I.	Introduction	90
II.	Background	93

III.	Electrode Materials	112
IV.	Analytical Techniques and Applications	120
V.	Indirect Methods	148
VI.	New Electrode Systems	154
VII.	On-line Monitoring and Control	162
	Glossary of Abbreviations	167
	References	167

ELECTROCHEMICAL ASPECTS OF LOW-DIMENSIONAL
MOLECULAR SOLIDS   181
Michael D. Ward

I.	Introduction	182
II.	Description of Low-Dimensional Solids	184
III.	Redox Properties of Molecular Solids	196
IV.	Electrochemical Preparation of Low-Dimensional Solids	220
V.	Low-Dimensional Solids as Electrode Materials	267
VI.	Summary	296
	Appendix: Definitions of Acronyms	297
	References	299

Author Index   313

Subject Index   335

Contents of Other Volumes

VOLUME 1

AC Polarography and Related Techniques: Theory and Practice, Donald E. Smith
Applications of Chronopotentiometry to Problems in Analytical Chemistry, Donald G. Davis
Photoelectrochemistry and Electroluminescence, Theodore Kuwana
The Electrical Double Layer, Part I: Elements of Double-Layer Theory, David M. Mohilner

VOLUME 2

Electrochemistry of Aromatic Hydrocarbons and Related Substances, Michael E. Peover
Stripping Voltammetry, Embrecht Barendrecht
The Anodic Film on Platinum Electrodes, S. Gilaman
Oscillographic Polarography at Controlled Alternating Current, Michael Heyrovsky and Karel Micka

VOLUME 3

Application of Controlled-Current Coulometry to Reaction Kinetics, Jiri Janata and Harry B. Mark, Jr.
Nonaqueous Solvents for Electrochemical Use, Charles K. Mann
Use of the Radioactive-Tracer Method for the Investigation of the Electric Double-Layer Structure, N. A. Balashova and V. E. Kazarinov
Digital Simulation: A General Method for Solving Electrochemical Diffusion-Kinetic Problems, Stephen W. Feldberg

VOLUME 4

Sine Wave Methods in the Study of Electrode Processes, Margaretha Sluyters-Rehbach and Jan H. Sluyters
The Theory and Practice of Electrochemistry with Thin Layer Cells, A. T. Hubbard and F. C. Anson
Application of Controlled Potential Coulometry to the Study of Electrode Reactions, Allen J. Bard and K. S. V. Santhanam

VOLUME 5

Hydrated Electrons and Electrochemistry, Geraldine A. Kenney and David C. Walker
The Fundamentals of Metal Deposition, J. A. Harrison and H. R. Thirsk
Chemical Reactions in Polarography, Rolando Guidelli

VOLUME 6

Electrochemistry of Biological Compounds, A. L. Underwood and Robert W. Burnett
Electrode Processes in Solid Electrolyte Systems, Douglas O. Raleigh
The Fundamental Principles of Current Distribution and Mass Transport in Electrochemical Cells, John Newman

CONTENTS OF OTHER VOLUMES

VOLUME 7

Spectroelectrochemistry at Optically Transparent Electrodes:
I. Electrode Under Semi-infinite Diffusion Conditions,
Theodore Kuwana and Nicholas Winograd
Organometallic Electrochemistry, Michael D. Morris
Faradaic Reactification Method and Its Applications in the Study of Electrode Processes, H. P. Agarwal

VOLUME 8

Techniques, Apparatus, and Analytical Applications of Controlled-Potential Coulometry, Jackson E. Harrar
Streaming Maxima in Polarography, Henry H. Bauer
Solute Behavior in Solvents and Melts, A Study by Use of Transfer Activity Coefficients, Denise Bauer and Mylene Breant

VOLUME 9

Chemisorption at Electrodes: Hydrogen and Oxygen on Noble Metals and Their Alloys, Ronald Woods
Pulse Radiolysis and Polarography: Electrode Reactions of Short-lived Free Radicals, Arnim Henglein

VOLUME 10

Techniques of Electrogenerated Chemiluminescence, Larry R. Faulkner and Allen J. Bard
Electron Spin Resonance and Electrochemistry, Ted M. McKinney

VOLUME 11

Charge Transfer Processes at Semiconductor Electrodes, R. Memming
Methods for Electroanalysis In Vivo, Kiri Koryta, Miroslav Březina, Jiři Pradáč, and Jarimila Pradáčová

Polarography and Related Electroanalytical Techniques in Pharmacy and Pharmacology, G. J. Patriarche, M. Chateau-Gosselin, J. L. Vandenbalck, and Petr Zuman

Polarography of Antibiotics and Antibacterial Agents, Howard Siegerman

VOLUME 12

Flow Electrolysis with Extended-Surface Electrodes, Roman E. Sioda and Kenneth B. Keating

Voltammetric Methods for the Study of Adsorbed Species, Etienne Laviron

Coulostatic Pulse Techniques, Herman P. van Leeuwen

VOLUME 13

Spectroelectrochemistry at Optically Transparent Electrodes. II. Electrodes Under Thin-Layer and Semi-infinite Diffusion Conditions and Indirect Coulometric Iterations, William H. Heineman, Fred M. Hawkridge, and Henry N. Blount

Polynomial Approximation Techniques for Differential Equations in Electrochemical Problems, Stanley Pons

Chemically Modified Electrodes, Royce W. Murray

VOLUME 14

Precision in Linear Sweep and Cyclic Voltammetry, Vernon D. Parker

Conformational Change and Isomerization Associated with Electrode Reactions, Dennis H. Evans and Kathleen M. O'Connell

Square-Wave Voltammetry, Janet Osteryoung and John J. O'Dea

Infrared Vibrational Spectroscopy of the Electron-Solution Interface, John K. Foley, Carol Korzeniewski, John L. Daschbach, and Stanley Pons

VOLUME 15

Electrochemistry of Liquid-Liquid Interfaces, H. H. J. Girault and
  D. J. Schiffrin
Ellipsometry: Principles and Recent Applications in Electrochemistry,
  Shimshon Gottesfeld
Voltammetry at Ultramicroelectrodes, R. Mark Wightman and
  David O. Wipf

ELECTROANALYTICAL CHEMISTRY

VOLUME 16

Voltammetry Following
Nonelectrolytic Preconcentration

Joseph Wang

*New Mexico State University
Las Cruces, New Mexico*

I. Introduction and Scope   2
II. Adsorptive Stripping Voltammetry   4
   A. Adsorption at Electrode/Solution Interfaces   5
   B. Voltammetric Response of Surface-Confined Analytes   7
   C. Method Development—Practical Considerations   15
   D. Adsorption of Complexes for Trace Measurements of Metals   19
   E. Trace Measurements of Organic Compounds   34
   F. Adsorptive Stripping Tensammetry   43
   G. Combination of Adsorptive Voltammetry with Catalytic Effects   47
   H. Problems and Solutions   48
III. Voltammetry Following Preconcentration at Chemically Modified Electrodes   53
   A. Requirements, Problems, and Solutions   55
   B. Ways to Introduce the Preconcentrating Agent   57
   C. Preconcentration Schemes   59
IV. Conclusion   80
   References   81

I. INTRODUCTION AND SCOPE

Modern analytical chemistry must satisfy very diverse demands. Among the most important tasks is the quantitation of trace and ultratrace components in complex samples of environmental, industrial, or biological origin. In the analysis of very dilute samples, it is often necessary to employ some type of preconcentration step prior to the actual quantitation. Besides its main (enrichment) objective, the preconcentration step may serve to isolate the analyte from the complex matrix. In the content of electrochemistry, this is usually accomplished by electrolytic deposition of trace metals [1]. Stripping analysis is probably the best-known analytical method that incorporates such electrolytic preconcentration step. Its most common version, anodic stripping voltammetry, involves reduction of a metal ion to the metal, which then dissolves in mercury (amalgam formation) as the preconcentration step,

$$M^{+n} + Hg + ne^- \longrightarrow M(Hg) \tag{1}$$

Preconcentration is followed by a positive-going potential scan; the resulting voltammogram corresponds to the anodic dissolution of the amalgam. Anodic stripping voltammetry has been used in trace analysis with relative ease and success in a variety of analytical applications [2]. Another common version of stripping analysis, cathodic stripping voltammetry, also involves preconcentration in connection with a faradaic process (formation, via oxidation, of an insoluble salt with the electrode material).

Although stripping analysis provides a highly sensitive route to the measurement of many electroactive species (particularly trace metals), its widespread utility for many analytical problems has been restricted. Numerous important analytes are not accessible to conventional stripping measurements because of the electrolytic nature of the preconcentration step. For example, stripping analysis is not generally applicable to the analysis of organics because, unlike metals, most organics cannot be electrodeposited. Nevertheless, in view of the inherent sensitivity afforded by analyte accumulation prior to voltammetric measurements, intense activity has been devoted to the development of alternative methods for the preconcentration step that do not include any faradaic process.

The area of voltammetry following nonelectrolytic preconcentration is a fascinating field of research. The two major nonfaradaic routes for effective preconcentration include adsorptive accumulation at ordinary electrodes and specific reactions at chemically modified electrodes (Fig. 1). As in conventional stripping voltammetry, the concentrated surface

VOLTAMMETRY FOLLOWING NON-ELECTROLYTIC PRECONCENTRATION

1. INTERFACIAL ACCUMULATION AT ORDINARY ELECTRODES
2. SPECIFIC (COORDINATION, COVALENT, ION-EXCHANGE) REACTIONS AT MODIFIED ELECTRODES

$O_{SOLN} \longrightarrow O_{SUR}$

CONVENTIONAL STRIPPING ANALYSIS ELECTROLYTIC DEPOSITION

$M^{n+} + ne^- + Hg \longrightarrow M(Hg)$

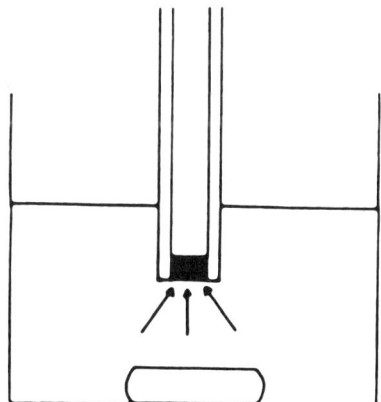
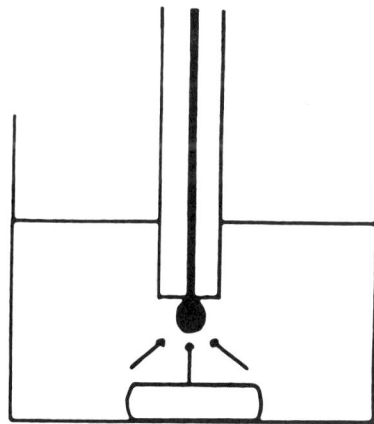

FIG. 1. Conventional and alternative (nonelectrolytic) approaches for the preconcentration step.

species retains its redox characteristics and can be quantified by applying a suitable voltammetric waveform. Over the last five years, adsorptive stripping voltammetry has evolved into a very versatile and powerful electroanalytical technique. Based on the wide range of applications reported during this period, the adsorptive concept appears to be the most universal stripping approach. Use of chemically modified electrodes as a means of selectively preconcentrating organic and inorganic analytes is also attracting intense interest. Rich chemistry can be utilized for preconcentration at these molecularly tailored surfaces, offering exciting prospects for trace electroanalysis. These nonelectrolytic preconcentration strategies add new dimensions to voltammetric stripping analysis by extending its scope of applications to a wide range of organic and inorganic analytes. The interest in these preconcentration avenues is further evidenced by the large number of publications on the subject that have appeared recently. These new developments are the subject of this chapter.

II. ADSORPTIVE STRIPPING VOLTAMMETRY

The development of adsorptive stripping voltammetry has been the result of exploiting the spontaneous adsorptive tendencies of reactants to enhance the sensitivity of voltammetric measurements. The adsorption phenomenon has traditionally been regarded as a problem and has been blamed for many unexplained results. On the other hand, adsorption-induced current enhancements have been observed in polarographic studies of numerous organic and inorganic species, thereby offering certain analytical advantages. The increased (five- to tenfold) response has been attributed to the presence of excess of electroactive species right at the surface of the dropping mercury electrode. The use of the dropping mercury electrode, however, does not allow full analytical exploitation of the sensitization possible with the interfacial accumulation. Stationary electrodes, in contrast, enable a drastic increase in the surface concentration, offering substantially lower detection limits. In adsorptive stripping voltammetry, a spontaneous adsorption process at

VOLTAMMETRY AFTER NONELECTROLYTIC PRECONCENTRATION

a stationary electrode is purposely utilized as an effective preconcentration step for trace measurements of important species that cannot be accumulated by electrolysis. Preconcentration is thus based solely on interfacial accumulation, that is, does not include any faradaic process. In addition to its growing analytical significance, adsorptive stripping voltammetry is a powerful research tool, particularly for elucidating the interfacial and redox properties of surface-active substances.

A. Adsorption at Electrode/Solution Interfaces

For proper application of adsorptive stripping voltammetry, it is important to understand the theory of adsorption. In electrochemistry, adsorption generally means the attachment of molecules or ions to the surface of electrodes. The electrode/electrolyte interface often exhibits properties that differ from those observed in the solution far from the interface or in the bulk of the electrode. The tendency to attract and retain reactants is one of the most important properties of the interface. The theory of adsorption from the solution phase to the electrode is a complicated field of electrochemistry. Our discussion will be limited to specific adsorption, which takes place in the compact double layer (unlike nonspecific adsorption, which involves long-range electrostatic forces).

Anson [3] summarized patterns of ionic and molecular adsorption at electrodes and divided these patterns into five classes. According to Anson, the equilibrium involved in adsorption at an electrode can be represented as in Eq. (2):

$$S\overset{S}{\underset{S}{(A)}}S + S\overset{S}{\underset{S}{\{E\}}} \rightleftharpoons (A)\overset{S}{\underset{S}{\{E\}}} + nS \qquad (2)$$

where A is the adsorbate, S the solvent, and E the electrode. Hence, adsorption of a solute is a displacement process in which molecules of solvent are replaced at the electrode surface by molecules or ions of solute. The latter have not usually displaced all the solvent molecules, as is described in the following. The amoung of adsorbate on a fully covered electrode surface (monolayer coverage) depends on the size of its molecules or ions and their orientation on the surface and is commonly 10^{-9} to 10^{-10} mol cm^{-2} for low-molecular-weight adsorbates. Large macromolecules are adsorbed with coverages of about 10^{-11} mol cm^{-2}.

The extent of adsorption is often related to the solubility of the reactant in the solvent concerned. Smaller solubilities tend to promote strong adsorption. Besides the hydrophobicity of adsorbate, other driving forces can lead to adsorption at electrode surfaces. These include electrostatic attraction between an ionic adsorbate and a charged electrode, field-dipole interaction between the electrode double-layer and functional groups of organic reactants, or chemisorption of certain electron or atomic groups on metallic electrode surfaces. The extent of these interactions and, hence, the resulting adsorptive stripping response depend on numerous experimental conditions, as will be discussed in the following sections.

The voltammetric response of surface-confined species is directly related to their surface concentration (see below). Hence, it is important to understand the relationship between the surface and bulk concentrations of the adsorbate. Adsorption isotherms provide such a relationship.

Adsorption isotherms are equilibrium relationships between concentrations of adsorbate on the surface and in the bulk of the solution. They are commonly derived from the formulations of the equilibrium constant for the adsorption reaction. Several isotherms have been proposed, corresponding to several approaches to the problem. The most frequently used is that of Langmuir,

$$\Gamma = \Gamma_m \left(\frac{BC}{1 + BC} \right) \qquad (3)$$

where Γ is the surface concentration of adsorbate (in moles cm^{-2}), Γ_m the surface concentration corresponding to a monolayer at the surface, C the bulk concentration of the adsorbate, and B the adsorption coefficient (depending on the interaction between the adsorbed species and the surface, thus relating to the free energy of adsorption). The

Langmuir isotherm can also be written in terms of the fractional coverage of the surface, $\theta = \Gamma/\Gamma_m$,

$$BC = \frac{\theta}{1-\theta} \tag{4}$$

When $1 \gg BC$, Eq. (3) can be linearized to yield

$$\Gamma = \Gamma_m BC \tag{5}$$

Hence, at very low adsorbate concentrations, the surface concentration is directly proportional to the bulk concentration.

The Langmuir adsorption isotherm is useful because it corresponds to a physical model in which the maximum surface coverage is that of a complete layer, generally a monolayer. Although it is the most widely used isotherm, many situations in which it is not applicable have been observed. This is primarily because the Langmuir isotherm assumes that there are no interactions between adsorbed species on the electrode surface. Characteristics of other adsorption isotherms, assuming interactions among the adsorbed species, are available, e.g., that of Frumkin,

$$BC = \left[\frac{\theta}{1-\theta}\right] \exp(-2g\theta) \tag{6}$$

where g is an interaction parameter (positive when the lateral interaction is one of attraction, and negative in the case of repulsion); g is often dependent on the electrode potential. When g is zero, the Frumkin isotherm reduces to the Langmuir isotherm. Forms of other isotherms (e.g., Temkin, Volmer, Virial, Mohilner, or Flory-Huggins) are available in the rich literature on the general topic of adsorption [4-6]. The interested reader is referred to these sources.

B. Voltammetric Response of Surface-Confined Analytes

Besides possible enhancement of the voltammetric response, the adsorption of electroactive species has complex effects on the voltammetric observations. These depend on whether the reactants, products, or both are adsorbed, on the nature of the adsorbed species, on the strength and potential dependence of the adsorption, on the supporting electrolyte, and on the voltammetric waveform employed. Detailed discussion of these effects is beyond the scope of this chapter. The reader is referred to earlier volumes in this series [7,8] in which the voltammetric response of immobilized redox species has been discussed in conjunction with the study of modified electrodes. The following discussion attempts to describe briefly the nature of current-potential curves for various voltammetric waveforms employed in adsorptive stripping voltammetry and to assess their analytical implications.

1. Linear-Scan Voltammetry

Ideal Nernstian behavior of surface-confined noninteracting species is manifested by symmetrical cyclic voltammetric peaks ($\Delta E_p = 0$), with peak half widths of $90.6/n$ mV. Under these conditions, the faradaic current arising from the presence of Γ mol cm^{-2} of an attached analyte is given by

$$i_p = \frac{n^2 F^2 \Gamma A v}{4RT} \tag{7}$$

where v is the potential scan rate, and n, F, A, R, and T have the conventional symbology. The double-layer chraging current i_{dl} in linear-scan experiments is given by

$$i_{dl} = C_{dl} A v \tag{8}$$

where C_{dl} is the double-layer capacitance. Because both faradaic and charging currents depend linearly on the potential scan rate, one would not expect to influence the ratio of faradaic-to-charging currents by changing the scan rate (unlike the case of measuring diffusing species, in which the faradaic current exhibits a square-root proportionality on the scan rate).

In most cases, the voltammetric behavior of surface-confined species is not ideal. Theoretical treatments of voltammograms of adsorbed species under different situations (nonidealities) that may be encountered in adsorptive stripping voltammetry have been derived, especially by Laviron; of particular analytical interest are the theories developed for (sub)monolayer strong adsorption of the oxidized (O) and reduced (R) forms of a redox couple [9,10]. Details of these are available elsewhere [7,11].

We will consider here only nonideality resulting from interactions between adsorbed molecules, for the case of a reversible surface voltammetric process. Under these conditions, the peak current obeys the following expression [12]:

$$i_p = \frac{n^2 F^2 \Gamma_T A v}{RT[4 - \Gamma_T(r_O + r_R)]} \tag{9}$$

where r_O and r_R are the interaction parameters of O and R, respectively, and Γ_T is the total amount present on the surface. [Equation (9) neglects the possibility that the interaction parameters depend on the potential.] Repulsive interactions within the adsorbed layer result in broadening of the peak (with peak half widths greater than 90.6/n mV). A more general expression for any degree of reversibility of the electrochemical reaction of a surface redox system, when interactions between the adsorbed molecules are taken into account, is also available [12].

In numerous adsorptive stripping applications, linear scan has been the waveform of choice, particularly because of its speed advantage over differential-pulse excitation. In addition, the large ratio of faradaic-to-charging currents characterizing conventional pulse voltammetry measurements is not always as apparent when the adsorption approach is concerned. The adsorptive accumulation may result in a substantial depression of double-layer capacitance (over a wide potential range), and hence in lower linear-scan background current [Eq. (8)]. The differential-pulse waveform may exhibit higher background response due to adsorptive/desorption "tensammetriclike" peaks associated with the adsorption of various species.

2. Differential-Pulse Voltammetry

The ability of the differential-pulse waveform to measure small quantities of surface-confined reactants has resulted in its widespread use in adsorptive stripping experiments. Brown and Anson [12] discussed the general properties of differential-pulse voltammetry of surface-bound species and compared its theoretical and experimental behaviors. By restricting measurements to sufficiently slow scan rates (1 mV sec^{-1} or less), peak potentials are obtained that fall close to the formal potentials of the attached reactants and are independent of the scan direction. The adsorption of a reactant introduces a large faradaic pseudocapacitance C_f into the cell impedance probed by the pulse polarograph. The presence of such large capacitance prolongs the effective time constant for decay of the charging current to the point at which it contributes to the net current sampled by most pulse polarographs (with ca. 50-msec delay between the pulse application and current measurement). Improved sensitivity can be achieved by introducing an uncompensated resistance in series with the working electrode [12]. Under these conditions, the maximum peak current occurs at the value of the uncompensated resistance R_u, is given by Eq. (10),

$$(R_u)_{max} = t\,(C_{dl} + C_f)^{-1} \qquad (10)$$

where t is the time at which differential-pulse current is sampled. Alternately, and often advantageously, solutions of increasing resistance (dilute electrolyte levels) can be employed [13]. For example, Fig. 2 shows the variation of the differential-pulse adsorptive stripping response for four organic compounds. Increased peak currents are observed at lower supporting electrolyte concentrations (unlike the peak current diminutions observed in conventional stripping procedures).

3. Other Waveforms

An analytically useful adsorptive stripping response can be obtained using potential-time waveforms other than linear-scan and differential-

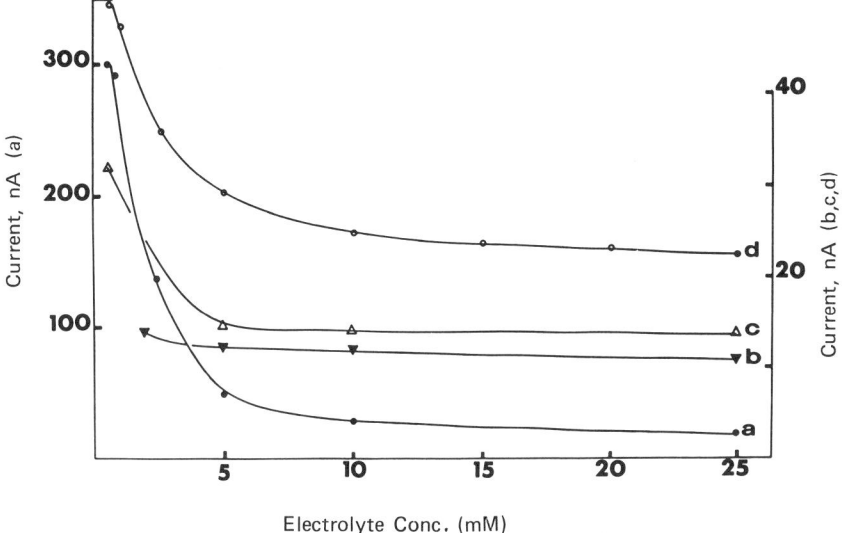

FIG. 2. Effect of supporting electrolyte preconcentration on the differential-pulse adsorptive stripping response for: riboflavin (a), testosterone (b), progesterone (c), and digoxin (d), all at 1×10^{-7} M. Electrolyte, borate buffer. Preconcentration for 90 (a,c), 120 (d), and 150 sec (b); at 0.0 (a), −0.8 (b,c) and −0.9 V (d). (From Ref. 13, with permission).

pulse voltammetry. Experimental and theoretical treatments are available for ac [14,15], square-wave [16,17], staircase [18], or normal-pulse [19] voltammetry of surface-confined redox species. Such excitations offer discrimination against the charging current background contribution, similar to that achieved by the differential-pulse mode. Because of its speed advantage, square-wave excitation is expected to play a major role in future analytical applications of the adsorptive approach. Further lowering of detection limits can be obtained with a background subtractive approach based on using different preconcentration periods and measuring the difference between the resulting stripping currents [20].

4. Analytical Calibration

Adsorptive stripping measurements require attention to the matter of analytical calibration. Independently of the voltammetric waveform employed, the adsorptive stripping voltammetric peak current is directly proportional to the surface concentration of adsorbate ($i_p = K\Gamma$). Hence, the current peak has the form of the Langmuir isotherm [Eq. (3)], when lateral interaction of adsorbed molecules is minimal. (Measurements of the peak area, which is also proportional to the quantity of analyte adsorbed, are not preferred for routine analytical work because the area is a physically cumbersome quantity to determine.) At the low analyte levels ($10^{-7} - 10^{-10}$ M) for which the method is usually applied, a linear adsorption isotherm is obeyed, with the slope of the resulting linear calibration plots proportional to $\Gamma_m B$ [Eq. (5)]. Deviations from linearity are observed at higher concentrations, as expected from Eq. (3). When approaching full surface coverage ($\theta \to 1$), the current response approaches its maximum value ($i \to i_{max}$). The concentration at which the electrode surface becomes saturated with the analyte is related to the area occupied by the adsorbed species and to the preconcentration time. The former depends on the size and orientation of the adsorbed species. Concentrations corresponding to a saturation coverage of a mercury drop typically range from 10^{-6} to 10^{-4} M. Means to extend the linear range include the use of shorter preconcentration times, lower rates of forced convection, or diluted samples. (With shortened preconcentration time and/or lowered convection rate, adsorption equilibrium may not be attained, but the quantity of adsorbed analyte is still related to the bulk concentration.) By adding, in the computer, several stripping curves, each obtained after a short accumulation time, the linear range can also be extended. The interdependence of the response on the bulk concentration and preconcentration time can be seen qualitatively with the aid of Fig. 3. This shows calibration plots for bilirubin obtained at different preconcentration times. For 30-sec preconcentration (a), the response is linear for the entire concentration range. At 60 and 120 sec [(b) and (c)],

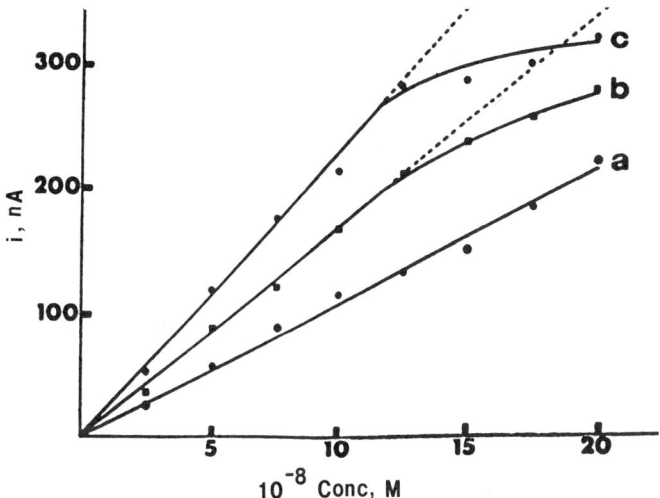

FIG. 3. Dependence of the differential-pulse adsorptive stripping response on bilirubin concentration at different preconcentration periods: (a) 30, (b) 60, (c) 120 sec. Preconcentration potential, −0.8 V. Electrolyte, 0.1 M sodium acetate. (From Ref. 21, with permission.)

the response is linear up to 1.25×10^{-7} M. The linear dynamic range can also be extended using the linear dependence of $1/i$ on $1/C$ for Langmuir-type systems:

$$\frac{1}{i} = \frac{1}{i_{max}} + \left(\frac{1}{i_{max} B}\right) \frac{1}{C} \qquad (11)$$

The slope of such reciprocal calibration plots is thus related to the adsorption coefficient. When lateral interactions exist between adsorbed molecules, deviation for a linear $1/i$ versus $1/C$ plot occurs. Similar deviations are expected for systems with a mixed control by mass-transport and adsorption isotherm. Other less common situations leading to deviation from Langmuir-type concentration dependence include concentration-induced orientional transitions (e.g., flat at low level, vertical at higher levels) [22], orientational transitions induced by another adsorbate [23], or a multilayer electrode process, with electron transfer proceeding through the first layer to layers formed on top [24]. Overall, because of the nature of adsorptive stripping measurements, calibration plots allow quantitation over the entire concentration range (as long as $i_{max} > i$). The method of standard additions is suitable for quantitation within the linear range; three additions are recommended to ensure that the response is within the linear portion.

5. Reproducibility

Reproducibility is assured by carefully controlling solution and instrumental conditions, such as the concentration of supporting electrolyte (and chelating ligand for metal analysis), the preconcentration time and potential, or convection rate, and by ensuring an analyte-free electrode surface prior to the preconcentration.

A great advantage of the hanging mercury drop electrode, particularly in the static mercury drop design, is the possibility of easily and reproducibly renewing the surface and thus allowing repetitive measurements at surfaces essentially free of adsorbed analyte. When this self-cleaning property is coupled with the automatic control of modern voltammetric instruments, highly reproducible data (with relative standard deivations of ca. 1-5%) can be achieved. In addition, such automation frees the operator from the manual control of the stripping procedure.

A lower degree of reproducibility (relative standard deviations of ca. 5-15%) is generally observed when oxidizable species are measured at solid electrodes. This is commonly attributed to changes in the surface behavior and activity during the adsorptive stripping cycle, including incomplete desorption of the analyte (and other coadsorbates). A cleaning step, usually electrochemical, is often required following the voltammetric scan to remove the remaining adsorbates from the surface. (In many situations, the product of the surface redox reaction is desorbed instantaneously, simplifying the procedure and improving the precision.) It is hoped that the recently developed solid-electrode activation procedures [24a] will be useful for improving the precision of adsorptive stripping measurements. Solid electrodes are also characterized

VOLTAMMETRY AFTER NONELECTROLYTIC PRECONCENTRATION 15

by a relatively large background current, and hence offer higher detection limits (about 1×10^{-9} M) compared to those obtained at the hanging mercury drop electrode ($10^{-10} - 10^{-11}$ M).

C. Method Development—Practical Considerations

To find the optimum conditions for the adsorptive preconcentration of the analyte of interest, the voltammetric response with and without accumulation should be compared over the entire accessible potential range of the working electrode. The resulting response and the peak current enhancement (over that without accumulation) should be evaluated as a function of numerous variables affecting the extent of adsorption. Among these are the electrolyte, pH, ionic strength, preconcentration potential, convection rate, or preconcentration period. The interdependence of the surface coverage (i.e., the response) on many of these parameters should be explored and considered in the optimization process. The effect of the ligand concentration should be considered in the metal-chelate adsorptive stripping procedure; ligand concentration higher than required may cause inhibition of the chelate adsorption via a competitive coverage of the free ligand. Figure 4 illustrates the dependence of the stripping peak current on the pH, ligand concentration, and preconcentration potential observed, during such parametric evaluation, for trace measurements of thorium. It is important also to compare possible voltammetric waveforms to be used for measuring the surface-bound analyte; once chosen, waveform parameters such as scan rate, amplitude, or frequency should be optimized. Some of these parameters may influence other characteristics of the response (e.g., baseline, peak shape, selectivity, and reproducibility). For example, a judicious choice of the preconcentration potential may minimize possible interferences (e.g., Ref. 26), thus requiring a trade-off between sensitivity and selectivity. Similarly, the pH of choice may lead not only to conditions of maximum adsorption (e.g., when acidic or basic analytes are concerned) but also to a more favorable hydrogen evolution background current. In

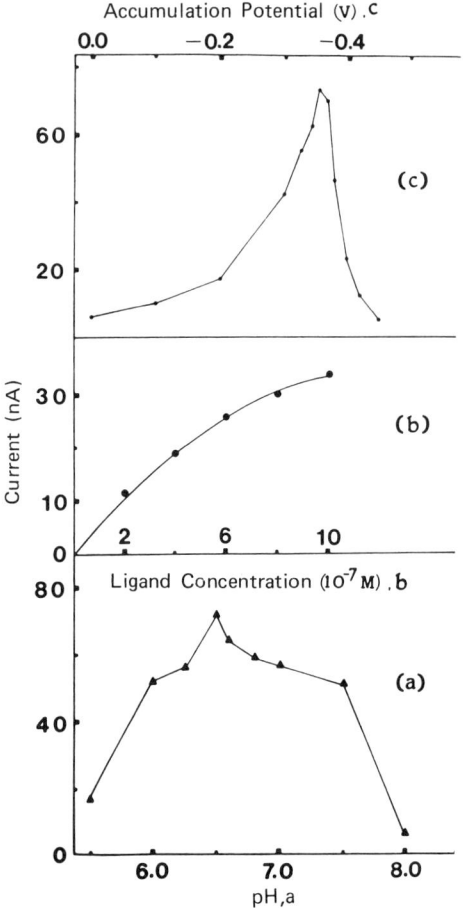

FIG. 4. Dependence of the adsorptive stripping response for thorium on the pH (a), ligand (mordant blue 9) concentration (b), and preconcentration potential (c). Thorium concentration, 20 μg liter^{-1}. Electrolyte, acetate buffer (pH 6.5). (From Ref. 25 with permission.)

experiments involving the medium-exchange procedure (often used to minimize matrix effects—Section II.H), the composition of the "exchange" solution should also be optimized to yield the best response characteristics.

The inherent rate of adsorption at the commonly used mercury electrodes is so fast that the overall process is usually governed by mass transport. Thus (and similar to conventional stripping measurements), forced convection during the preconcentration step increases the sensitivity by increasing the rate of transport of analytes to the electrode surface. On the basis of the concept of the Nernst diffusion layer, Delahay and Trachtenbert [27] obtained, for the linearized isotherm,

$$\frac{\Gamma_t}{\Gamma_{eq}} = 1 - \exp \frac{Dt}{\delta b} \qquad (12)$$

where Γ_t is the surface concentration at time t, Γ_{eq} the equilibrium surface concentration, δ the thickness of the diffusion layer, and b the adsorption constant (b = $B\Gamma_m$). Because diffusion coefficients of small molecules are very similar, the rate of adsorption under given experimental conditions depends primarily on the adsorption constant b.

Forced convection during the preconcentration step must be reproducible, because of the dependence of the surface coverage on δ. Common forms of convective transport (stirring, rotation, flow) can be employed, depending on the specific system employed. The convective transport is stopped at the end of the preconcentration step and, after a 15-sec rest period, the voltammetric scan is initiated in a quiescent solution.

Special consideration should be given to the choice of preconcentration time. As in conventional stripping measurements, such choice requires a compromise between sensitivity and speed. The time for establishing adsorption equilibrium depends strongly on the reactant concentration (with the rate of attaining equilibrium increasing with concentration). For example, Fig. 5 shows the dependence of the surface concentration of the nickel-dimethylglyoxime complex on the preconcentration

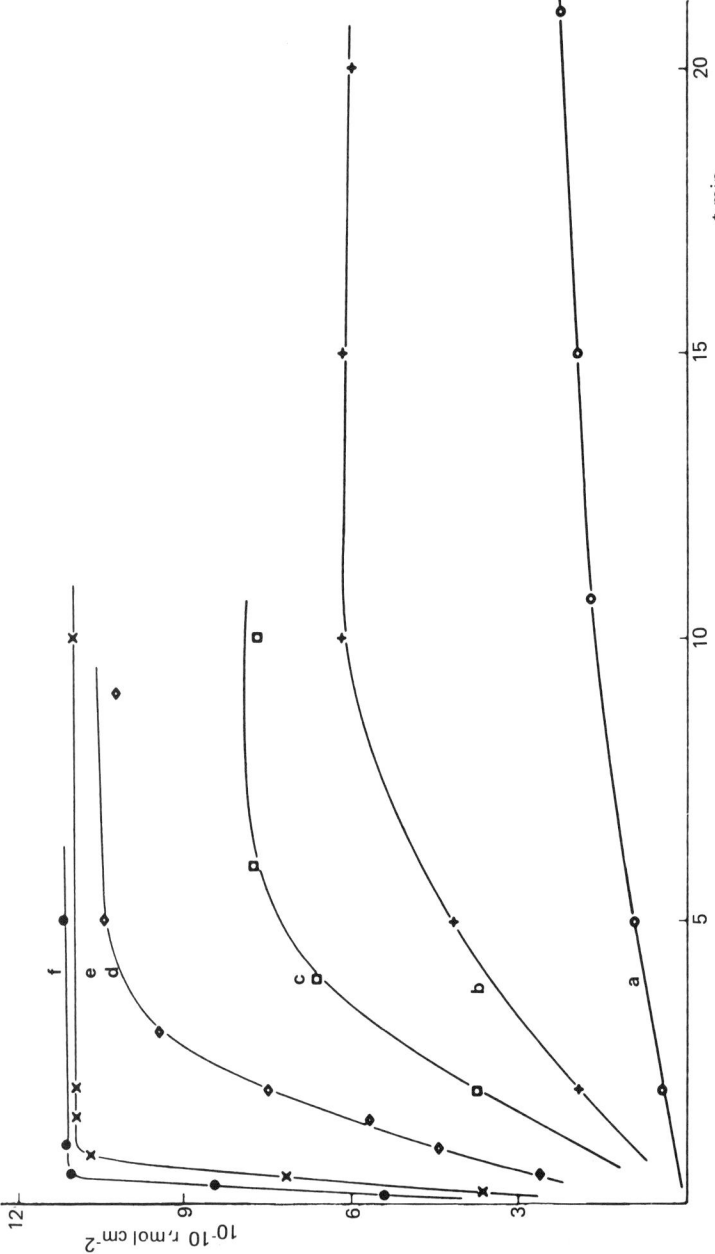

FIG. 5. Dependence of the surface concentration of nickel-dimethylglyoxime on the preconcentration time at different levels of nickel: 0.5 (a), 3 (b), 5 (c), 13 (d), 38 (e) and 62.5 (f) μg liter^{-1}. Preconcentration potential, −0.7 V. Electrolyte, 0.1 M NH_3/NH_4Cl. (From Ref. 22, with permission.)

time at different levels of nickel. Equilibrium surface concentrations are reached after ca. 20 min and 30 sec at 0.5 and 62.5 µg liter^{-1} nickel, respectively. Only for very low concentrations, corresponding to the linearized isotherm, the rate of attainment of equilibrium coverage is independent of the bulk concentration [Eq. (12)]. In general, the length of the preconcentration period depends on the analyte, its concentration, and the working electrode. Typical preconcentration times for convenient measurement of reducible species at the hanging mercury drop electrode range from 0.5 to 2 min at the 5×10^{-8} M level to 10-15 min at 5×10^{-10} M; measurements of oxidizable species at solid electrodes (with higher background current) require longer preconcentration periods, typically 1-3 min at the 1×10^{-7} M level and 10-15 min at 1×10^{-9} M. In a few situations, short preconcentration times and/or nonconvective systems are employed to extend the linear range via lowering of the surface coverage (see Section II.B).

Finally, it is essential to perform the adsorptive stripping cycle also with the blank (supporting electrolyte) solution of choice to detect impurities that may not contribute to the response without accumulation. Nonelectroactive adsorbates present in the supporting electrolyte are also regarded as interferences in adsorptive stripping measurements because of their competition for adsorption sites. Thus, whenever possible, reagents used for the preparation of the supporting electorlyte and standards should be of the highest purity possible. As in other trace analytical methods, all the principles of good laboratory practice (glassware cleanliness, storage, etc.) must be observed to obtain high accuracy and low detection limits.

D. Adsorption of Complexes for Trace Measurements of Metals

Electroanalytical methods in general and stripping voltammetry in particular provide very sensitive routes for the quantitation of many trace metals [2].

However, the electrochemical behavior of numerous metal ions of analytical interest makes their quantitation by conventional stripping voltammetry difficult or impossible. Alternative procedures based primarily on metal-chelate adsorptive accumulation have been developed in recent years and have proved extremely useful for ultratrace measurements of such metals. These developments are the direct result of early observations of substantial enhancement of the polarographic response of several metals in the presence of different ligands [28,29]. Controlled adsorptive accumulation at stationary electrodes (i.e., longer adsorption times) permits substantial enhancements of the surface concentration of the chelates, and hence lower detection limits.

The steps involved in such measurements of metal ions are shown in Fig. 6. The formation of an appropriate metal chelate is followed by its controlled interfacial accumulation (for a given time, potential, convection rate, etc.) onto the working electrode, commonly the hanging mercury drop electrode. The extensive chemistry of metal-ligand complexes

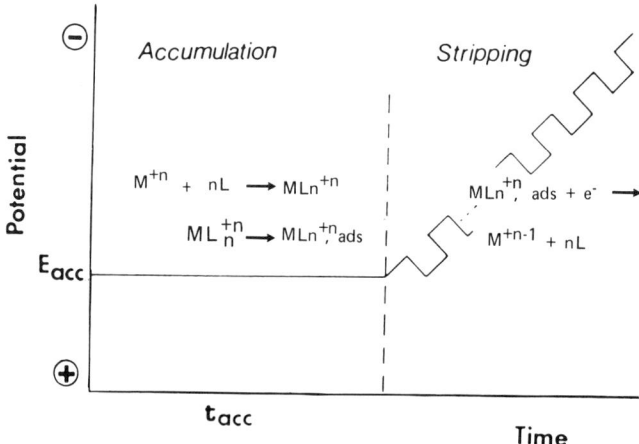

FIG. 6. Steps in voltammetric measurement of a metal ion, based on the formation, adsorptive preconcentration, and reduction of its surface-active complex.

developed during the past two decades provides a rich and versatile source for the judicious choice of the ligand (see discussion below). The selectivity of the chemical step (complex formation) can be used to increase the overall selectivity of the analysis; a ligand capable of binding only a few metals should make possible the formulation of a highly selective scheme. Since all complex equilibria in the sample solution are adjusted, as are the adsorption equilibria of the metal complex, the amount of metal adsorbed remains proportional to the bulk concentration of the metal (as long as one works sufficiently below full surface coverage). In contrast to conventional anodic stripping voltammetry, the accumulated complex is measured during a negative-going potential scan; linear-scan, differential-pulse, or square-wave waveforms can be employed. In addition to the commonly used voltammetric quantification of the surface-bound complex, a potentiometric approach based on applying a constant reducing current and monitoring the potential-time behavior has recently been suggested [30].

Figure 7 is a periodic table of elements indicating those that have been determined by adsorptive stripping voltammetry. Detection limits at the subnanomolar concentration level have been estimated for these elements, following short (3-5 min) accumulation periods. Note, in particular, that important transition metals, actinides, or rare-earth elements (e.g., Fe, Ti, Mo, V, Cr, Pd, V, Th, La) can be measured at trace levels by voltammetry. Combined with approximately fifteen additional elements, commonly quantified by conventional stripping voltammetry, the scope of stripping voltammetry is rapidly approaching that of relevant atomic absorption spectroscopic procedures. Apart from extending the scope of voltammetry, the metal-chelate adsorption approach offers alternative, and often improved, schemes for measuring metals such as Ni, Mn, Ga, Cu, Co, or Sn. The quantitation of these elements by conventional stripping voltammetry often suffers from difficulties such as the formation of intermetallic compounds, extreme redox potentials, or poor selectivity. The improvements observed in adsorptive stripping

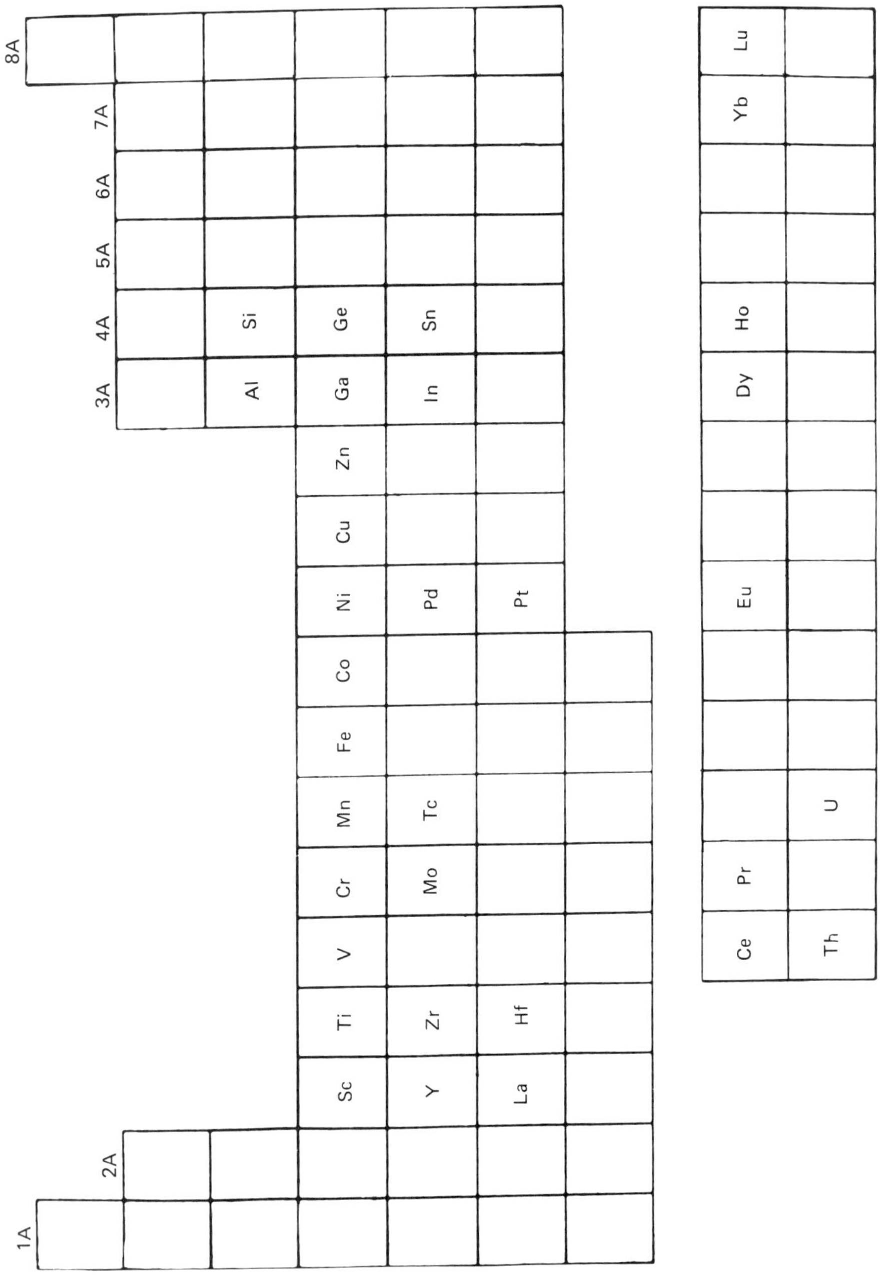

FIG. 7. Metals determined by adsorptive stripping voltammetry.

measurements are attributed to the different nature of the response (associated with the different detection principles).

1. Complexes of Dimethylglyoxime

Dimethylglyoxime (DMG) is commonly used as a highly specific organic precipitating agent for gravimetric determinations of nickel (particularly in steel samples). Nonbonding electron pairs on the nitrogen atoms form a five-membered chelate ring to nickel with a 2:1 stoichiometry as follows:

$$Ni^{2+} + 2 \begin{array}{c} H_3C \\ \diagdown \\ C=N-OH \\ | \\ C=N-OH \\ \diagup \\ H_3C \end{array} \longrightarrow \begin{array}{c} \text{Ni(DMG)}_2 \text{ chelate} \end{array} + 2H^+ \quad (13)$$

Various investigators [28,31] observed substantial enhancements of the polarographic response for nickel in the presence of DMG. Such sensitization, associated with the adsorption of the Ni(DMG)$_2$ chelate, was exploited by Pihlar et al. [32] for developing a highly sensitive and reliable adsorptive stripping voltammetric procedure for the determination of trace and ultratrace levels of nickel. For this purpose, the sample solution is adjusted to pH 9.2 with ammoniacal buffer, and enough DMG is spiked to give a concentration of about 10^{-4} M. The resulting Ni(DMG)$_2$ complex is surface-active and adsorbs at the surface of the hanging mercury drop electrode. To attain optimal adsorption conditions, the electrode is held during the accumulation period at a potential of −0.7 V (vs. Ag/AgCl). Subsequently, the adsorbed complex is reduced during a negative-going potential scan. A flat baseline facilitates peak-current measurement, particularly using a differential-pulse waveform. Hence, remarkably low detection limits—of a few ng liter^{-1}— are obtained following short preconcentration periods. The only interferences is cobalt, which forms a surface-active complex with DMG, thus accumulating at the electrode surface in a way similar to nickel. However, the potential for reduction of Co(DMG)$_2$ is 0.15 V more negative than for Ni(DMG)$_2$, and interference is only a problem when cobalt is present in large excess. Indeed, such separation in peak potentials allows simultaneous determination of both elements. Bond and co-workers [33] assessed conditions for the determination of varying concentration ratios of cobalt and nickel in selected biological materials. Figure 8 illustrates the utility of the method for the analysis of orchard leaves. Other "real-life" applications based on the accumulation of DMG complexes include the determination of nickel in a human nail [34] or rainwater [35]

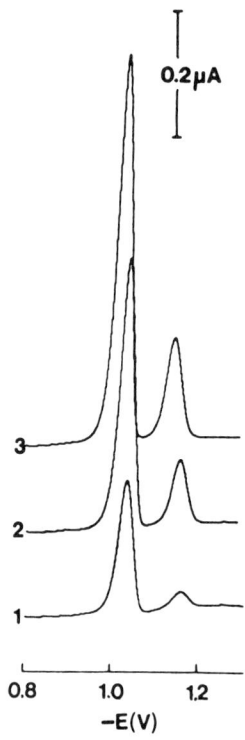

FIG. 8. Simultaneous determination of nickel and cobalt in orchard leaves by the standard addition method. Preconcentration for 240 sec at -0.8 V in NH_3/NH_4Cl containing 0.8-mM DMG. Standards added (2 and 3): 1.25 µg liter^{-1} Ni and 0.25 µg liter^{-1} Co. (From Ref. 33, with permission.)

and of cobalt and nickel in natural waters [30] or simulated PWR coolant [36]. Other versions of the DMG-sensitized adsorption approach, based on the use of a mercury film working electrode [37], flow system [38], or chronopotentiometric stripping mode [30], have been described. The mechanism of the redox process of the adsorbed $Ni(DMG)_2$ complex was explored recently [22]. Palladium(II), which forms a complex with DMG in dilute acid solutions, can also be measured in a similar way down to 2×10^{-10} M [39]. It is expected that other dioximes used as precipitants (e.g., methylbenzoyldioxime, 1,2-cyclohexanedione—dioxime) would find similar use in trace-metal analysis.

2. *Complexes of Di-O-Hydroxyazo Dyes*

One class of complexing agents that offers great promise for adsorptive stripping measurements of trace metals is that of di-o-hydroxyazo dyes. (The common structural unit is shown in I.)

I

The ability of certain di-o-hydroxyazo dyes to form discrete polarographic reduction waves in the presence of metal ions was first discovered by Willard and Dean [40], who used the dye solochrome violet RS for the determination of aluminum. Since this work, other di-o-hydroxyazo dyes have been shown to behave similarly but selectively with various metal ions. The polarographic behavior of metal chelates of di-o-hydroxyazo dyes was reviewed by Latimer [41]. Because of the surface-active properties of these complexes, their polarographic response is enhanced by adsorption at the dropping mercury electrode. However, the short lifetime of the drop does not allow full analytical exploitation of the enhancement possible with chealte adsorption, and measurements are limited to the micromolar concentration level. In contrast, stripping measurements, following chelate adsorption over longer periods, allows substantial lowering of detection limits. Recent studies in the author's laboratory have been directed at developing new adsorptive stripping measurement schemes based on chelation with di-o-hydroxyazo dyes. These studies resulted in new procedures for measuring titanium [42], aluminum [43], gallium [44], thorium [45], uranium [46], iron [47], zirconium [48], manganese [49], or yttrium [50], over the 10^{-7}–10^{-10} M concentration range, using solochrome violet RS [42–44, 47, 48, 50], mordant blue 9 [45, 46], or eriochrome black T [49] as chelating dyes. Detection limits at the 10^{-10} M level can be obtained following short preconcentration periods (3–5 min). Successful use of di-o-hydroxyazo dyes for adsorptive stripping measurements depends on the resolution between the free dye and metal-chelate voltammetric peaks (ΔE_p). Several theories have been proposed to explain why the metal chelate of a di-o-hydroxyazo dye is reduced at a more negative potential than the free dye [41, 51]. Because the redox process involves the reduction of the azo-group, which is involved in the coordination, the magnitude of ΔE_p depends on the stability of the metal-dye chelate (actually on the ratio of the stability constants of the metal ion with the azo- and hydrazo- derivatives of the dye [51]). An example of the spontaneous accumulation and redox behavior, using the manganese chelate with eriochrome black T, is shown in Fig. 9. The first scan following accumulation exhibits large cathodic peaks, at -0.82 and -0.99 V, related to the reduction of the adsorbed dye and chelate, respectively. Rapid desorption of the products is indicated from subsequent scans. Adsorptive stripping measurements based on di-o-hydroxyazo dyes require a careful choice of the preconcentration potential—usually to a point between the dye and chelate peaks—for minimizing coadsorption of the free dye and improving the baseline current associated with the preceding free-dye peak. Depending on the metal-chealte peak potentials, simultaneous measurements of two metals (e.g., Ti-Fe, Th-Ni, U-Fe) present at diverse concentration ratios is feasible

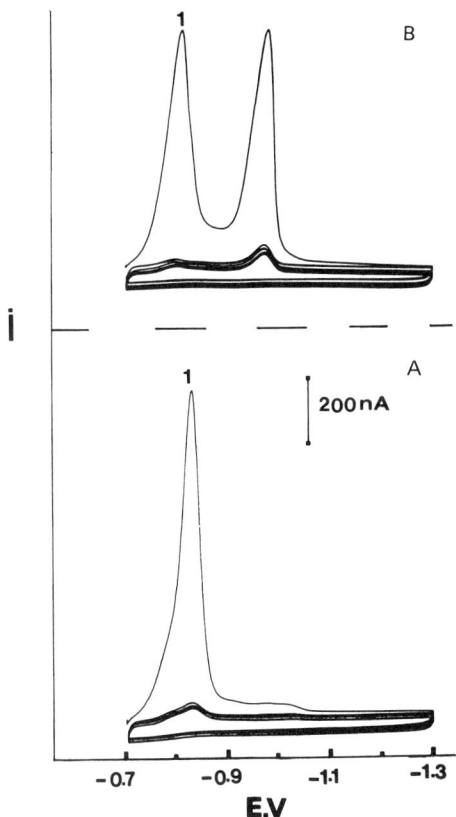

FIG. 9. Repetitive cyclic voltammograms recorded after 1-min accumulation at −0.70 V: (A) 2×10^{-6} M eriochrome black T; (B) same as (A) but after addition of 40 µg liter^{-1} manganese(II). (From Ref. 49, with permission.)

[52]. In contrast, a separation step is required to differentiate between metals, such as rare earths, yielding chelate peaks at similar potentials [50]. The new procedures, based on chelation with di-o-hydroxyazo dyes, have been applied successfully for the determination of metals such as titanium, iron, or gallium in environmental matrices such as seawater, river water, or rainwater. Typical voltammograms for traces of titanium in such samples are shown in Fig. 10. The quantitation of gallium in the presence of solochrome violet RS [44] provides an example to an adsorptive stripping procedure that offers certain advantages over analogous measurements by conventional stripping voltammetry (e.g., absence of intermetallic compounds with copper and zinc). Unlike most chelates of di-o-hydroxyazo dyes, which yield one reduction peak, two peaks are observed for the gallium chelate.

3. *Complexes of Catechol, Oxine, and Other Ligands*

Several other chelating agents, forming slightly soluble chealtes with certain metal ions, have been used successfully in the development of highly sensitive adsorptive stripping schemes. In particular, van den Berg's group illustrated the utility of catechol(II) and oxine(III) for the determination of various trace metals in seawater. These agents form highly stable complex ions with several metal ions. In particular, highly sensitive procedures have been reported for measuring vanadium-(V) [53], copper(II) [54] and iron [55] in the presence of catechol, and for measuring molybdenum(VI) [56] and uranium(VI) [57] in the presence of oxine. Direct measurements in seawater were performed following pH adjustment by the addition of a PIPES/NaOH buffer. For example, Fig. 11 illustrates the use of oxine for the determination of uranium in seawater, with a simultaneous response of the copper and nickel oxinates

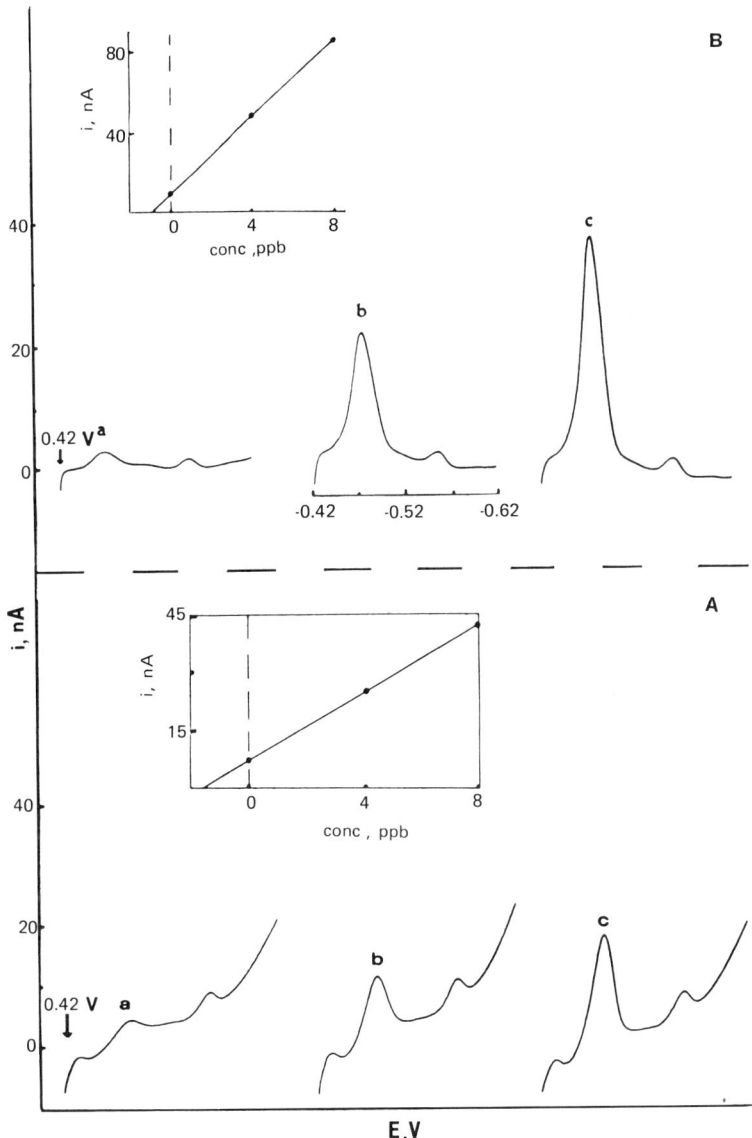

FIG. 10. Measurement of titanium in river water (A) and seawater (B) (8-ml natural water + 2-ml acetate buffer + 1.5×10^{-6} M solochrome violet RS). 30-sec preconcentration at -0.40 V. (a) Sample; (b) same as (a) but after adding 2 µg liter^{-1} Ti; (c) same as (b) but after adding 2 µg liter^{-1} Ti. Also shown, the resulting standard addition plots. (From Ref. 42, with permission.)

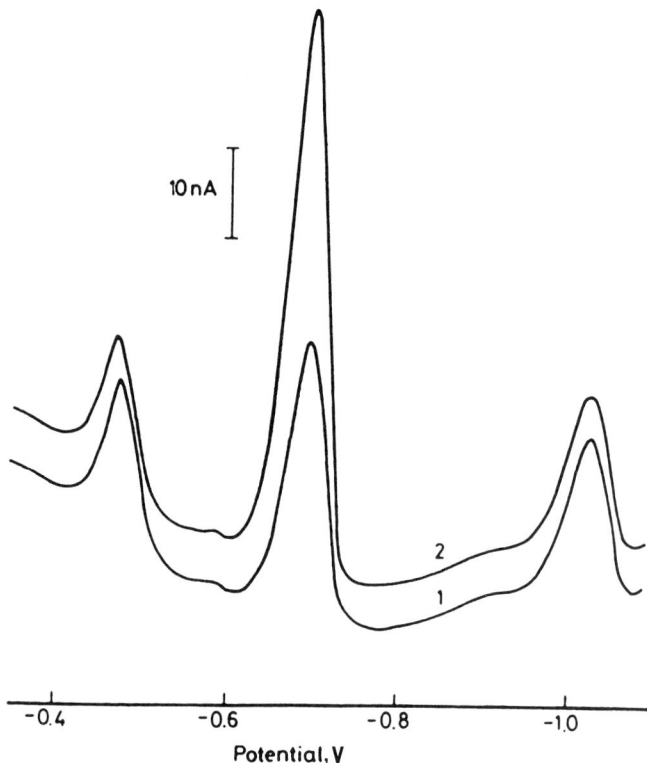

FIG. 11. Adsorptive stripping voltammetric scans for uranium in seawater: (1) sample; (2) same as (1) but after adding 10 nM U (VI). Preconcentration for 60 sec at −0.35 V. Sample containing 0.01 M PIPES and 2×10^{-5} M oxine. (From Ref. 57, with permission.)

(at −0.47 and −1.02 V, respectively). Because the desired complex of catechol or oxine is usually not formed selectively, interferences from other reducible complexes (with similar peak potentials) should be minimized, for example, by addition of EDTA to eliminate copper and lead interferences in measurements of iron [55]. Detection limits usually lie at the 10^{-10} M level; a lower value of ca. 10^{-11} was reported for copper [54]. The adsorptive collection of the copper-catechol complex offers improvement in sensitivity and/or speed compared to analogous measurements by conventional stripping analysis. The cathodic peak is associated with the reduction of Cu(II) in the adsorbed complex to Cu(0):

$$Cu(C_6H_4O_2) + 2e^- + 2H^+ \longrightarrow Cu^\circ + C_6H_4(OH)_2 \qquad (14)$$

The saturation adsorption density of this and other complexes of catechol or oxine is of the order of 10^{-10} mol cm^{-2}.

Effective metal-chelate adsorptive stripping procedures have been developed recently utilizing other common (or less common) complexing agents. Thus, low concentrations of chromium(III) [58], aluminum [59], lanthanum and cerium [60], tin [61], or zinc [62] can be determined by complexation with diethylenetriaminepentaacetic acid [58], 1,2-dihydroxy-anthraquinone-3-sulphonic acid [59], o-cresolphthalexon [60], tropolone [61], and aminopyrrolidine dithiocarbamate [62], respectively. These schemes involve a controlled adsorptive collection of the chelate at the hanging mercury drop electrode and offer great potential for environmental surveillance. For example, various natural waters were analyzed reliably for trace and ultratrace levels of chromium [58] and aluminum [59]. Adsorptive voltammetry of tin was shown to improve the selectivity (in the presence of lead and cadmium) compared to conventional stripping measurements and to be applicable for analysis of fruit juices [61]. Details of these and other metal-chealte adsorptive stripping schemes are given in Table 1.

4. *Metal Speciation in Natural Waters Based on Adsorptive Stripping Voltammetry*

In recent years, increasing efforts have been made to obtain detailed information on the various chemical forms of trace metals in natural waters. Conventional stripping voltammetry (with electrolytic deposition) has been widely used for these speciation studies [2,66]. Because adsorptive stripping voltammetric methods are becoming widely available for the determination of many metals, they are expected to play a major role in future speciation investigations. Adsorptive stripping voltammetry, which is based on detection principles that are different from

TABLE 1

Measurement of Trace Metals Based on Stripping Voltammetry with Complex Adsorption

Metal	Complexing agent	Supporting electrolyte	Detection limit, M	Refs.
Al	Solochrome violet RS	Acetate buffer (pH 4.5)	5×10^{-9}	43
	Dihydroxyanthraquinone-sulfonic acid	BES (pH 7.1)	1×10^{-9}	59
Co	Dimethylglyoxime	NH_3/NH_4Cl (pH 9.2)	1×10^{-10}	33
Cr	Diethylenetriamine-pentaacetic acid	Sodium acetate	4×10^{-10}	58
Cu	Catechol	HEPES (pH 7.7)	1×10^{-11}	54
Fe	Catechol	PIPES (pH 6.8)	6×10^{-10}	55
	Solochrome violet RS	Acetate buffer (pH 5.1)	7×10^{-10}	47
Ga	Solochrome violet RS	Acetate buffer (pH 4.8)	1×10^{-9}	44
La	Cresolphthalexon	NH_3/NH_4Cl (pH 9.4)	2×10^{-10}	60
Mn	Eriochrome black T	PIPES (pH 12)	6×10^{-10}	49
Mo	Oxine	Hydrochloric acid (pH 2.6)	1×10^{-10}	56
Ni	Dimethylglyoxime	NH_3/NH_4Cl (pH 9.2)	1×10^{-10}	32
Pd	Dimethylglyoxime	Acetate buffer (pH 5.15)	2×10^{-10}	39
Pt	Formazone	Sulfuric acid	1×10^{-12}	63
Sn	Tropolone	Acetate buffer (pH 4.0)	2.3×10^{-10}	61
Tc	Thiocyanate	Sulfuric acid	5×10^{-10}	64
Th	Mordant blue 9	Acetate buffer (pH 6.5)	4×10^{-10}	45
Ti	Solochrome violet RS	Acetate buffer (pH 5.1)	7×10^{-10}	42

TABLE 1 (continued)

Metal	Complexing agent	Supporting electrolyte	Detection limit, M	Refs.
U	Catechol	PIPES (pH 6.8)	3×10^{-10}	65
	Mordant blue 9	Acetate buffer (pH 6.5)	2×10^{-10}	46
	Oxine	PIPES (pH 6.8)	2×10^{-10}	57
V	Catechol	PIPES (pH 9)	1×10^{-10}	53
Zn	Ammonium pyrrolidine dithiocarbamate	BES (pH 7.3)	3×10^{-11}	62
Zr	Solochrome violet RS	Acetate buffer	2.3×10^{-10}	48
Y	Solochrome violet RS	PIPES (pH 11)	1.4×10^{-9}	50

those of conventional stripping voltammetry, may provide fundamentally different speciation information.

For example, in various situations, adsorptive stripping voltammetry has been shown to be extremely useful for oxidation state discrimination. This ability is attributed to the specificity of some of the coordination reactions employed toward a certain oxidation state of the metal. For example, chelation with solochrome violet RS has proved useful for selective measurements of iron(III) in the presence of iron(II) [47].

Van den Berg [67] described a ligand-exchange adsorptive stripping procedure for measuring complexing capacity, based on the competition between natural ligands and catechol for free copper ions, following by voltammetric measurements of the adsorbed copper-catechol complex. This method also allows estimates of conditional stability constants for complexes of copper(II) with dissolved organic ligands in seawater. This adsorptive approach greatly minimizes problems, such as dissociation of the copper complex or surfactant adsorption, that characterize analogous measurements by conventional stripping voltammetry.

The extent to which the metal chelate of interest is formed from other species in the equilibrium system (and thus the fraction of the metal contributing to the adsorptive response) is determined by its thermodynamic stability and the concentration of competing natural ligands. Hence, the adsorptive approach measures the free (hydrated) ion and metal displaced from complexes (weaker than the metal chelate of interest). Conventional stripping voltammetry, in contrast, measures the labile metal fraction. It has been shown recently [68] that under the concentration excess of a strong chelating agent, for example, dimethylglyoxime, used for adsorptive stripping measurements of nickel and cobalt, essentially the *total* metal content is determined. In contrast, with solochrome violet RS as the chelating agent, the iron response is affected by the presence of EDTA (but not by tannic acid, glycine, NTA, or cysteine), whereas the titanium response is affected by EDTA, NTA, and tannic acid (but not by glycine and cysteine) [69]. Hence, it is stressed here that all adsorptive stripping speciation work on aquatic matrices be assessed critically to establish which forms of a given metal are actually determined.

E. Trace Measurements of Organic Compounds

Conventional stripping analysis has been traditionally concerned with the determination of trace metals. Trace levels of organic compounds have been measured in only a few cases involving film formation cathodic stripping schemes (based on reactions with mercury ions formed electrolytically). Adsorptive stripping voltammetry has been shown, over the last five years, to be highly suitable for measuring organic compounds that exhibit surface-active properties.

Organic molecules adsorb on the electrode surface from aqueous solutions, primarily because of their hydrophobicity. In general, the less soluble an organic compound, the stronger its adsorption. Neutral organic compounds exhibit their greatest adsorption on an uncharged electrode. Adsorption decreases as the charge of the electrode is increased in both the positive and negative directions because of displacement by oriented water molecules. Cationic and anionic organic compounds exhibit strong adsorption at potentials negative or positive, respectively, to the potential of zero charge. Besides its pronounced effect on hydrophobicity, molecular structure has other important effects. For example, the strength of adsorption of various organic analytes on carbon electrodes, and hence the sensitivity, appears to correlate with the number of aromatic rings in the adsorbing molecule. Such behavior is attributed to the interaction of molecules having extended pi-electron systems with carbon surfaces. For example, although o-dianisidine (with two aromatic rings) yields a detectable response, no such response is obtained for N,N,N',N'-tetramethyl-p-phenylenediamine (with one aromatic ring) [70]. The adsorptive stripping response of organic compounds can also be affected by the presence of various substituents that may affect the molecular orientation, and hence the surface concentration [71]. The systematic work of Hubbard (e.g., Refs. 72 and 73) is particularly useful for understanding the various factors affecting the orientation of adsorbed organic molecules.

Overall, the interfacial activity of numerous compounds of pharmaceutical, biological, and environmental significance has been exploited for their adsorptive stripping measurements at trace levels. For example, Fig. 12 shows voltammograms at the hanging mercury drop electrode that had been immersed in a 5×10^{-8} M diazepam solution for increasing periods of time. Careful optimization of the preconcentration conditions permits convenient measurement of reducible and oxidizable organic compounds down to the 10^{-10} M and 10^{-9} M levels, respectively. The lowest detection limit, 2.5×10^{-11} M, was reported for riboflavin (30-min preconcentration) [75]. Nonelectroactive organic compounds can also be quantified using tensammetric procedures (see Section F) or following an appropriate derivatization reaction [76]. Whereas most early applications have been concerned with low molecular weight compounds, large biological macromolecules have been the focus of recent investigations.

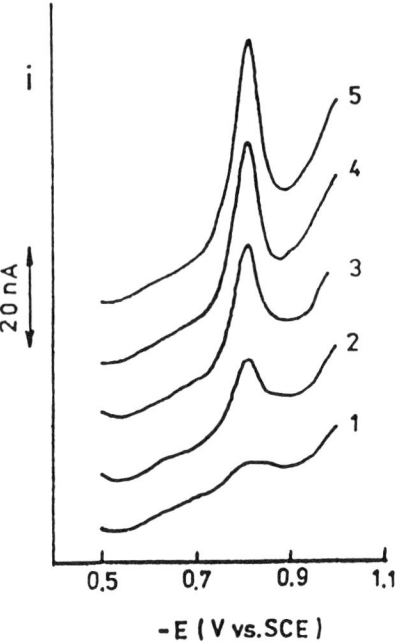

FIG. 12. Differential-pulse voltammograms for 5×10^{-8} M diazepam after different preconcentration times: (1) 0, (2) 30, (3) 60, (4) 120, (5) 180 sec. Preconcentration potential, -0.5 V. (From Ref. 74, with permission.)

1. Measurement of Low Molecular Weight Compounds

Compounds of pharmaceutical significance have attracted most of the attention because of the clinical needs for their quantitation at trace levels. In particular, Kalvoda [74,77] reported effective procedures for measuring alkaloids and anesthetics such as diazepam, papaverine, or codeine, while work in the author's laboratory [78-82] has been aimed at a variety of cardiac drugs (e.g., digoxin, reserpine, diltiazem) and anticancer agents (e.g., mitomycin C, methotrexate). The clinical utility of many cardiac and anticancer drugs is often hampered by their toxicity. Adsorptive stripping voltammetry can provide the desired high sensitivity, as indicated, for example, from voltammograms for 5×10^{-9} M digoxin obtained with and without accumulation (Fig. 13). Although quantitation at this level is not feasible without accumulation, a well-defined peak is observed after accumulation; a detection limit near 3×10^{-10} M can be estimated from this voltammogram. Recent adsorptive stripping schemes for measuring anticancer drugs are summarized in Table 2. Other highly sensitive schemes have been developed for pharmaceuticals, such as tetracycline antibiotics [88], streptomycin and related antibiotics [89], phenothiazine compounds [90,91], tricyclic antidepressants [71], and cimetidine [17], as well as compounds of biological significance, such as bilirubin [21], riboflavin [75], folic acid [92], or sex hormones [93]. The clinical utility of these procedures is often hampered by various matrix effects, particularly the presence of

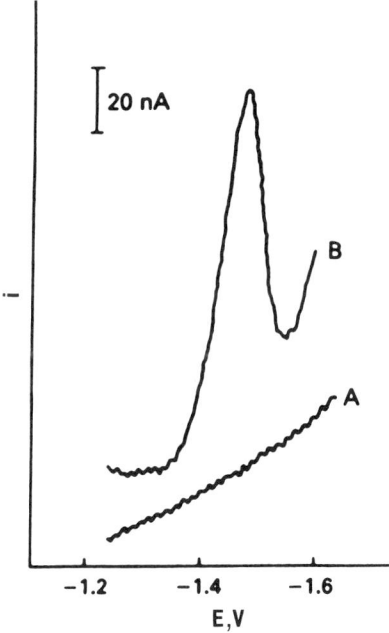

FIG. 13. Voltammograms obtained for 5×10^{-9} M digoxin using preconcentration for 0(A) and 15(B) min at -0.9 V; 0.005 M NaOH solution. (From Ref. 78, with permission.)

TABLE 2

Trace Measurements of Anticancer Drugs

Compound	Supporting electrolyte	Working electrode	E_p, V	Detection limit, M	Refs.
Methotrexate	Phosphate buffer (pH 2.5)	HMDE	−0.48	2×10^{-9}	82
Mitomycin C	Borate buffer (pH 10.2)	HMDE	−0.50	2×10^{-9}	81
Adriamycin	Acetate buffer (pH 4.5)	Carbon paste	+0.50	1×10^{-8}	83
Daunorubicin	Acetate buffer (pH 4.4)	HMDE	−0.57	1×10^{-9}	84
Cis-platin	Potassium chloride	HMDE	−0.49	2×10^{-8}	85
Chlorambucil	Acetate buffer (pH 5.4)	HMDE	−1.30	3×10^{-8}	86
Vinblastine	Ammonia buffer (pH 9.3)	HMDE	−1.70	1×10^{-8}	63, 87
5-fluorouracil	Borate buffer (pH 10)	HMDE	−0.40	3×10^{-9}	86

coadsorbing surfactants. As a result, assays of biological samples usually require a separation step; this and other approaches for minimizing possible interferences are discussed in Section H.

Adsorptive stripping procedures for measuring environmentally important compounds were developed, particularly in Czechoslovakia. Such schemes allowed measurements of trichlorobiphenyl [94] or nitro-containing pesticides [95] in waste and natural waters. Fogg et al. [96] reported effective adsorptive stripping procedures for 16 coloring substances and for their determination in tablet coatings and cosmetics. Similar procedures were reported for the antioxidant butylated hydroxy-anisole [97], thiourea and thiourea derivatives [26], and phenanthrene-quinone-like compounds [98]. New schemes for measuring other low molecular weight compounds are expected in the near future.

2. *Adsorptive Stripping Measurements of Biological Macromolecules*

Whereas early adsorptive stripping investigations were focused at low molecular weight compounds, recent activity has shifted to the measurement of macromolecules of biological significance. Biomacromolecules almost always possess surface activity and, thus, they are very suitable for adsorptive stripping measurements. Both the adsorption and redox processes of biomacromolecules are considerably more complex than those of small molecules. Adsorption usually takes place only with segments of the whole chain molecule, while the connecting parts (in between) form loops into the bulk of the solution. In various situations, the adsorption process has a pronounced (favorable) effect on the kinetics of the electron-transfer reaction. In addition to electroactive macromolecules, nonelectroactive ones can be determined by tensammetric procedures (Section II.F). Besides possible bioanalytical applications, the adsorptive stripping experiment may provide new insights into the physiological behavior of the macromolecule. For example, by using the medium-exchange concept, the biomacromolecule is adsorbed independently of the conditions under which the redox process occurs, thereby allowing evaluation of the interfacial behavior of biomacromolecules under more versatile conditions. Early studies of Stankovich and Bard [98a, 98b] demonstrated the utility of the adsorptive voltammetric scheme for obtaining useful knowledge of the interfacial and redox properties of biomacromolecules.

Palecek's group pioneered the use of adsorptive stripping voltammetry for nucleic acid research [99-102]. In neutral and weakly acidic media, single-stranded deoxyribonucleic acid (DNA) produces a cathodic peak due to the reduction of the adenine and cytosine residues and an anodic peak due to the guanine residues. The adsorptive accumulation of DNA has allowed an enhancement of the sensitivity by two orders of magnitude compared to pulse polarographic procedures. In particular, by monitoring the anodic guanine peak, single-stranded DNA can be quantified at concentrations as low as 0.1 mg l^{-1} (10-min preconcentration) [99]. Because double-helical DNA yields a substantially lower adsorptive stripping response, it does not interfere in the measurement of single-stranded DNA. As a result of the strong adsorption of DNA, the electrode (with the adsorbed DNA) can be transferred from the sample to an electrolyte blank solution to yield voltammetric peaks similar to those observed in the sample. This phenomenon can be exploited for a substantial reduction of the sample volume and for studying the interaction of the adsorbed DNA with various substances (mutagens, enzymes, etc.) [100]. It is possible also to modify chemically the DNA with an electroactive marker prior to the adsorptive stripping measurement. In particular, nanogram quantities of osmium-labeled DNA can be quantified following short preconcentration periods (1-3 min) [101]. The formation of such stable electroactive markers is illustrated in Eq. (15). The high sensitivity of adsorptive stripping DNA measurements thus shows great promise in the analysis of recombinant-DNA molecules of viruses and plasmids, which play an important role in the current development of molecular biology and genetics. A very recent paper [102] reported the determination of submicrogram quantities of duplex

$$\text{(structure)} \xrightarrow[\text{pyridine}]{O_s O_4} \text{(structure)} \qquad (15)$$

DNA in plasmid samples. The technique has also been applied to the in vitro investigation of damage in native DNA by small γ-radiation doses [103].

The adsorptive stripping quantitation of other important biomacromolecules has recently attracted increased attention. Submicromolar levels of the important heme protein cytochrome C can be measured following controlled adsorption and reduction at the hanging mercury drop electrode [24]. Binding of this protein is a prerequisite for the achievement of direct electrode transfer, because it overcomes the large activation energy for the process. Hence, well-defined increasing peaks are observed following preconcentration for longer periods (Fig. 14). The electron transfer is not restricted to the first monolayer but can proceed through this layer to the layers formed on top. The consequence of this behavior is a favorable adsorptive stripping response because saturation effects are minimized. Adsorptive stripping measurements of ferritin [103] are another example of establishing direct electrical communication between the redox center of a macromolecule and the mercury surface.

Measurements of other proteins have been reported recently. For example, the disulfide proteins trypsin, insulin, and chymotrypsin yield well-defined cathodic peaks following preconcentration from acidic solutions [104,105]. Optimal conditions (0.1 M HCl, preconcentration at +0.05 V versus Ag/AgCl, and a differential-pulse waveform) permit convenient quantitation down to the nanomolar concentration level using short preconcentration periods. The similar response of human serum albumin and immunoglobulin G has been exploited recently by Smyth's group for monitoring their immunochemical reactions with antihuman

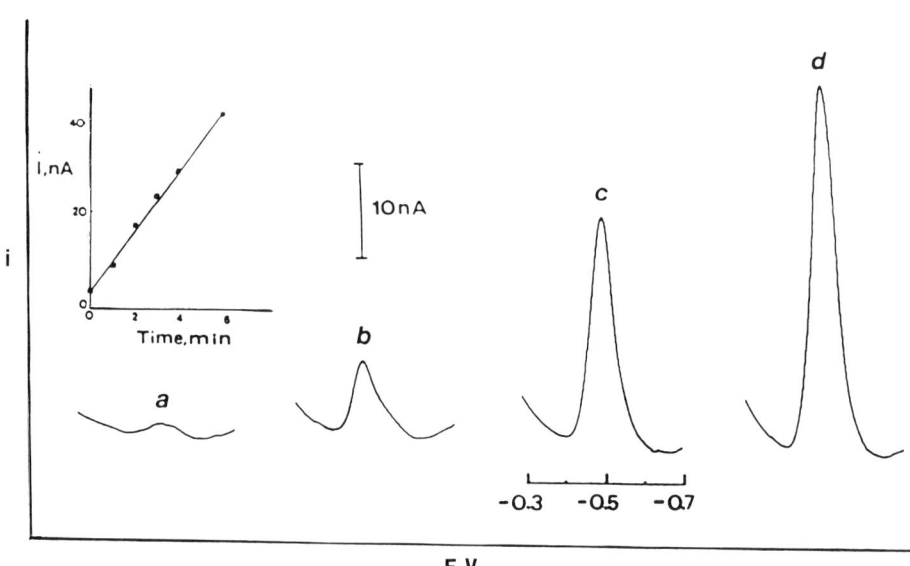

FIG. 14. Effect of preconcentration time on the voltammetric response for 4×10^{-6} M cytochrome C in 0.01 M Tris-HCl (pH 6.5). Preconcentration time: (a) 0, (b) 60, (c) 180, (d) 360 sec. Preconcentration potential, -0.1 V, with 400-rpm stirring. (From Ref. 24, with permission.)

serum albumin and anti-immunoglubolin G, respectively [106]. Nanomolar levels of chlorophyll and ferredoxin can be measured following their controlled adsorptive accumulation [106a]. Palecek and co-workers [100] are currently evaluating the properties of lipid layers by means of adsorptive stripping voltammetry. Such applications will greatly enhance the capabilities of voltammetry in biological membrane research.

3. *Adsorptive/Extractive Accumulation at Carbon Paste Electrodes*

Use of carbon paste electrodes for voltammetric measurements following interfacial accumulation represents a unique case. Carbon paste electrodes, originated by Adams, consist of a mixture of graphite and organic liquid. Because of their unique composition, various organic compounds can be extracted into the paste. Evidence of such extractive accumulation was obtained using a radioactive tracer [107] or a piston-like electrode configuration [108]. Penetration to an appreciate depth below the surface was reported for compounds such as chlorpromazine or butylated hydroxyanisole. Kuwana and French [108a] dissolved organic compounds in the pasting liquid and suggested that the carbon paste electrode functions for such compounds in a manner analogous to a mercury electrode for metals. Wang et al. [109] systematically investigated trends and changes in the accumulation behavior of organic compounds at carbon paste electrodes. The correlation between distribution experiments and analogous voltammetric measurements was explored. It is clear from their data that both extraction and adsorption occur simultaneously at different degrees, depending on the nature of the pasting liquid, graphite, and liquid-to-graphite ratio. For example, Fig. 15 shows distribution and current profiles as a function of the liquid-to-graphite ratio. Such behavior indicates that adsorption dominates in graphite-rich compositions, whereas extraction occurs mainly in pasting-liquid-rich electrodes. In the latter case, coverage of the graphite particles blocks adsorption sites. Penetration of organic analytes into the electrode interior was also reported at wax-impregnated graphite electrodes [110].

F. Adsorptive Stripping Tensammetry

Nonfaradaic processes resulting from an adsorption-desorption mechanism can be employed for measuring trace levels of nonelectroactive surfactants, following their controlled interfacial accumulation. Such processes often yield sharp current spikes (at very negative or positive potentials), as well as depression in the double-layer capacitance, which can be monitored primarily by ac (fundamental and second harmonic) or differential-pulse waveforms. The current spikes are attributed to sudden changes in the structure of the double layer when a molecule adsorbs or desorbs; the depression in capacitance occurs, as expected, over the

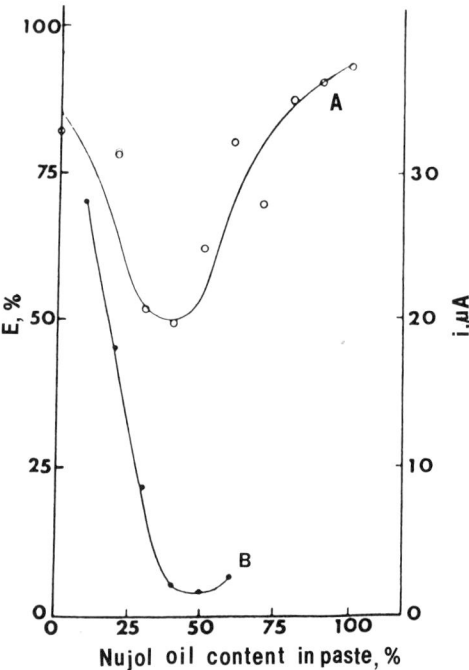

FIG. 15. Dependence of percent extraction (A) and voltammetric peak current (B) on the carbon paste composition. Initial concentration of chlorpromazine in the aqueous phase, 2.5×10^{-5} M. (From Ref. 109, with permission.)

potential region in which the substance is adsorbed. Empirically, the measured current change Δi is proportional to the change in the differential capacitance ΔC,

$$\Delta i \sim \frac{\partial q}{\partial t} \sim \Delta C \tag{16}$$

where q is the charge density on the electrode surface. The interested reader is referred to the theoretical treatment of Jehring [111]. Whereas early studies used tensammetric procedures in conjunction with the dropping mercury electrode, recent work has demonstrated that a marked sensitivity enhancement can be obtained at the hanging mercury drop electrode, following preconcentration from a stirred solution. Thermodynamic adsorption isotherms (discussed in Section II.A) are commonly obtained for low concentrations of surfactants, with mass transport usually controlling the rate of the adsorption process.

Bednarkiewicz and Kublik [112] compared the utility of various tensammetric techniques (ac fundamental and second-harmonic or differential-pulse) for trace measurement of surfactants at the hanging mercury drop electrode. Figure 16 illustrates the response characteristics obtained by these techniques for the desorption peak of Triton X-100, following 100-sec preconcentration from a neutral solution. Each tensammetric mode was shown to possess its own advantages and disadvantages. For example, ac second-harmonic tensammetry offers improved sensitivity

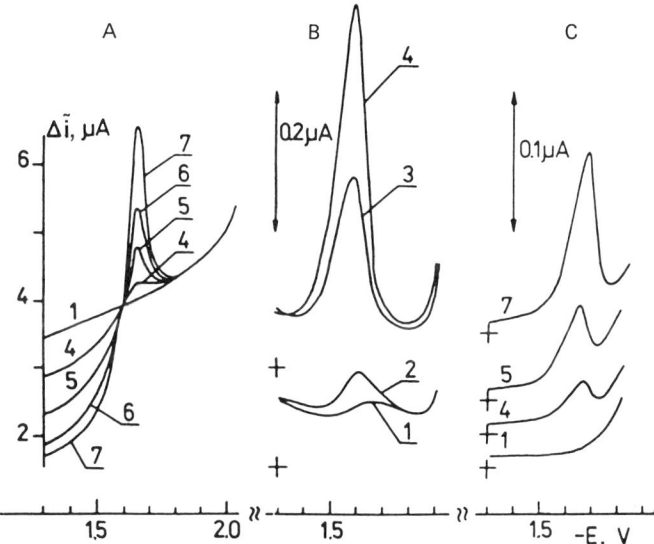

FIG. 16. Tensammetric peaks obtained for Triton X-100 after 100-sec accumulation at −1.3 V, using ac fundamental (A), ac second-harmonic (B), and differential-pulse (C) waveforms. Triton X-100 concentration: (1) 0, (2) 10, (3) 20, (4) 50, (5) 100, (6) 150, (7) 250 g dm^{-3}. (From Ref. 112, with permission.)

without improving the precision and requires sophisticated instrumentation. All adsorptive stripping tensammetric modes allowed convenient detection of surfactants down to the parts-per-billion (ppb) level. Measurements at this level required removal of surface-active contaminants from the supporting electrolyte, for example, by passage through active charcoal. Batycka and Lukaszewski [113] investigated the adsorptive stripping tensammetric response for mixture of surfactants. Three different types of mixtures were employed. The first consisted of mixtures of components that have very similar properties and behaved additively (i.e., yielding one peak, the height of which depends linearly on the total concentration). The second group consisted of mixtures of components with rather different properties; in this case, one component was measured in the presence of excess of a second one by a proper choice of the preconcentration potential. The third group consisted of mixtures of components that have similar properties but behave non-additively and thus cannot be quantified. In a related study [114], the same authors explored the influence of the preconcentration potential on the cathodic tensammetric peak height of several poly(ethylene glycols). In accordance with the theoretical prediction of Damaskin and Tedoradze [115], it was shown that the tensammetric peak potential depends linearly on the logarithm of the surfactant concentration.

In an interesting application, Kalvoda and Novotny [116] explored the adsorptive stripping tensammetric behavior of petroleum components in aqueous solutions. Their data indicate that water pollution with petroleum can be determined by this approach, with the sensitivity depending on the quality of the pollutant and the composition of the supporting electrolyte. Well-defined calibration curves were obtained for diesel oil at the sub-ppm level. In a different study, Kalvoda [117] demonstrated the utility of adsorptive stripping tensammetry for measuring the polyether antibiotic monensin over the 0.1 to 1.5-ppm concentration range.

VOLTAMMETRY AFTER NONELECTROLYTIC PRECONCENTRATION 47

The adsorptive stripping tensammetric approach is expected to play an important role in future investigations of the interfacial behavior of nonelectroactive biomacromolecules. Analysis of complex examples, however, would require prior separation of the surfactants, as coadsorption remains the major obstacle to direct assays of surfactant mixtures.

G. Combination of Adsorptive Voltammetry with Catalytic Effects

A novel combination of adsorptive voltammetry with catalytic hydrogen processes has recently been proposed to yield extremely low detection limits. Surface-catalyze proton reduction processes have been used for many years for polarographic measurements of platinum, platinum-group metals, and organics such as amines or heterocycles [118]. Adsorption of the catalyst has been observed often in many of these systems, yielding enhanced polarographic response. Recent work in the author's laboratory [63,87] has illustrated that coupling such hydrogen catalytic systems with controlled interfacial accumulation of the catalyst at the hanging mercury drop electrode can be used for ultratrace voltammetric measurements. A possible mechanism for such preconcentration and measurement steps is as follows:

$$CatH^+ \longrightarrow CatH^+_{ads} \tag{17}$$

$$CatH^+_{ads} + e^- \longrightarrow Cat_{ads} + \tfrac{1}{2}H_2 \tag{18}$$

where Cat is the catalyst and the subscript ads designated adsorption at the electrode. Alternately, for Mairanovskii-type processes [118], protonation of the catalyst commonly occurs after its adsorption,

$$Cat_{ads} + H^+ \longrightarrow CatH^+_{ads} \tag{19}$$

This concept has been applied recently for ultratrace measurements of platinum in the presence of formazone (detection limit of ca. 1×10^{-12} M!)

[63], as well as for the quantification of the vinca alkaloids vinblastine and vincristine [87]. Many new applications of this dual-amplification scheme are expected in the near future.

H. Problems and Solutions

The major type of interference in adsorptive stripping measurements is the presence of other surface-active species in the sample solution. Such species compete for adsorption sites, resulting in diminished sensitivity. For example, 1 ppm of gelatin results in 25 and 45% depressions of the bilirubin and riboflavin stripping peaks, respectively [21,75]. The surface concentration of the analyte i in the presence of interfering adsorbate j can be described in terms of the appropriate Langmuir isotherm,

$$\Gamma_i = \Gamma_{m,i} \frac{B_i C_i}{1 + B_i C_i + B_j C_j} \tag{20}$$

The extent of this interference thus depends on the relative affinities of the analyte and interferent toward the surface (B_i and B_j) and their bulk concentrations (C_i and C_j). The affinity toward the surface depends not only on the species involved but also on numerous experimental conditions (e.g., solution, preconcentration time, and potential). For example, nonionic surfactants exhibit more pronounced depression following accumulation around the potential of zero charge. The presence of halide ions that exhibit specific adsorption can result in similar effects. In addition to suppressive effects, coadsorbing electroactive substances may yield overlapping responses if they have a redox potential similar to that of the analyte.

The susceptibility of the method to the presence of surfactants may pose serious problems, especially when biological samples are concerned. Knowledge of the coadsorption effects is required for their minimization. Various approaches have been suggested to address the coadsorption problem. In situations involving modest (10–50%) peak depressions, untreated samples may be employed. Since the depression of the peak current depends on the fraction of the surface covered by the interfering surfactant (independent of the analyte concentration), calibration curves or standard addition measurements made uniformly, using the sample control, should correct for these effects [21,83]. Obviously, the linear range is altered by the presence of other surfactants. Surfactant interferences can also be alleviated by using shorter accumulation times [74,21] or by a proper choice of the accumulation potential [26]. For example, Fig. 17 illustrates the effect of gelatin on the nitrazepam peak current at different accumulation times. Such different profiles may be obtained when adsorption equilibrium for the analyte and interferent is attained after different periods.

More severe peak depressions usually require a separatory technique to isolate the analyte from coadsorbing interferents. Among the proposed treatments suggested for this purpose are separation of interfering surfactants by molecular exclusion chromatography (on Sephadex) [74], by high-performance liquid chromatography [17,94], or by using a Sep-Pak extraction cartridge [118a]. Alternatively, it is possible to perform

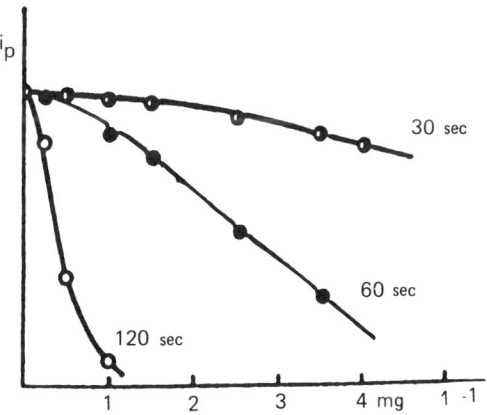

FIG. 17. Effect of gelatin on the 1×10^{-7} M nitrazepam peak height at different preconcentration times. Preconcentration potential, -0.50 V. (From Ref. 74, with permission.)

the separation step in situ, at the electrode surface, using a permselective polymeric coating such as cellulose acetate [119]. Fine control
of the film permeability can also be used for minimizing interferences
from adsorbable electroactive species. Chemical treatment of the sample
may also be useful in various situations [94]. Effective correction of
surfactant interferences can be achieved when using the metal-chelate
approach to measure metal ions. In this case, interfering surfactants
are destroyed by ultraviolet (UV) irradiation prior to the addition of
the chelating ligand of interest [53,56,57].

It is possible also to exploit the suppressive effect observed in the
presence of various surfactants for their indirect measurement. Pihlar
et al. [120] demonstrated recently that trace (ppb) levels of Triton
X-100 can be measured on the basis of the linear decrease of the
$Ni(DMG)_2$ peak height in the presence of increasing concentrations of
the surfactant.

Nonadsorbable (solution-phase) constituents, with redox potentials
similar to the analyte, represent a second type of potential interference.
The extent of this interference depends on the peak potential and concentration of these solution-phase species in comparison to those of the
adsorbable analyte. In many practical situations, a large response associated with a macrosolution constituent may mask almost completely the
adsorptive stripping peak of interest (for example, measurements of
chlorpromazine in the presence of ascorbic acid, Fig. 18A). Wang and
Freiha [121] proposed the use of the medium-exchange procedure in
conjunction with adsorptive stripping voltammetry to eliminate this type
of interference. After adsorption has proceeded under controlled conditions, the electrode is transferred into a suitable electrolyte solution,
where the voltammetric measurement is carried out. Hence, currents
arising from significant concentration of solution-phase species are
eliminated (Fig. 18B). Batch operations require washing the electrode
during the transfer to remove the sample solution adhering to the surface. Because of the mechanical instability of the hanging mercury drop

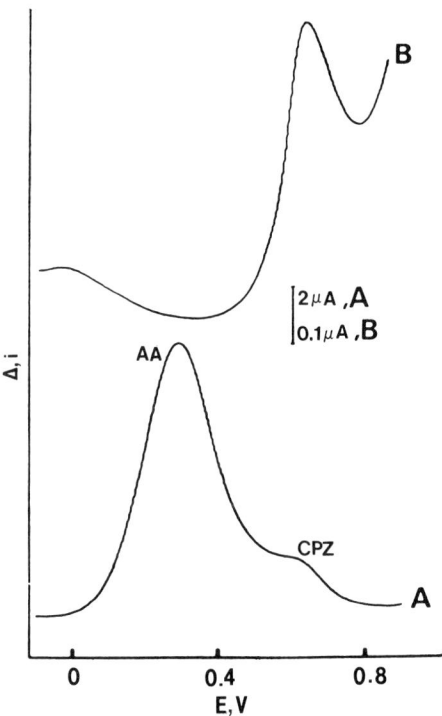

FIG. 18. Voltammograms for 2.5×10^{-7} M chlorpromazine in the presence of 5×10^{-4} M ascorbic acid: (A) without medium exchange, (B) after medium exchange. Preconcentration for 3 min at -0.2 V. (From Ref. 121, with permission.)

electrode, most applications of the medium-exchange approach have involved measurement of oxidizable species at solid electrodes. Of particular bioanalytical interest is the possibility of measuring dopamine in the presence of a huge concentration excess of ascorbic acid following accumulation at a platinum electrode [122]. A more elegant, effective, and reproducible approach to performing medium exchange in conjunction with adsorptive accumulation utilizes the manifold of flow-injection systems [123]. In a flow-injection system, the preconcentration period can be started as the sample plug arrives in the detector and terminated on the arrival of the carrier (blank) solution. The effective preconcentration period is thus determined by the sample volume and flow rate. The advantages of such flow-injection/adsorptive stripping voltammetric schemes have been illustrated recently in direct measurements of drugs in various body fluids, including assays of urine for chlorpromazine [123] and adriamycin [124] or of plasma for cyadox [125], with no preliminary cleanup steps. For example, Fig. 19 shows flow-injection/voltammetric measurement of 95 µg liter^{-1} cyadox in blood. This study has illustrated that flow systems also permit a convenient use of mercury electrodes in conjunction with medium exchange.

Trace metal measurements by the metal-chelate approach may be affected by coexisting metal ions that form reducible and adsorbable

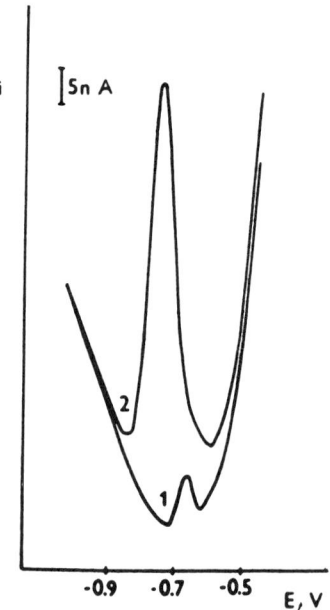

FIG. 19. Voltammograms for injections of blood plasma containing (1) 0 and (2) 95 g liter^{-1} cyadox. Sample volume, 1.0 ml; preconcentration potential, −0.1 V; flow rate, 0.2 ml min^{-1}. (From Ref. 125, with permission.)

chelates with the ligand of interest. These can interfere via an overlapping response and/or competition on adsorption sites. Addition of a suitable masking agent or adjustment of the preconcentration potential or the pH are often sufficient to eliminate this source of interference. Samples "rich" in interfering metals would also require a prior separation step. As was discussed in Section II.D, it is possible to measure simultaneously several metal ions, providing that their metal-chelate peaks are well separated. Competition by other ligands on the metal of interest should also be taken into account in the metal-chelate approach. The extent of this competitive effect is determined by the concentration of natural ligands (vs. that of the added chelator) and by the stability of the complexes involved; such behavior provides the basis for the exploitation of the metal-chelate approach for speciation work (as was discussed in Section II.D 4).

Finally, the inherent sensitivity of adsorptive stripping voltammetry permits significant dilution of various samples, thus potentially minimizing possible interferences.

III. VOLTAMMETRY FOLLOWING PRECONCENTRATION AT CHEMICALLY MODIFIED ELECTRODES

The field of chemically modified electrodes (CMEs) has grown dramatically since its inception in the mid-1970s and continues to generate considerable attention [3,126]. Besides their importance in electrocatalysis, electrosynthesis, and energy conversion, these electrodes are used increasingly for electroanalysis [127]. Three major advantages might be realized in this direction: determination of species with slow electron-transfer kinetics by use of electrodes modified with electrocatalytic moieties; improvements in selectivity and stability at electrodes coated with permselective polymeric layers; and preferential accumulation of analytes onto bound surface functionalities. The objective of the following discussion is to review the utility of voltammetric methods based on preconcentration at chemically modified electrodes.

The use of properly designed CMEs adds a powerful dimension to preconcentration/voltammetric schemes. The basic concept is to perform the desired chemistry at the surface of the electrode via a judicious choice of reagents for preconcentrating various analytes. By deliberately attaching such reagents to the surface, one hopes that the electrode surface would take on the binding properties of the attached reagents. The precise nature of the preconcentration mechanism is determined by the reactivity of the electrode-modifying group. Most popular schemes used to trap analytes on the electrode surface are based on complexation and electrostatic attractions; covalent linkages can also be used. The modifying agent (ligand, ion exchanger) is commonly introduced into the surface as part of an appropriate polymeric coating or directly into the matrix of a carbon paste electrode. Hence, preconcentration is accomplished by a purely nonelectrolytic step (with forced convection being used to facilitate mass transport, thus shortening the deposition period). Quantitation of the analyte accumulated at the surface is subsequently carried out by the usual voltammetric waveforms, with the electrode residing either in the sample solution itself or in an electrolyte solution possessing more ideal background characteristics (i.e., medium exchange). The analyte concentration in the original sample is determined through the use of working curves.

This approach enjoys an inherently high sensitivity because it is a preconcentration technique. A second major advantage lies in the added dimension of selectivity, which is provided by the chemical requirement of the modifier-analyte interactions. Such coupling of chemical selectivity with the inherent sensitivity of preconcentration/voltammetric schemes is attractive for practical analytical situations. Other advantages involve the ability to measure numerous analytes that cannot be accumulated electrolytically, the use of "mercury-free" electrodes or "reagent-free" solutions, the convenience of medium exchange between the preconcentration and voltammetric steps, and (in various situations) the elimination of oxygen interference. Although methods for analyte preconcentration on modified electrode surfaces will continue to evolve, this is already a powerful concept, as described below.

A. Requirements, Problems, and Solutions

There are several requirements for a successful use of modified electrodes as preconcentrating surfaces. These requirements differ from those expected for conventional stripping measurements because these techniques are based on fundamentally different detection principles.

First (and usually ideally), the preconcentration process should be selective for the analyte species of interest. If not, the analyte must be voltammetrically discriminated from other collected species. This is not always a simple task; for example, measurements of ionic analytes at ion-exchanger modified electrodes are subject to interference from other ionic species (whether electroactive or not) that occupy exchange sites in the electrode, thus denying access to the analyte. Interferences from noncollected (solution-phase) electroactive species can be easily eliminated using the medium-exchange approach described in Section II.H. The nonelectroactive nature of the preconcentration step simplifies the medium-exchange procedure.

A second requirement for a successful use of CMEs is that the surface not be easily saturated during the preconcentration step. Saturation is a particularly serious problem because, once it is reached, the voltammetric response no longer bears any concentration dependence. Hence, the capacity of the electrode should be sufficient to minimize saturation by the analyte.

Finally, there should be a convenient way to regenerate a fresh and reproducible surface following the voltammetric scan. It is desirable that the product of the redox reaction be rapidly removed (stripped) from the surface on completion of the voltammetric scan, so that a fresh surface is available for immediate reuse. In numerous situations, however, the electrode can be used for only a single determination and, thus, a new surface has to be regenerated. This procedure is not only tedious but also requires careful normalization of the data. Some novel strategies, suggested recently to circumvent the regeneration problem, may be suitable in specific situations. For example, Abruna and co-workers [128] described a multiple-use polymer modified electrode based on 2,9-dimethylsulfonated bathophenanthroline for voltammetric measurements of copper. The stability constant of the resulting copper complex depends strongly on the oxidation state of the metal; thus, after the redox reaction, the metal/ligand complex dissociates (while the ligand remains intact), and the electrode is ready to be reused. Martin's group [129] described an analogous approach for ion-exchanger modified electrodes. This concept allows electrochemical modulation of the ion-exchange characteristics of certain cation exchangers; as a result, counterions can be expelled and reincorporated electrochemically. This electrorelease concept is illustrated schematically in Fig. 20 for an electrode modified with a vinylferrocene-styrenesulfonate copolymer. Active research on electrically controlled release of chemicals, especially drugs, will undoubtedly lead to similar elegant schemes for regenerating fresh

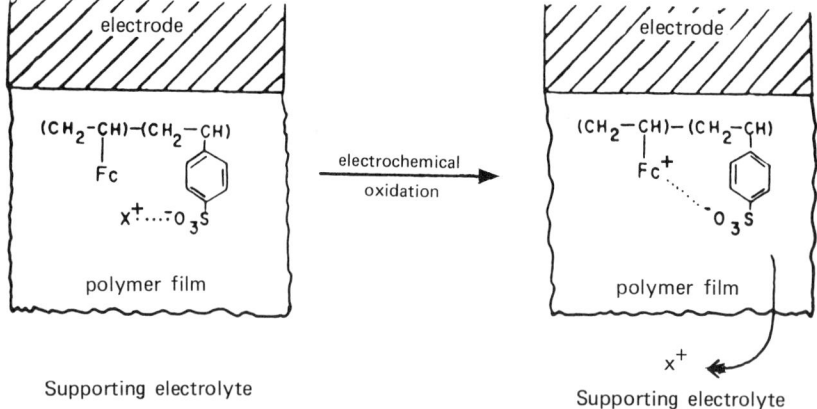

FIG. 20. Model for electrorelease of counterions from a vinylferrocene-styrenesulfonate copolymer film modified electrode. (From Ref. 129, with permission.)

("analyte-free") surfaces. Chemical regeneration, involving dipping the electrode in an appropriate solution (e.g., an acidic solution for cation-exchanger modified electrodes) has been useful in numerous situations. In addition, the ease by which carbon paste electrodes can be mechanically renewed for subsequent reuse makes these electrodes extremely useful for the preconcentration-CMEs approach. Specific problems and solutions for various preconcentration-CMEs systems are described in the following sections.

It should be stressed that all the steps in the procedure should be carried out with the greatest care, and always in the same manner, in order to achieve the utmost reproducibility of results. Nonequivalent treatment of electrodes (e.g., modification, regenerating, stirring, rinsing) may result in irreproducible amounts of analyte being accumulated, and hence errors in the measured concentration.

B. Ways to Introduce the Preconcentrating Agent
1. Polymeric Coatings

Polymeric coatings have attracted great interest for their role in the modification of electrode surface properties. Advantages offered by these systems over directly chemically bound monolayers include ease of preparation, chemical and physical stability, and complete surface coverage. Polymer layers are commonly produced by spin-coating or dip-coating methods. The film is then held on the surface by a combination of chemisorptive and solubility effects. It is possible also to produce the polymer layer by electrodeposition or by inducing the polymerization of monomers by other means. The general properties and methods of preparing polymer-coated electrodes have been reviewed earlier in this series [8].

Several types of polymer electrodes have been evaluated in conjunction with electrostatic or coordinative preconcentration. The preparation of such electrodes can proceed along two major avenues: the preconcentration moiety may be a part of the polymeric backbone itself, or the polymeric phase can be functionalized with the desired modifier. In both cases, a high concentration of the preconcentration moiety can be obtained, and hence a large voltammetric response (compared to an analogous work with monolayers). To achieve high sensitivity due to the analyte preconcentration, the polymer should permit efficient charge and mass transport. Hence, the film structure and thickness have a pronounced effect on the analytical performance. Improved performance has been observed recently at composite polymers that couple the chemical reactivity of one polymeric domain with a structural/mechanical function of a second one. The behavior, response characteristics, and applications of preconcentration/voltammetric procedures at polymer-coated (and other modified) electrodes are described in the following sections.

2. Modified Carbon Paste Electrodes

"Mixed carbon paste electrodes," made of carbon paste that is doped during preparation with the modifying molecule have gained popularity for the preconcentration/voltammetric approach. Numerous preconcentration agents can easily be inserted into the carbon paste, with no need to devise individualized attachment schemes for each modifier. Several studies [130,131] demonstrated the versatility, stability, and ease of operation of mixed carbon paste electrodes; details are given in the following sections. Such electrodes offer several advantages as preconcentration surfaces. First, the effective surface coverage can be altered by varying the weight of the modifier added to the paste mixture. In addition, the surface can be removed and renewed for subsequent reuse with a reproducibility of 5 to 10%. Such reproducibility requires a uniform surface coverage, which can be achieved by a thorough mixing of the modifier and the organic pasting liquid. Hand-mixing or ultrasonic dispersion can be employed. A chemical regeneration step, leading to an "analyte-free" surface, can also be used to minimize the errors and inconvenience associated with manual resurfacing. An important requirement for the success of this modification approach is that the solubility of the modifier be sufficiently slight in the solvents employed

VOLTAMMETRY AFTER NONELECTROLYTIC PRECONCENTRATION 59

to keep it stably incorporated in the carbon paste matrix. More robust and polishable preconcentrating electrodes, based on other carbon composites, are being explored in various laboratories.

C. Preconcentration Schemes

1. Preconcentration at Ligand-Containing Complexing Electrodes

The use of complexing electrodes is a particularly exciting prospect for analytical electrochemistry. Because the metal-ligand complex formation is used as the preconcentration step, an appropriate modifier with very high affinity for a particular metal ion (i.e., large formation constant) should be used. While this assures high sensitivity, selectivity trends in solution can be used to obtain good specificity. Such trends can be extrapolated to the electrode surface, hence obtaining a rich surface coordination chemistry. Various ligands and modification strategies have been explored for the coordinative uptake and subsequent measurement of metal ions. Complexing electrodes based on ligand-containing polymeric coatings or "mixed carbon paste" have been employed, as illustrated below. Important speciation information may be obtained when the ligand chosen possesses high affinity for a particular oxidation state of the metal.

Abruna and co-workers [132,133] proposed and demonstrated the utility of electrodes modified with functionalized polymer films that carry a coordinating group as well as an internal redox couple (Fig. 21). The former is chosen to exhibit high affinity for the particular metal ion, whereas the latter serves in the deposition of the polymer and as an internal standard. For example, by knowing the ratio of internal redox sites to ligand centers, one can predict the maximum electrode response and thus determine when the surface is saturated. The ligand can be part of the polymer backbone itself, for example, using a vinylbipyridine/vinylferrocene copolymer (Fig. 21, case I). It is possible also to incorporate the ligand via ion exchange, for example, by attaching a negatively charged ligand to a pyridinium group (Fig. 21, case II). Judicious choice of the ligand can attain the desired selectivity. For

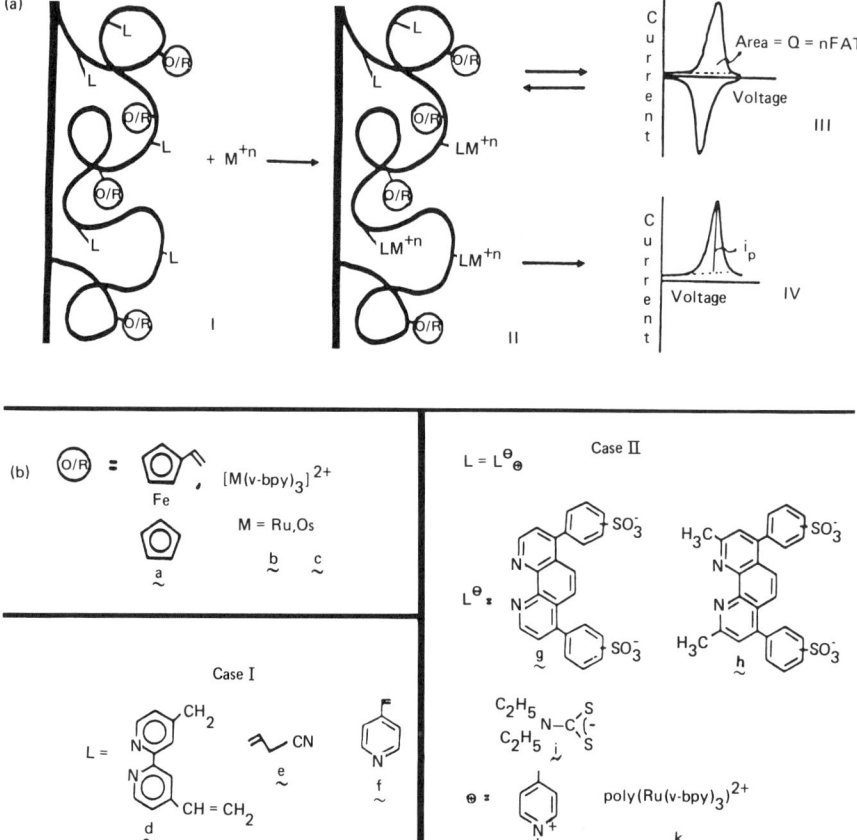

FIG. 21. Schematic description of preconcentration/voltammetric analysis of metal ions at electrodes modified with functionalized polymer films. (From Ref. 133, with permission.)

example, sulfonated 2,9-dimethylbathophenanthroline was employed for selective measurements of copper in the presence of iron. Because of steric hindrance (associated with the two methyl groups), this ligand has a greatly diminished affinity for iron, whereas the affinity for copper remains essentially the same (compared to that for unmodified sulfonated bathophenanthroline). High sensitivity has also been reported; for example, an electrode modified with the copolymer [Os(vinylbipyridine)$_3$]$^{+2}$/vinylbipyridine yielded a well-defined response following the uptake of iron (II) from a 4×10^{-7} M solution of the ion.

O'Riordan and Wallace [134] described the utility of poly(pyrrole-N-carbodithioate) electrode for trapping and quantitating copper ions. The dithiocarbamate ligand (IV) possesses several properties that make it

IV $\quad \begin{matrix} R \\ \\ R \end{matrix} N = C \begin{matrix} S \\ \\ S \end{matrix}$

attractive for the preconcentration/CME approach. It forms stable complexes with a wide range of metals, with selectivity achieved via an appropriate pH control. Such complexes are amenable to both oxidation and reduction, for example,

$$[Cu(II)(dtc)_2] + e^- \rightleftharpoons [Cu(I)(dtc)_2]^- \qquad (21)$$

$$[Cu(II)(dtc)_2] \rightleftharpoons [Cu(III)(dtc)_2]^+ + e^- \qquad (22)$$

In particular, the oxidative response permits measurements without oxygen interference, although very long accumulation periods are required for measuring micromolar levels of copper. The electropolymerized polypyrrole used in the above study can be employed as a general matrix into which other preconcentration moieties can be incorporated. This and other conducting polymers may also offer a convenient way to regenerate the surface via an electrically controlled release of the analyte.

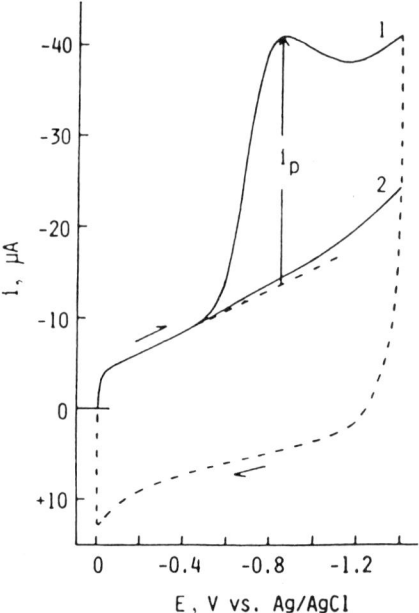

FIG. 22. Voltammograms at a glassy carbon electrode coated with TOPO following 10-min preconcentration at 0.0 V: (1) in the presence of 1×10^{-7} M UO_2^{2+}; (2) in the absence of UO_2^{2+}. (From Ref. 136, with permission.)

Sensitive measurements of uranium(VI) based on accumulation at electrodes modified by trioctylphosphine oxide (TOPO) have attracted considerable attention [135,136]. The uranyl ions are effectively concentrated into the TOPO layer on a glassy carbon electrode held at 0 V (versus Ag/AgCl) and, subsequently, give a reduction peak (at about −0.75 V) when the potential is scanned in the negative direction (Fig. 22). The voltammetric scan regenerates a "fresh" surface, suitable for repetitive measurements. The method is very sensitive (the detection limit is 2×10^{-9} M); it is also very selective, hence allowing the determination of uranium(VI) in seawater following a simple sample treatment [137]. The redox reaction can be considered as mediated electron transfer, with the preconcentrated uranium(VI) amplifying the reduction current of glassy-carbon surface functionalities [138]. A schematic description of this process is given in Fig. 23. TOPO-based modified electrodes can be used for measuring other metal ions. For example, mercury film electrodes prepared on silver metal and modified with a layer of TOPO in a poly(vinyl chloride) matrix have been shown effective for a preconcentration of Bi(III) [139]. Regeneration of a fresh surface is obtained by immersing the electrode in a dilute acid to remove any remaining bismuth ions.

Baldwin et al. [130] described the characteristics and practical analytical utility of nickel-sensitive CMEs based on dimethylglyoxime-containing carbon paste. Measurements were performed using ammonia buffer (pH 8); regeneration of a fresh ("nickel-free") surface was accomplished by a 5-sec treatment in 1 M nitric acid. The detection limit was 5×10^{-8} M (4-min preconcentration), the response was linear over the 5×10^{-8} to 5×10^{-6} M range, and the relative standard deviation

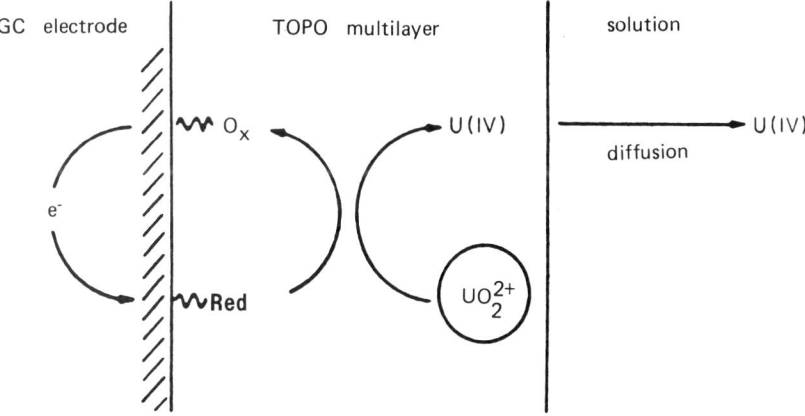

FIG. 23. Schematic description of the mediated electron transfer of a glassy carbon electrode coated with TOPO. Red and Ox are the reduced and oxidized forms of the surface functionalities. (From Ref. 138, with permission.)

(for 10 deposition/measurement/regeneration cycles) was 3%. Satisfactory results were reported for nickel in a variety of NBS reference materials, including water, coals, coal fly ash, and sediments. A typical set of voltammograms obtained for a digested coal fly ash are shown in Fig. 24. Caron paste electrode containing 2,9-dimethyl-1,10-phenanthroline was employed by the same group for the preconcentration and determination of copper [140]. Uniform dispersion of the modifier was obtained by forming an ethanol slurry of the graphite/ligand mixture, placing the slurry in a ultrasonicator for 10 min, and allowing the ethanol to evaporate before mixing with Nujol oil. A 1-sec accumulation time and voltammetric measurement of the anodic peak gave a detection limit of 0.3 μM and a linear response up to 10 μM. Coordination chemistry can be applied for measuring other metal ions at ligand-containing carbon paste electrodes. For example, gold(III) can be preconcentrated from hydrochloric acid solution on a surface of a carbon paste electrode that has been modified with dithizone [141]. A linear relationship between the

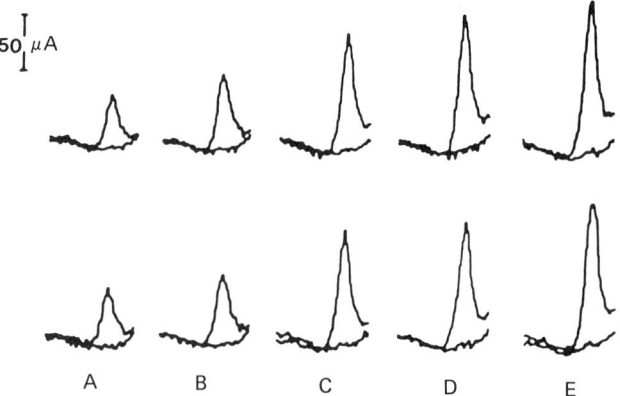

FIG. 24. Determination of nickel in decomposed dissolved coal fly ash using a dimethylglyoxime-containing carbon paste electrode. Voltammograms (A) are those obtained for the unspiked digest; the remaining signals were obtained after addition of (B) 0.5, (C) 1.0, (D) 1.5, and (E) 2.0 μM nickel (II). (From Ref. 130, with permission.)

current and concentration was observed for 10 to 1000 µg liter^{-1} gold. Severe interferences were observed in the presence of metal ions, for example, Ag(I), Pt(II), Pd(II), commonly reacting with dithizone. Carbon paste electrode-containing tropolone was used for the uptake and voltammetric quantitation of tin [142]. One of the early preconcentration/CME experiments was carried out by Cheek and Nelson [143], who complexed Ag(I) from aqueous solutions at carbon paste electrodes made from graphite powder, to which diethylenetriamine was bonded by an amidization process. Coordination of Ag(I) by the surface amine resulted in a remarkably low detection limit of 1×10^{-11} M. Finally, a recent study in the author's laboratory illustrated the utility of crown ether modified carbon paste electrodes for preconcentrating mercury from dilute solutions [144].

2. Ion-Exchange Voltammetry

Electrostatic binding in an ion-exchange process may provide another useful and versatile avenue to preconcentration/voltammetric analysis. A wide range of applications in the determination of ionic analytes can be achieved by modification of working electrodes with ion exchangers. The modification can be accomplished by coating the electrode with an appropriate polyelectrolyte or by introducing the ion exchanger into the carbon paste matrix. During a controlled preconcentration step, counterionic analytes are extracted from the solution and held at the surface by electrostatic binding. The distribution coefficient D_c for the ion of interest between the electrode (exchanger) and the solution governs the resulting voltammetric peak current. For the hypothetical partition reaction,

$$M^{+n}_{sol} \rightleftharpoons M^{+n}_{elec} \tag{23}$$

D_c is given by

$$D_c = \frac{[M^{+n}]_{elec}}{[M^{+n}]_{sol}} \tag{24}$$

Large distribution coefficients (ca. 10^3 to 10^5) have been reported [145,146]; the resulting enhancement of the ionic-analyte concentration in the electrode surface permits very sensitive analysis when combined with an advanced voltammetric method. A detection limit of 1×10^{-9} M has been reported in conjunction with differential pulse detection [146a]. In general, electrostatic trapping follows a reactivity pattern consistent with its essential ion-exchange character (although the kinetics of ion-exchange reactions on electrodes and in beads is not necessarily the same). Good correlation between cyclic voltammetric data for trapping various ions and the known binding strength of a similar ion-exchange resin has been reported (e.g., Ref. 147). It is thus possible to deploy the substantial available body of ion-exchange selectivity data to design more optimally the electrostatic preconcentration of redox ions at the electrode surface. The extent of the ion-exchange reaction may be conveniently visualized by plotting the partition isotherm ($[M^{+n}]_{elec}$ versus $[M^{+n}]_{sol}$) [146].

The selectivity between various pairs of counter ions can be expressed by the ratio of distribution coefficients:

$$\alpha_j^i = \frac{D_{c,i}}{D_{c,j}} \tag{25}$$

This can be measured by a pairwise competitive partitioning. Actually, competitive partitioning represents the major drawback of the electrostatic binding/preconcentration scheme. The voltammetric response may be affected by both electroactive and nonelectroactive counterions with high affinity for the ion exchanger. For example, a supporting electrolyte anion would be expected to compete with the anionic analyte for surface-charged sites. Figure 25 illustrates the effect of ionic strength on the peak height of the $Fe(CN)_6^{3-/4-}$ voltammograms at an anion-exchanger modified (+) and unmodified (•) electrode. An ionic-strength-independent response is observed at the unmodified electrode. In contrast, at the modified electrode, the peak increases with increasing

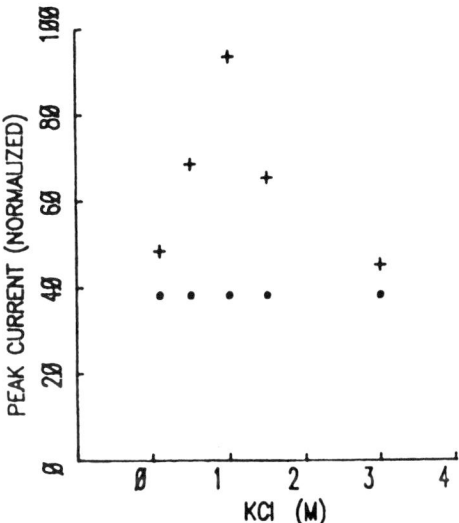

FIG. 25. Dependence of normalized $K_3Fe(CN)_6$ peak current on the KCl concentration at a (+) poly(dimethyldiallylammonium chloride) platinum-coated electrode and (•) unmodified graphite electrode. (From Ref. 147, with permission.)

concentration of KCl up to 1.0 M, above which a decrease is observed. This profile is the net result of two opposing effects: competition between $\overline{Cl}$ and ferri-/ferrocyanide for surface sites and the tendency of polyelectrolytes to shrink with increasing ionic strength (thus forming a more compact film with a greater density of active groups). For various polyelectrolytes, the solution pH often exhibits similar opposing effects between structural (swollen) changes and protonation of ion-exchange moieties [148]. It should be stressed, however, that changes in the response owing to competitive partitioning of nonelectoractive ions do not imply that the determination is impossible. In most cases, well-defined calibration plots are obtained and only the sensitivity is altered. The situation is more serious when the counterionic electroactive interferent undergoes redox reaction at a potential adjacent to that of the analyte. This is, for example, the situation when the cationic neurotransmitter dopamine is measured, in the presence of epinephrine and norepinephrine, at Nafion-coated electrodes [149]. Sufficient peak separation would allow simultaneous measurement of several electroactive counterions, although competitive partitioning should be taken into account in the quantification process. Various preconcentration strategies based on electrostatic binding at ion-exchange modified electrodes are described in the following sections.

a. Polyelectrolyte-Coated Electrodes as
   Preconcentration Surfaces

Use of adherent polyelectrolytes to bind counterionic reactants to electrode surfaces has grown in popularity since this concept was first demonstrated by Oyama and Anson [150]. It is desirable that the polyelectrolyte coating possess large ion-exchange capacity, chemical and mechanical stability with strong adhesion to the electrode surface, good swelling in aqueous solutions, water insolubility, prolonged retention of counterions, and rapid charge propagation rates. The number of useful polyelectrolyte systems presently available is limited because common polyelectrolytes usually lack one or more of the above requirements.

With the cationic polyelectrolyte coating, poly(4-vinylpyridine) (PVP), used in the original work of Oyama and Anson [150], anionic species can be attached to the surface from solutions of low pH (at which the pyridine groups are protonated). The cyclic voltammetric response of the incorporated counterion remains essentially the same on transferring the electrode to a pure supporting electrolyte solution and repeating the cycling for periods up to 2 hr. Similarly to what is commonly observed in homogeneous solutions, the protonated PVP film exhibits preferential incorporation of multiply charged counter ions over singly charged anions. These observations have suggested that PVP-coated electrodes possess an analytically useful potential for trace measurements of anionic species. This was illustrated by Cox and Kulesza [151], who described the determination of trace levels of Cr(VI) following preconcentration at a PVP-coated platinum electrode. Preconcentration conditions included a 0.15-M

NaF solution (pH 3.5) and a potential of +0.9 V (vs. SCE). A defined peak, due to the reduction of the surface-bound Cr(VI) and Cr(III) was observed at +0.2 V (Fig. 26); this peak was proportional to the Cr(VI) concentration over the $10^{-6}-10^{-8}$ M range. A 1000-fold excess of Cr(III) did not affect the response.

Improved analytical performance, including protection against organic surfactants, prolonged retention of counterions, or large ion-exchange capacity, can be achieved by means of composite electrode coatings prepared by mixing PVP with other polymers [152,153]. The analytical potential of these new coatings (in which a polyelectrolyte establishes the electrostatic and chemical environments while a templating polymer defines the overall morphology) seems broad and appealing. An analogous approach, yielding controllable microstructure and improved performance, has been reported using aluminum oxide film impregnated with PVP [154]. Gamma-irradiation cross-linking can also be used to obtain size selectivity and protection [147].

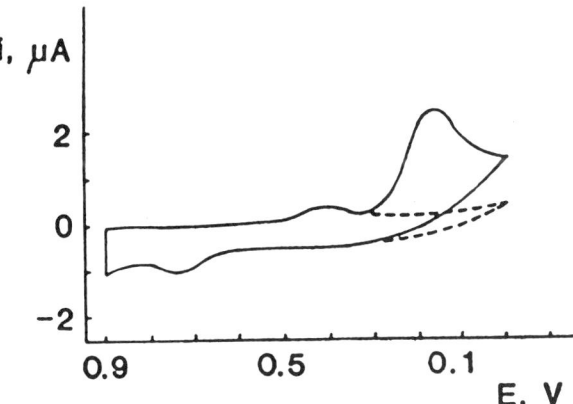

FIG. 26. Cyclic voltammogram for 6×10^{-7} M dichromate at a PVP-coated platinum electrode. Preconcentration for 60 sec at +0.9 V. Dotted line represents the background current. (From Ref. 151, with permission.)

$$-[(CF_2-CF_2)_x-(CF_2-CF)]_y-$$
V
$$O-CF_2-CF-O-CF_2-CF_2-SO_3H$$
$$|$$
$$CF_3$$

Perfluorinated Nafion films (V) have been extensively examined because of their utility in entrapping positively charged analytes [146,155]. Nafion polymer is believed to be composed of a network of interconnected hydrophilic ionic clusters in a bulk of hydrophobic fluorocarbon phase. The resulting coating can extract cationic analytes from highly dilute solutions [146]. For example, Fig. 27 shows voltammograms for Nafion-coated electrodes after equilibration with very dilute ($<5 \times 10^{-8}$ M) solutions of various counterions. The ion-exchange accumulation reaction is given by

$$M^{+n}_{soln} + n(SO_3^-Na^+)_{elec} \rightleftharpoons [(SO_3^-)_nM^{+n}]_{elec} + nNa^+_{soln} \quad (26)$$

Because of its unique structural characteristics, Nafion exhibits tremendous ion-exchange affinity for large hydrophobic cations (e.g., Fig. 27A, C, D) relative to simple inorganic cations. It is even possible to use the hydrophobic effect of Nafion to preconcentrate neutral species, as was illustrated recently with α-cyclodextrin [156]. Although, thermodynamically, Nafion appears to be ideal for preconcentration of hydrophobic ions, the dynamics of these exchange processes is relatively slow. Such slow attainment of equilibria is a result of the low ionic diffusion coefficient of large organic cations in Nafion. Hence, a compromise between analysis time and detection limit has to be reached. Nafion-based composite polymers that can transport ions at much higher rates have

---

FIG. 27. Steady-state cyclic voltammograms for a Nafion-coated electrode after equilibration with (A) 1.51×10^{-8} M Ru(2,2-'bipyridine)$_3^{2+}$; (B) 2.78×10^{-8} M Ru(NH$_3$)$_6^{3+}$; (C) 4.36×10^{-8} M methyl viologen; (D) 3.4×10^{-8} M ferrocenylmethyl trimethylammonium hexafluorophosphate. (From Ref. 146, with permission.)

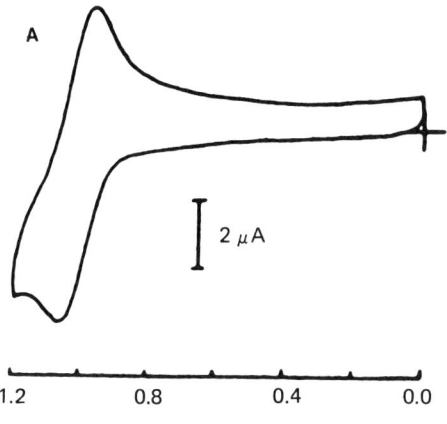
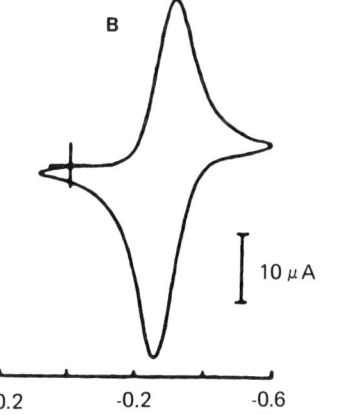
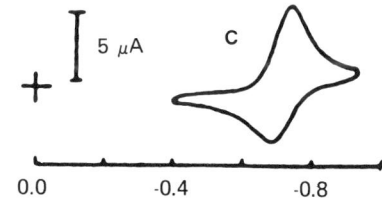
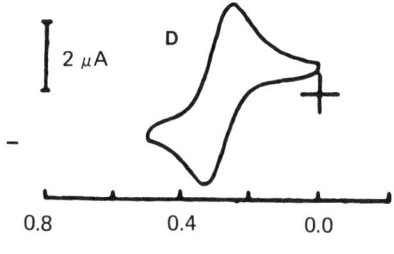

E (V vs. SCE)

been described recently. For example, Nafion-impregnated Gore-Tex membranes produce apparent diffusion coefficients for $Ru(NH_3)_6^{3+}$ that are over an order of magnitude larger than the corresponding diffusion coefficients in Nafion [157]. Similar, and additional, advantages have been reported for Nafion/Nuclepore or Nafion/polypyrrole composite polymeric coatings [158]. Some practical applications of Nafion-coated electrodes include measurements of cationic primary neurotransmitters in brain tissue and fluids, without interference from anionic species (particularly ascorbic acid) [159], or metal speciation studies in aquatic media [160]. A novel bilayer electrode configuration, with a cellulose acetate layer atop the Nafion film (Fig. 28), can improve the disciminative properties of Nafion-based sensors by allowing transport and preconcentration of cationic analytes in the presence of larger cations with

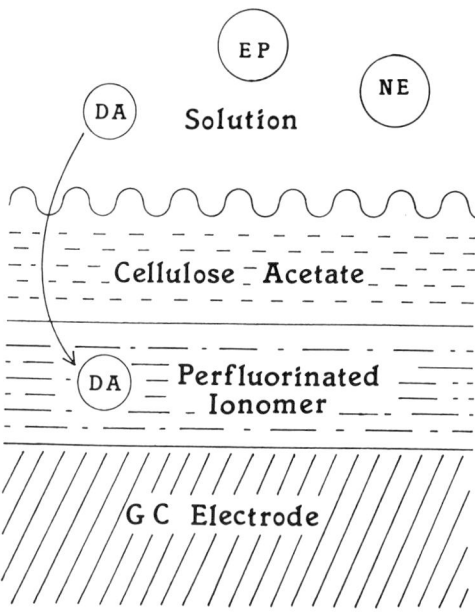

FIG. 28. Schematic description of a glassy carbon electrode modified with a Nafion/cellulose acetate bilayer coating. (From Ref. 149, with permission.)

similar redox potentials, e.g., detection of dopamine in the presence of epinephrine or norepinephrine [149].

Organic electroanalysis can also be accomplished at other ion-exchanger-based CMEs. For example, preconcentration and voltammetric measurement of aromatic amines can be carried out at electrodes modified with a functionalized polymer film that contains styrene sulfonate groups [161]. Good sensitivity and reproducibility have been reported in measurements of diphenylamine. The presence of an aromatic group is essential to ensure large partition coefficients at this electrode. Hence, aliphatic amines do not interfere, even when present in 100-fold concentration excess.

b. Electrostatic Preconcentration at Inorganic Coatings

Although organic polyelectrolytes have been used extensively for electrostatic binding and preconcentration of ionic analytes, inorganic materials may play an important role in future studies. Recent work by Bard, Anson, and their co-workers [162-164] has indicated that modification of electrode surfaces by coverage with a thin layer of a treated clay possesses a great potential for the electrostatic preconcentration of ionic analytes. Clay and other inorganic layers on electrode surfaces have the advantages of high chemical and mechanical stability and special structural (chemical and physical) features. For example, the coating permeability is advantageous in electroanalytical applications where high fluxes of analytes are desirable to facilitate the preconcentration step. Figure 29 illustrates cyclic voltammograms for the incorporation of $Os(2,2\text{-bipyridine})_3^{2+}$ at a graphite electrolyte coated with sodium montmorillonite, including comparison to an unmodified electrode (B) and the response after transfer to a pure supporting electrolyte solution (C). Although most work in this direction has used montmorillonite layers, other types of materials—particularly zeolites—can be used. The pores of zeolites contain counterions that are mobile and may be ion-exchanged. The unique cagelike pore structure may lead to new selectivity dimensions based on size and shape. For example, the preferential uptake and

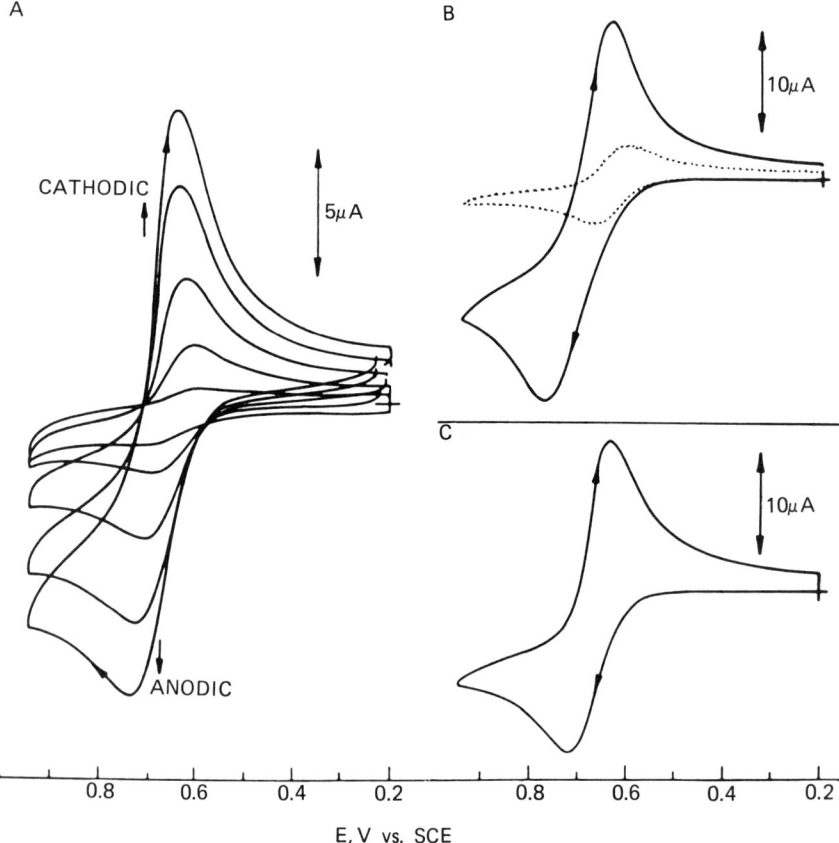

FIG. 29. Cyclic voltammetry of 2×10^{-4} M Os(2,2-bipyridine)$_3^{2+}$ at a montmorillonite-coated graphite electrode. (A) continuous scanning; curves shown were obtained 0 (innermost), 5, 10, 15, and 20 min after immersion. (B) Steady-state voltammogram obtained after 50 min, compared to the response at the uncoated electrode (dotted lines). (C) After transfer of the coated electrode from B to pure supporting electrolyte. (From Ref. 164, with permission.)

voltammetric measurement of silver at zeolite-containing carbon paste electrodes was demonstrated recently [164a]. It is expected, therefore, that inorganic coatings will offer new patterns of reactivity that may be exploited in a variety of analytical (and other) applications.

c. Ion-Exchanger Modified Carbon Paste Electrodes

Several studies have demonstrated the versatility, stability, and ease of use of carbon paste electrodes containing an appropriate ion exchanger for preconcentration/voltammetric measurements. When carbon paste electrodes are used, the exchanger is usually added to the pasting liquid as finely ground resins or as liquids. The strategy of immobilizing an ion exchanger into the carbon paste matrix was introduced by Wang et al. [131]. In this study, an acidic cation-exchange resin was employed for trace measurements of copper(II) ions. The response was shown to increase linearly with preconcentration time, copper concentration, or the amount of resin in the paste. Typical voltammograms obtained for solutions containing increasing levels of copper are shown in Fig. 30. Renewal of the surface was accomplished by immersing the electrode in a hydrochloric acid solution. Loading of counterions onto carbon paste electrodes modified with cation-exchange resins was also described by Kutner et al. [165].

Liquid ion exchangers are preferable to resins for the modification of carbon paste electrodes. Such exchangers can easily be added to the pasting liquid to give a high degree of homogeneity. In addition, the liquid phase exposed many more functional groups to the solution than do particles, resulting in enhanced sensitivity. The use of carbon-paste electrodes modified with liquid ion exchangers has been explored by Kalcher; voltammetric procedures for measuring nitrite [166] and gold [167] at pastes modified with the commercial exchangers tricaprylmethylammonium chloride and Amberlite LA2, respectively, have been reported. The ion-exchange (ion-pair) accumulation process of nitrite can be described by

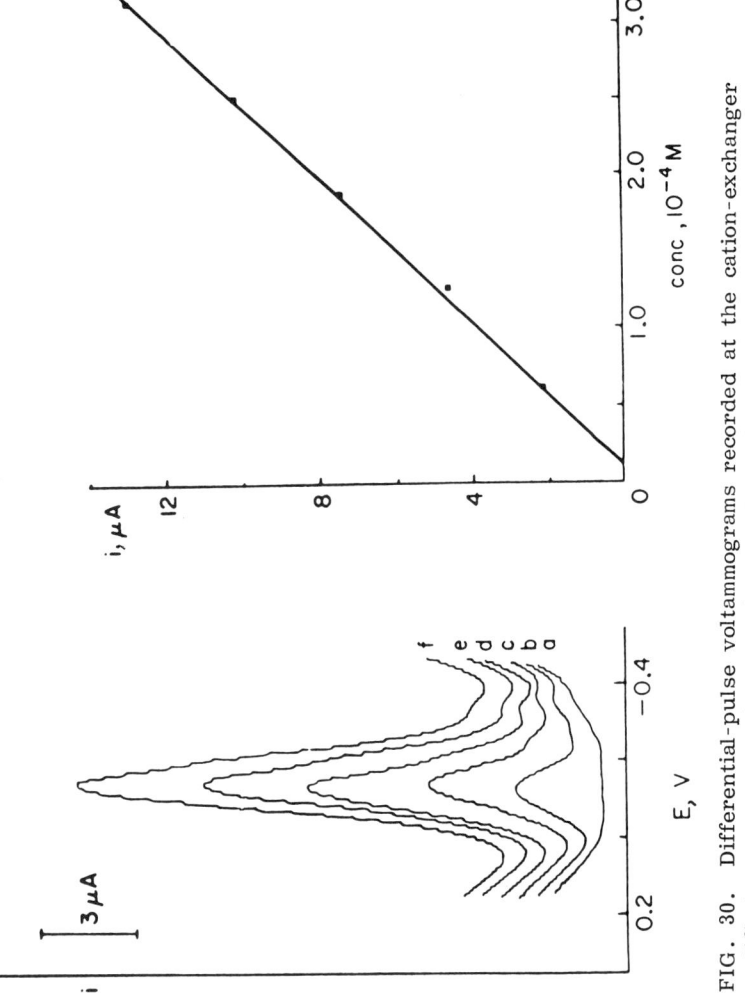

FIG. 30. Differential-pulse voltammograms recorded at the cation-exchanger modified carbon paste electrode for solutions of increasing copper levels (6.25×10^{-5} M steps). Preconcentration for 4 min. (From Ref. 131, with permission.)

$$[R_4N^+Cl^-]_{sur} + [NO_2^-]_{aq} \rightleftharpoons [R_4N^+NO_2^-]_{sur} + [Cl^-]_{aq} \qquad (27)$$

The irreversible oxidation of the accumulated nitrite to nitrate yields a well-defined peak, as indicated from the cyclic voltammogram shown in Fig. 31. A low detection limit (1 µg liter^{-1}) has been reported using the differential-pulse waveform; the peak, however, is prone to serious interference from coexisting halide ions. Polyelectrolytes can also be admixed into the carbon paste matrix and used for preconcentration/voltammetric measurements. For example, the incorporation of poly-(4-vinylpyridine) into the paste matrix can be used for electrostatic binding and voltammetric measurement of ferrocyanide [168]. Bioaccumulation of heavy metals via electrostatic binding to an algae-modified carbon paste electrode is another promising avenue, already yielding very useful analytical data [169].

3. *Preconcentration via Covalent Reactions*

Covalent reactions may offer another promising avenue for selective and sensitive electroanalysis at CMEs. Effective preconcentration of target analytes can be accomplished via their specific reaction with a surface moiety. For example, electrodes modified with primary amines can yield a surface selective to carbonyl-containing compounds; hence, the following condensation-type preconcentration reaction can be employed:

$$\Big|\!-NH_2 + R-\overset{O}{\underset{\|}{C}}-R' \rightleftharpoons \Big|\!-N = C\!\!\begin{array}{c} R \\ R' \end{array} + H_2O \qquad (28)$$

The redox reaction of the resulting imine can be used as the analytical signal and related to the bulk concentration of the carbonyl compound [170]. Alternately, a redox moiety of the preconcentrated analyte can be monitored. Price and Baldwin [171] prepared a suitable CME by adsorption of allylamine onto the surface of a platinum wire. For the model analyte ferrocencarboxaldehyde, a detection limit of 1×10^{-7} M

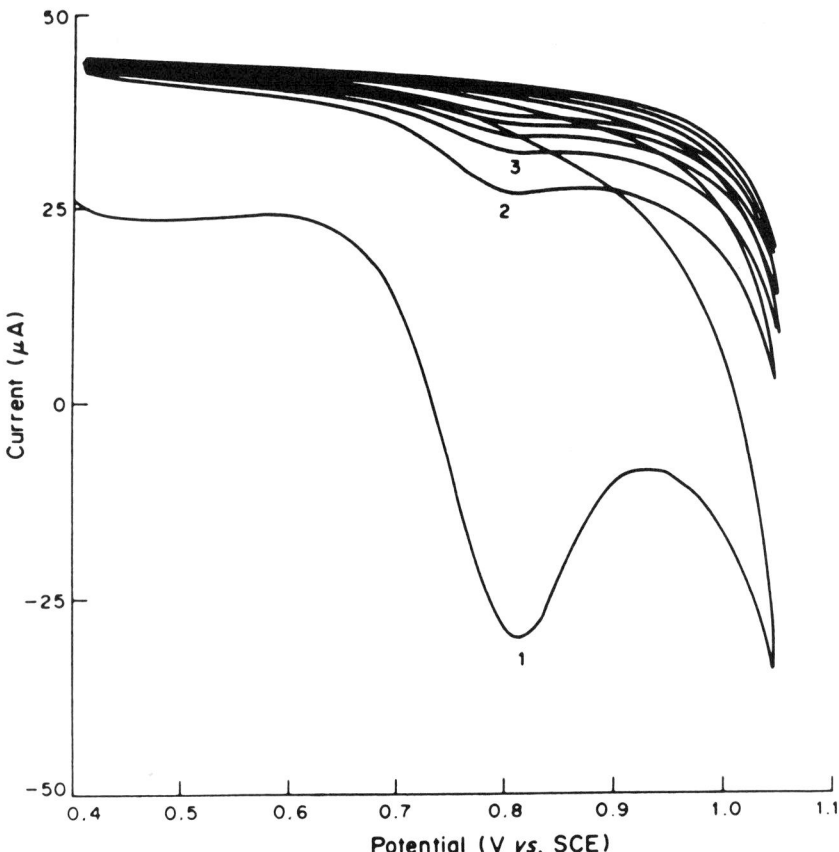

FIG. 31. Repetitive cyclic voltammograms recorded at an anion-exchanger modified carbon paste electrode for 5 mg liter^{-1} NO_2^-, following 1-min preconcentration. (From Ref. 166, with permission.)

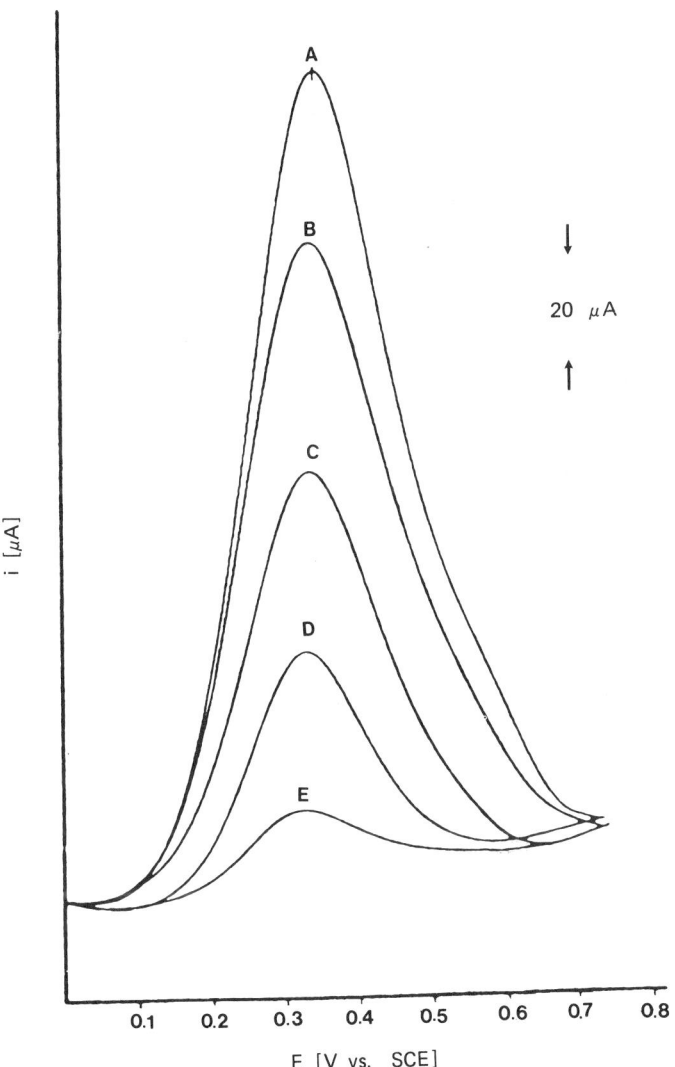

FIG. 32. Pulse voltammograms for solutions of increasing ferrocene-carboxaldehyde concentration, following 5-min accumulation at allylamine modified platinum electrode. Analyte concentration: (A) 4.5×10^{-3} M; (B) 9.0×10^{-4} M; (C) 4.5×10^{-4} M; (D) 9.0×10^{-5} M; (E) 1.1×10^{-5} M. (From Ref. 171, with permission.)

was obtained (5-min accumulation) by monitoring the ferrocene-ferrocenium oxidation reaction in a positive-going differential-pulse scan. Typical voltammograms obtained at different levels of ferrocenecarboxaldehyde are shown in Fig. 32. One can reverse the role of the surface functionality and analyte, in the same imine-formation condensation reaction, for measuring primary amines. For example, Abruna and coworkers [172] demonstrated recently the utility of carbon paste electrodes modified with a polymer functionalized with an aldehyde for the measurement of primary amines. The oxidation of the produced imine was related to the solution concentration of the amine. This scheme permits convenient measurements of submillimolar concentrations, coupled with excellent selectivity (e.g., in the presence of a large excess of dimethylaniline).

The covalent-attachment preconcentration strategy can be easily extended to the measurement of other organic analytes, based on the reactivity of various surface functionalities. Despite its enormous potential and synthetic diversity, research in this direction has progressed very slowly (compared to schemes based on electrostatic or coordinative attachments).

Current work in various laboratories involves the preparation of many new types of preconcentrating surfaces and studies of the fundamental and practical aspects of their behavior. Given the versatile tailor-made appeal of chemically modified electrodes, they will surely become much more widely used as preconcentration/voltammetric-based sensors.

IV. CONCLUSION

Voltammetry following nonelectrolytic preconcentration has greatly improved the application of voltammetric techniques to the determination of numerous analytes (metals and nonmetals) that cannot be preconcentrated by electrolysis. Such appealing coupling of specific chemical reactions (at the surface or in the solution phase) with preconcentration/voltammetric measurements offers high sensitivity, selectivity, versatility, and simplicity. An enormous potential remains for using modified surfaces, in conjunction with the preconcentration/voltammetric concept, as highly sensitive and selective sensing devices. Hence, many new exciting developments and applications lie on the horizon.

ACKNOWLEDGMENTS

Preparation of this chapter was assisted by a grant from the National Institutes of Health (GM 30913-04).

REFERENCES

1. R. E. Sioda, G. E. Batley, W. Lund, J. Wang, and S. C. Leach, Talanta, 33:421 (1986).
2. J. Wang, "Stripping Analysis: Principles, Instrumentation, and Applications," VCH Publishers, Deerfield Beach, FL, 1985.
2a. E. Barendrecht, "Electroanalytical Chemistry," A. J. Bard, Ed., Vol. 1, Marcel Dekker, New York, 1967.
3. F. C. Anson, Acc. Chem. Res. 8:400 (1985).
4. B. B. Damaskin, O. A. Petrii, and V. V. Batrakov, *Adsorption of Organic Compounds on Electrodes*, Plenum, New York, 1971.
5. S. Trasatti, J. Electroanal. Chem. 53:335 (1974).
6. A. N. Frumkin and B. B. Damaskin, *Modern Aspects of Electrochemistry*, Vol. 3, Chap. 3 (1964).
7. E. Laviron, in *Electroanalytical Chemistry*, A. J. Bard, Ed., Vol. 12, Marcel Dekker, New York, 1983.
8. R. W. Murray, in *Electroanalytical Chemistry*, A. J. Bard, Ed., Vol. 13, Marcel Dekker, New York, 1984.
9. E. Laviron, J. Electroanal. Chem. 52:395 (1974).
10. E. Laviron, J. Electroanal. Chem. 101:19 (1979).
11. A. J. Bard and L. R. Faulkner, *Electrochemical Methods: Fundamentals and Applications*, Wiley, New York, 1980.
12. A. P. Brown and F. C. Anson, Anal. Chem. 49:1589 (1977).
13. J. Wang and P. A. M. Farias, J. Electroanal. Chem. 182:211 (1985).
14. T. Kakutani and M. Senda, Bull. Chem. Soc. Jap. 52:3236 (1979).
15. E. Laviron, J. Electroanal. Chem. 100:263 (1979).

16. L. Ramaley, J. A. Palziel, and W. T. Tan, Can. J. Chem. 59: 3334 (1981).
17. A. Webber, M. Shah, and J. Osteryoung, Anal. Chim. Acta 154: 105 (1983).
18. W. J. Blaedel and G. A. Mabbott, Anal. Chem. 53:2270 (1981).
19. M. Lovric, J. Electroanal. Chem. 170:143 (1984).
20. J. Wang and B. A. Freiha, Talanta 30:837 (1983).
21. J. Wang, D. B. Luo, and P. A. M. Farias, J. Electroanal. Chem. 185:61 (1985).
22. B. Pihlar, P. Valenta, and H. W. Nürnberg, J. Electroanal. Chem. 214:157 (1986).
23. M. P. Soriaga and A. T. Hubbard, J. Am. Chem. Soc. 104:2742 (1982).
24. J. Wang and M. S. Lin, J. Electroanal. Chem. 221:257 (1987).
24a. I. F. Hu, D. Karweik, and T. Kuwana, J. Electroanal. Chem. 188: 59 (1985).
25. J. Wang and J. Zadeii, Anal. Chim. Acta 188:187 (1986).
26. V. Stara and M. Kopanica, Anal. Chim. Acta 159:105 (1984).
27. P. Delahay and I. Trachtenberg, J. Am. Chem. Soc. 79:2355 (1957).
28. C. J. Flora and E. Nieboer, Anal. Chem. 52:1013 (1980).
29. G. Weber, Anal. Chim. Acta 186:49 (1986).
30. H. Eskilsson, C. Haraldsson, and D. Jagner, Anal. Chim. Acta 175:79 (1985).
31. V. V. Astafeva, G. V. Prokhorova, and R. M. F. Salikhdzhanova, Zh. Anal. Khim. 32:260 (1976).
32. B. Pihlar, P. Valenta, and H. W. Nürnberg, Z. Anal. Chem. 307: 337 (1981).
33. S. B. Adeloju, A. M. Bond, and M. H. Briggs, Anal. Chim. Acta 164:181 (1984).
34. B. Gammelgaard and J. R. Andersen, Analyst 110:1197 (1985).
35. L. Vos, Z. Komy, G. Reggers, E. Roekens, and R. Van Grieken, Anal. Chim. Acta 184:271 (1986).
36. K. Torrance and C. Gatford, Talanta 32:273 (1985).
37. H. Braun and M. Metzgar, Z. Anal. Chem. 318:321 (1984).
38. F. Wahdat and R. Neeb, Z. Anal. Chem. 320:334 (1985).
39. J. Wang and K. Varughese, Anal. Chim. Acta, 199:185 (1987).

40. H. H. Willard and J. A. Dean, Anal. Chem. 22:1264 (1950).
41. G. W. Latimer, Talanta 15:1 (1968).
42. J. Wang and J. S. Mahmoud, J. Electroanal. Chem. 208:383 (1986).
43. J. Wang, P. A. Farias, and J. S. Mahmoud, Anal. Chim. Acta 172:57 (1985).
44. J. Wang and J. Zadeii, Anal. Chim. Acta 185:229 (1986).
45. J. Wang and J. Zadeii, Anal. Chim. Acta 188:187 (1987).
46. J. Wang and J. Zadeii, Talanta 34:247 (1987).
47. J. Wang and J. S. Mahmoud, Z. Anal. Chem. 327:789 (1987).
48. J. Wang, P. Tuzhi, and K. Varughese, Talanta 34:561 (1987).
49. J. Wang and J. S. Mahmoud, Anal. Chim. Acta 182:147 (1986).
50. J. Wang and J. Zadeii, Talanta 33:321 (1986).
51. T. M. Florence and W. L. Belew, J. Electroanal. Chem. 21:157 (1967).
52. J. Wang, J. S. Mahmoud, and J. M. Zadeii, unpublished work.
53. C. M. G. van den Berg, Anal. Chem. 56:2383 (1984).
54. C. M. G. van den Berg, Anal. Chim. Acta 164:195 (1984).
55. C. M. G. van den Berg, J. Electroanal. Chem. 177:269 (1984).
56. C. M. G. van den Berg, Anal. Chem. 57:1532 (1985).
57. C. M. G. van den Berg and M. Nimmo, Anal. Chem. 59:269 (1987).
58. J. Golimowski, P. Valenta, and H. W. Nürnberg, Z. Anal. Chem. 322:315 (1985).
59. C. M. G. van den Berg, K. Murphy, and J. P. Riley, Anal. Chim. Acta 188:177 (1986).
60. J. Wang, P. A. M. Farias, and J. S. Mahmoud, Anal. Chim. Acta 171:215 (1985).
61. J. Wang and J. Zadeii, Talanta, 34:909 (1987).
62. C. M. G. van den Berg, Talanta 31:1069 (1984).
63. J. Wang, J. Zadeii, and M. S. Lin, J. Electroanal. Chem., 237:281 (1987).
64. M. Friedrich and H. Ruf, J. Electroanal. Chem. 198:261 (1986).
65. C. M. G. van den Berg, Anal. Chim. Acta 164:209 (1984).
66. T. M. Florence, Analyst 111:489 (1986).
67. C. M. G. van den Berg, Mar. Chem. 15:1 (1984).

68. A. M. Bond, J. Electroanal. Chem. 214:21 (1986).
69. J. Wang, P. Tuzhi, and T. Martinez, Anal. Chim. Acta, 201:43 (1987).
70. A. P. Brown, C. Koval, and F. C. Anson, J. Electroanal. Chem. 72:379 (1976).
71. J. Wang, M. Bonakdar, and C. Morgan, Anal. Chem. 58:1024 (1986).
72. M. P. Soriaga, P. H. Wilson, A. T. Hubbard, and C. S. Benton, J. Electroanal. Chem. 142:317 (1982).
73. M. P. Soriaga, V. K. F. Chia, J. H. White, D. Song, and H. T. Hubbard, J. Electroanal. Chem. 162:143 (1984).
74. R. Kalvoda, Anal. Chim. Acta 162:197 (1984).
75. J. Wang, D. B. Luo, P. A. M. Farias, and J. S. Mahmoud, Anal. Chem. 57:158 (1985).
76. A. G. Fogg and J. M. Lewis, Analyst 111:1443 (1986).
77. R. Kalvoda, Anal. Chim. Acta 138:11 (1982).
78. J. Wang, J. S. Mahmoud, and P. A. M. Farias, Analyst 110:855 (1985).
79. J. Wang, T. Tapia, and M. Bonakdar, Analyst 111:1245 (1986).
80. J. Wang, P. A. M. Farias, and J. S. Mahmoud, Analyst 111:837 (1986).
81. J. Wang, M. S. Lin, and V. Villa, Anal. Lett. 19:2293 (1986).
82. J. Wang, P. Tuzhi, M. S. Lin, and T. Tapia, Talanta 33:707 (1986).
83. E. N. Chaney and R. P. Baldwin, Anal. Chem. 54:2556 (1982).
84. J. Wang, M. S. Lin, and V. Villa, Analyst 112:1303 (1987).
85. J. Wang, T. Peng, and M. S. Lin, Bioelectrochem. Bioenerg. 16:395 (1986).
86. J. Wang, M. S. Lin, and V. Villa, Analyst 112:247 (1987).
87. J. Wang and M. S. Lin, submitted.
88. J. Wang, T. Peng, and M. S. Lin, Bioelectrochem. Bioenerg. 15:147 (1986).
89. J. Wang and J. S. Mahmoud, Anal. Chim. Acta 186:31 (1986).
90. T. B. Jarbawi and W. R. Heineman, Anal. Chim. Acta 186:11 (1986).
91. J. Wang, B. A. Freiha, and B. K. Deshmukh, Bioelectrochem. Bioenerg. 14:457 (1985).

92. D. B. Luo, Anal. Chim. Acta 189:277 (1986).
93. J. Wang, P. A. M. Farias, and J. S. Mahmoud, Anal. Chim. Acta 171:195 (1985).
94. N. K. Lam and M. Kopanica, Anal. Chim. Acta 161:315 (1984).
95. H. Benadikova and R. Kalvoda, Anal. Lett. 17:1519 (1984).
96. A. G. Fogg, A. A. Barros, and J. O. Cabral, Analyst 111:831 (1986).
97. J. Wang and B. A. Freiha, Anal. Chim. Acta 154:87 (1983).
98. H. Y. Cheng, L. Falat, and R. L. Li, Anal. Chem. 54:1384 (1982).
98a. M. T. Stankovich and A. J. Bard, J. Electroanal. Chem. 85:173 (1977).
98b. M. T. Stankovich and A. J. Bard, J. Electroanal. Chem. 86:189 (1978).
99. E. Palecek, P. Boublikova, and F. Jelen, Anal. Chim. Acta 187: 99 (1986).
100. E. Palecek and I. Postbieglova, J. Electroanal. Chem. 214:359 (1986).
101. E. Palecek and M. A. Hung, Anal. Biochem. 132:236 (1983).
102. P. Boublikova, M. Vojtiskova, and E. Palecek, Anal. Lett. 20:275 (1987).
103. J. M. Sequaris, P. Valenta, and H. W. Nürnberg, Int. J. Radiat. Biol. 42:407 (1982).
103a. J. Wang and V. Villa, unpublished work.
104. J. Wang, V. Villa, and T. Tapia, Bioelectrochem. Bioenerg., in press.
105. J. Wang and V. Villa, unpublished work.
106. J. R. Flores and M. R. Smyth, J. Electroanal. Chem., 235:317 (1987).
106a. J. Wang, V. Villa, and A. Bernsteiner, unpublished work.
107. C. A. Chambers and J. K. Lee, J. Electroanal. Chem. 14:309 (1967).
108. J. Wang and B. A. Freiha, Anal. Chem. 56:849 (1984).
108a. T. Kuwana and W. G. French, Anal. Chem. 36:38 (1964).
109. J. Wang, B. K. Deshmukh, and M. Bonakdar, J. Electroanal. Chem. 194:339 (1985).
110. T. B. Jarbawi and W. R. Heineman, Anal. Chim. Acta 135:359 (1982).

111. H. Jehring, *Elektrosorptionanalyse mit der Wecheselstrompolarographie*, Akademi Verlag, Berlin, 1974.

112. E. Bednarkiewicz and Z. Kublik, Anal. Chim. Acta *176*:133 (1985).

113. H. Batycka and Z. Lukaszewski, Anal. Chim. Acta *162*:215 (1984).

114. H. Batycka and Z. Lukaszewski, Anal. Chim. Acta *162*:207 (1984).

115. B. B. Damaskin and G. A. Tedoradze, Electrochim. Acta *10*:529 (1965).

116. R. Kalvoda and L. Novotny, Collection Czechoslavak Chem. Commun. *51*:1587 (1986).

117. R. Kalvoda, J. Electroanal. Chem. *180*:307 (1984).

118. S. G. Mairanovskii, *Catalytic and Kinetic Waves in Polarography*, Plenum, New York, 1968.

118a. L. Hernandez, A. Zapardiel, J. A. P. Lopez, and E. Bermejo, Analyst *112*:1149 (1987).

119. J. Wang, M. Bonadkar, and M. Pack, Anal. Chim. Acta. *192*:215 (1987).

120. B. Pihlar, B. Gorenc, and D. Petric, Anal. Chim. Acta *189*:229 (1986).

121. J. Wang and B. A. Freiha, Anal. Chim. Acta *148*:79 (1983).

122. J. Wang and B. A. Freiha, J. Electroanal. Chem. *151*:273 (1983).

123. J. Wang and B. A. Freiha, Anal. Chem. *55*:1285 (1983).

124. E. N. Chaney and R. P. Baldwin, Anal. Chim. Acta *176*:105 (1985).

125. M. Kopanica and V. Stara, J. Electroanal. Chem. *214*:115 (1986).

126. L. R. Faulkner, Chem. Eng. News, February *27*:28 (1984).

127. R. W. Murray, A. G. Ewing, and R. A. Durst, Anal. Chem. *59*:379A (1987).

128. L. M. Wier, A. R. Guadalupe, and H. D. Abruna, Anal. Chem. *57*:2009 (1985).

129. M. W. Espenschied and C. R. Martin, J. Electroanal. Chem. *188*:73 (1985).

130. R. P. Baldwin, J. K. Christensen, and L. Kryger, Anal. Chem. *58*:1790 (1986).

131. J. Wang, B. Greene, and C. Morgan, Anal. Chim. Acta *158*:15 (1984).

132. A. R. Guadalupe and H. D. Abruna, Anal. Chem. *57*:142 (1985).

133. A. R. Guadalupe, L. M. Wier, and H. D. Abruna, Am. Lab. 18(8):102 (1986).
134. D. M. T. O'Riordan and G. G. Wallace, Anal. Chem. 58:128 (1986).
135. K. H. Lubert, M. Schnurrbusch, and A. Thomas, Anal. Chim. Acta 144:123 (1982).
136. K. Izutsu, T. Nakamura, R. Takizawa, and H. Hanawa, Anal. Chim. Acta 149:147 (1983).
137. K. Izutsu, T. Nakamura, and T. Ando, Anal. Chim. Acta 152:285 (1983).
138. K. H. Lubert and M. Schnurrbusch, Anal. Chim. Acta 186:57 (1986).
139. J. Lexa and K. Stulik, Talanta 32:1027 (1985).
140. S. V. Prabhu, R. P. Baldwin, and L. Kryger, Anal. Chem. 59:1074 (1987).
141. K. Kalcher, Z. Anal. Chem. 325:181 (1986).
142. J. Wang and P. Tuzhi, unpublished results, 1987.
143. G. T. Cheek and R. F. Nelson, Anal. Lett. 11:393 (1978).
144. J. Wang and M. Bonakdar, Talanta 35:277 (1988).
145. J. Schneider and R. W. Murray, Anal. Chem. 54:1508 (1982).
146. M. N. Szentirmay and C. R. Martin, Anal. Chem. 56:1898 (1984).
146a. L. D. Whiteley and C. R. Martin, Anal. Chem. 59:1746 (1987).
147. E. S. De Castro, E. W. Huber, D. Villarroel, C. Galiatsatos, J. E. Mark, W. R. Heineman, and P. T. Murray, Anal. Chem. 59:134 (1987).
148. M. J. Gehron and A. Brajter-Toth, Anal. Chem. 58:1488 (1986).
149. J. Wang and P. Tuzhi, Anal. Chem. 58:3257 (1986).
150. N. Oyama and F. C. Anson, J. Electrochem. Soc. 127:247 (1980).
151. J. A. Cox and P. J. Kulesza, Anal. Chim. Acta 154:71 (1983).
152. J. Wang and P. Tuzhi, J. Electrochem. Soc. 134:589 (1987).
153. D. D. Montgomery and F. Anson, J. Am. Chem. Soc. 107:3431 (1985).
154. C. J. Miller and M. Majda, J. Am. Chem. Soc. 107:1419 (1985).
155. C. R. Martin, I. Rubinstein, and A. J. Bard, J. Am. Chem. Soc. 104:4817 (1982).

156. T. Matsue, V. Akiba, and T. Osa, Anal. Chem. 58:2096 (1986).
157. R. M. Penner and C. R. Martin, J. Electrochem. Soc. 132:514 (1985).
158. F. R. F. Fan and A. J. Bard, J. Electrochem. Soc. 133:301 (1986).
159. G. A. Gerhardt, A. F. Oke, F. Nagy, B. Moghaddam, and R. N. Adams, Brain Res. 290:390 (1984).
160. R. D. Guy and S. Namaratne, Can. J. Chem., in press.
161. A. R. Guadalupe and H. D. Abruna, Anal. Lett. 19:1613 (1986).
162. P. K. Ghosh and A. J. Bard, J. Am. Chem. Soc. 105:5691 (1983).
163. D. Ege, P. K. Ghosh, J. R. White, J. F. Equey, and A. J. Bard, J. Am. Chem. Soc. 107:5644 (1985).
164. H. Y. Liu and F. C. Anson, J. Electroanal. Chem. 184:411 (1985).
164a. J. Wang and T. Martinez, Anal. Chim. Acta, in press.
165. W. Kutner, T. J. Meyer, and R. W. Murray, J. Electroanal. Chem. 195:375 (1985).
166. K. Kalcher, Talanta 33:489 (1986).
167. K. Kalcher, Anal. Chim. Acta 177:175 (1985).
168. P. W. Geno, K. Ravichandran, and R. P. Baldwin, J. Electroanal. Chem. 183:155 (1985).
169. J. Gardea, D. Darnall, and J. Wang, Anal. Chem., 60:72 (1988).
170. R. P. Baldwin, J. F. Price, K. Ravichandran, and D. L. Packett, 1981 LCEC Symposium, Indianapolis, IN, May 1981, Paper No. 8.
171. J. F. Price and R. D. Baldwin, Anal. Chem. 52:1940 (1980).
172. A. R. Guadalupe, S. S. Jhaveri, K. E. Liu, and H. D. Abruna, Anal. Chem. 59:2346 (1987).

Hydrodynamic Voltammetry

in Continuous-Flow Analysis

Hari Gunasingham

*National University of Singapore
Kent Ridge, Singapore*

Bernard Fleet

*Scada Systems Inc.
Rexdale, Ontario, Canada*

I. Introduction 90

II. Background 93

    A. Theoretical Considerations 93
    B. Wall-Jet and Thin-Layer Detectors 98
    C. Practical Considerations 103

III. Electrode Materials 112

    A. Mercury Electrodes 112
    B. Metal Electrodes 115
    C. Carbon Electrodes 117

IV. Analytical Techniques and Applications 120

    A. Flow-Injection Electrochemical Detection 120
    B. Stripping Voltammetry 121
    C. Liquid Chromatography-Electrochemical Detection 127
    D. Pulse and Potential-Scan Techniques 140

V. Indirect Methods 148

    A. Photoexcitation 148
    B. Use of an Electroactive Intermediary 149
    C. Indirect Electrode Reactions 152
    D. Chemical Derivatization 153

VI. New Electrode Systems   154
   A. Ultramicroelectrodes in Continuous-
      Flow Analysis   154
   B. Chemically Modified Electrodes   156
   C. Enzyme Electrodes   160

VII. On-Line Monitoring Applications   162
   A. Process Monitoring   162
   B. Patient Monitoring   164
   C. Computer Control and Automation   166

   Glossary of Abbreviations   167

   References   167

I. INTRODUCTION

Hydrodynamic voltammetry (HDV) defines the use of voltammetric techniques, including potential-scan techniques and amperometric measurements at a fixed potential, where the flux of electroactive analyte to the working electrode is maintained under well-developed convection diffusion conditions. The most common routes to establishing these hydrodynamic conditions are by stirring the solution or, in the case of discrete samples, by rotating the electrode or by a continuous flow of electrolyte for dynamic analyses.

Hydrodynamic voltammetry followed close on the heels of the earliest developments in electroanalytical chemistry. In 1947, Muller described a platinum electrode-based flow-through cell design [1]. And, in the following two decades, several works on the use of mercury electrode-based detectors were reported [2]. Practical demonstrations of continuous-flow HDV were few until the mid-1970s, when the situation changed dramatically as a result of a number of developments. The most important of these were the techniques of liquid chromatography-electrochemical detection (LC-EC) and flow injection-electrochemical detection (FI-EC).

This chapter deals with the theory and applications of continuous-flow HDV in flow-through detectors. Section II gives an overview of the theoretical framework for deriving the current response of flow-through electrochemical detectors under steady-state as well as FI and LC conditions. Special attention is given to the wall-jet (WJ) and thin-layer (TL) detectors that appear to have become the most widely used flow-through cell designs. A critical assessment is made of the practical requirements for detector design, including a comparison of the WJ and TL detectors.

The major problem of voltammetry that has hindered its practical development remains the working electrode. In the selection of a suitable electrode material, long-term stability and high electrocatalytic activity appear to be contradictory goals. Section III reviews various electrode materials that have been employed in continuous-flow HDV. Most modern work supports the conclusion that glassy carbon is the most useful electrode material. Current work has been directed at developing procedures for electrode conditioning as a means of improving performance. Whereas conditioning can be used to bring about dramatic improvements, the result is often unpredictable and, interestingly, dependent on the analyte being determined.

Although continuous-flow techniques are subject to the limitations of electrode performance, because the analytical response is continuously monitored in a well-defined hydrodynamic system, it is possible to obviate some of the problems. For example, changes in the electrode response can often be compensated for by periodic calibration with a standard. Also, data-handling techniques, such as background subtraction, can be used to good effect.

Section IV is concerned with the practical applications of LC-EC and FI-EC. These techniques enable analysis of extremely small and discrete sample volumes (10–1000 µl) with high sample throughput (typically, 1–10 samples/sec). Coupled with the cost-effectiveness of voltammetric instrumentation, LC-EC and FI-EC have become attractive alternatives to spectrometric methods for routine usage in analytical laboratories. Applications are widespread in clinical analysis, environmental monitoring, and food analysis, among others.

In recent years, work has been directed at extending the basic techniques of LC-EC and FI-EC. Improvement of the analytical information content can be achieved by employing dual-electrode or multiple electrode schemes and by using pulsed waveforms and potential-scan techniques. These are also reviewed.

The use of indirect techniques to measure electroinactive analyte is the subject of Section V. Two general approaches have been taken. The first is to convert the analyte into an electroactive one. This conversion may be effected by photoexcitation or chemical derivatization. The second approach is to use an electroactive mediator that can interact with the analyte.

The development of new electrode systems is generating considerable interest. Section VI examines current work, which includes chemically modified electrodes, microelectrodes, and enzyme electrodes.

The prospect now exists for synthesizing tailor-made organic microstructures at the working electrode surface to respond selectively and catalytically to the specific analyte of interest while ensuring long-term stability and reproducibility. Of particular interest are the development of novel sensing mechanisms and the use of microfabrication techniques derived from semiconductor technology.

Electrode miniaturization is attracting widespread attention. The last three years have seen a surge in activity in the development of microelectrodes less than 10 μ in diameter used singly or in arrays. A number of practical advantages arise with these electrode systems, the most significant of which is the avoidance of solution resistance effects and the ability to do very fast potential scans.

Section VI also reviews the theory and application of enzyme electrodes. Although these devices have been around for some time, their systematic use in continuous-flow monitoring is quite recent. Apart from enzyme electrodes, there are tremendous prospects for a wide range of biosensors that seek to couple specific biochemical reactions, other than enzymatic reactions, with electrochemical sensing systems.

Use of HDV techniques in on-line monitoring of chemical processes and systems and in biomedical monitoring is going to be an area of immense practical importance. Section VII examines the potential applications. In biomedical monitoring, particular interest is centered around the development of on-line patient monitoring, which includes extracorporeal and implantable systems.

With increasing demands for automation, use of on-line computer control is becoming more and more important, enabling subtle approaches to automation that could not otherwise be realized. Not only can the computer be programmed to control the voltammetric cell and data handling, it can also be used to control precisely the timing and synchronization of sample handling, as well as the solution flow parameters. The advent of the microprocessor has made possible dedicated computer control, where previously it would have been ruled out on grounds of cost. Whereas the use of computer control is mentioned throughout the chapter, Section VII specifically examines its implications for on-line process monitoring.

The integration of microvoltammetric electrodes with the associated control electronics appears to be an inevitable step. The concept of the intelligent sensor, in which microprocessor-controlled electronics is combined with the actual electrode system on a single silicon chip, is fast becoming a reality. Developments in this area are also briefly examined in Section VII.

II. BACKGROUND
A. Theoretical Considerations

One of the first approaches to the treatment of mass transfer in electrode processes was that of Nernst [3]. Nernst postulated the existence of a motionless thin layer of solution adjacent to the electrode and also predicted a linear concentration gradient of the electroactive species in this layer. Nernst's theory, which introduces the important concept of the diffusion layer, is now recognized as an oversimplification in the case of both stirred and unstirred solutions.

In more recent work, Nernst's approximate approach has been replaced by more rigorous treatments, which take into account the hydrodynamic characteristics of the flowing solution. Equations that have been thus deduced relate the diffusion-layer characteristics to the hydrodynamic parameters.

A wide variety of designs and cell geometries have been used in continuous-flow applications. In general, the theory developed for the rotating disk electrode (RDE) by Levich and co-workers [4] has found application in the treatment of convective diffusion currents arising in flow-through electrochemical detectors. However, a complete solution has been obtained for only a limited number of cell geometries. The mass transfer of an electroactive species by convective diffusion may be described by the general equation in cartesian coordinates.

$$\frac{\partial c}{\partial t} = D\left(\frac{\partial c}{\partial x} + \frac{\partial c}{\partial y} + \frac{\partial c}{\partial z}\right) - \left(V_x \frac{\partial c}{\partial x} + V_y \frac{\partial c}{\partial y} + V_z \frac{\partial c}{\partial z}\right) \tag{1}$$

The first group of terms on the right-hand side arise from diffusive mass transfer and the second group of terms from convective mass-transfer processes. The hydrodynamic characteristics of the flowing solution are embodied in the velocity distribution functions V_x, V_y, and V_z, and these are often quite complex. Equation (1) ignores the influence of migration, which is negligible if the electroactive species contribute negligibly to the ionic strength.

Solutions to Eq. (1) may be arrived at by setting appropriate initial and boundary conditions, depending on the particular electrode geometry. However, the mathematical treatment is usually quite difficult.

Analogy Between Mass and Momentum Transfer

Another route to solving the problem of convective mass transfer is through the analogy between mass transfer and momentum transfer. This approach, described by Levich [4], can be applied to obtain the limiting-current equation for HDV systems. Whereas the flux of mass is given by the equation

$$j = D \frac{\partial c}{\partial y} \tag{2}$$

the flux of momentum is given by

$$\tau = \mu \frac{\partial V_x}{\partial y} \tag{3}$$

where y is the direction normal to the electrode, τ = skin friction, and μ = kinematic viscosity.

In terms of mass transfer, the equation for the limiting current is given by

$$i_{lim} = nFj_{max} \tag{4}$$

Matsuda [5] derived an analogous expression for the limiting current for momentum transfer for axisymmetrical flow past a body given by

$$I = knFC^0 D^{2/3} \mu^{-1/3} \left(\int_{X_0}^{X_1} [r(x)^3 \tau(x)]^{1/2} \, dx \right)^{2/3} \tag{5}$$

where $\tau(x)$ = skin friction, μ = solution viscosity, $r(x)$ is the distance from the axis of the body, x is the direction parallel to the surface of the body, and k is a numerical constant. The other terms have their usual meanings.

Rigorous solutions for the limiting diffusion current have been derived for a number of electrode geometries. These include the conical electrode [6], planar electrode with solution flowing parallel over the surface with zero incidence [7], tubular electrode [8], flat plate perpendicular to solution flow [9], wall-jet electrode [10–12], and impinging-jet electrode [13,14]. Table 1 sets out the limiting-current equations under steady-state conditions for these geometries, and Fig. 1 shows the different electrode configurations.

TABLE 1

Limiting Current Equations for Various Electrode Geometries (Adapted from Ref. 22)

Electrode geometry	Limiting current equation
Tubular	$i = 1.61\ nFC(DA/r)^{2/3}V^{1/3}$
Planar (parallel flow)	$i = 0.68\ nFCD^{2/3}v^{-1/6}(A/b)^{1/2}V^{1/2}$
TLD	$i = 1.47\ nFC(DA/b)^{2/3}V^{1/3}$
Planar (perpendicular flow)	$i = 0.903\ nFCD^{2/3}v^{-1/6}A^{3/4}u^{1/2}$
WJD	$i = 0.898\ nFCD^{2/3}v^{-5/12}a^{-1/2}A^{3/8}V^{3/4}$

Definition of terms: a = diameter of inlet, A = electrode area, b = channel height, C = concentration in mM/liter, F = Faraday constant, D = diffusion coefficient, v = kinematic viscosity, r = radius of tubular electrode, V = average volume flow rate, u = velocity, cm/sec, n = number of electrons.

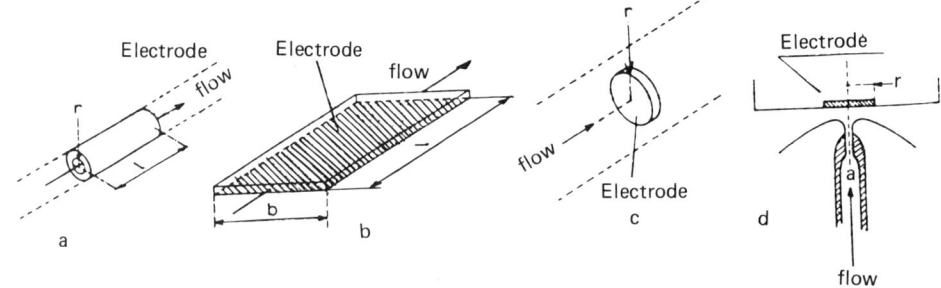

FIG. 1. Electrode geometries for continuous-flow HDV: (a) tubular, (b) planar with parallel flow, (c) planar with perpendicular flow, (d) wall-jet. (From Ref. 28, with permission of Elsevier Scientific Publishing.)

HYDRODYNAMIC VOLTAMMETRY IN CONTINUOUS-FLOW ANALYSIS 97

Diffusion Layer and Hydrodynamic Boundary Layer

When a fluid flows over a surface, a very thin layer is formed at the surface in which the velocity gradient normal to the surface is very large. This thin layer, termed the boundary layer, is not, however, strictly defined. For example, for parallel laminar flow over a flat plate with zero incidence, the limit of the boundary layer is taken as the point at which the velocity approaches 90% of the velocity of the mainstream [4]. Gunasingham and Fleet have similarly defined the boundary layer formed for a wall jet as the layer where most of the radial velocity profile is found [12]. Knowledge of the boundary-layer thickness is important in optimizing detector performance, as is shown in the discussion on the wall-jet and thin-layer detectors.

The diffusion layer immediately adjacent to the electrode, which defines the concentration gradient during the electrode process, is considered to be stationary. The thickness of the diffusion layer is a few percent of the boundary layer. Levich [4] has described an approximate relationship given by

$$\delta_{dl} = \left(\frac{D}{\nu}\right)^{1/3} \delta_{bl} \tag{6}$$

Turbulent Flow

The above treatment of continuous-flow HDV requires the assumption that the flow is laminar. It does not apply to turbulent flow, where the motion of the solution is best described as chaotic with only an average direction. The transition from laminar to turbulent flow is usually characterized by the dimensionless Reynolds number Re, which is the ratio of inertial to viscous forces characteristic of the electrode geometry under consideration. The Reynolds number is given by

$$Re = Ul/\nu \tag{7}$$

where U is the characteristic velocity, l the characteristic length, and ν the kinematic viscosity.

For a given cell geometry, the onset of turbulent flow is marked by a critical Reynolds number, R_{Cr}. The solution flow remains laminar as long as $Re < Re_{Cr}$. For the rotating disk electrode (RDE), R_{Cr} is about 10^4 [4] whereas, for the WJ, it is about 10^3 [12].

Theory of FI-EC and LC-EC

In FI-EC applications, the analyte is injected into the detector as a discrete sample. The flux of analyte to the electrode is thus a function of the concentration profile of the sample as it is swept through the detector by the carrier stream. Meschi et al., however, have shown that under amperometric conditions, at a constant applied potential, the assumption that the analyte flux is a steady-state function can be used to describe the current response [14]. Also, the total charge obtained by integrating the current profile is found to be independent of the sample dispersion.

In the case of potential-scan techniques under FI conditions, if the time scale of the potential change is significantly faster than the concentration change of the sample passing through the cell, the concentration can be assumed to be a steady state for the duration of the scan.

The direct use of steady-state theory to describe the current response in LC-EC without taking into account the influence of chromatographic conditions is, however, inappropriate. Prabhu and Anderson have described a modified approach that couples the Van Deemter model for chromatographic dispersion with the assumption of a steady-state response of the detector [15]. A good correlation between theory and experiment was found. An interesting outcome of this work was the finding that optimum conditions, where the detector response is independent of flow rate, can be predicted.

B. Wall-Jet and Thin-Layer Detectors

The two most widely used electrode geometries in practical applications appear to be detectors based on the wall-jet and thin-layer configurations.

The principle of the wall jet (WJ) was defined by Glauert to describe the flow of a jet of fluid that strikes a wall perpendicularly and flows radially over its surface [16]. A rigorous solution for the limiting-current equation has been obtained by Albery and Brett, who used the correlations derived by Glauert to solve the three-dimensional convective diffusion equation [11]. Yamada and Matsuda earlier obtained a solution through the skin-friction concept [10]. Also, Gunasingham and Fleet have derived the equation for the boundary layer and diffusion layer that was used to obtain the limiting-current equation [12]. The limiting-current equation for the wall-jet electrode (WJE) is given in Table 1. The validity of the equation has been adequately confirmed [10-12].

An interesting characteristic of the velocity component normal to the electrode is that, near the electrode, the flow is away from the electrode whereas, away from the electrode, the flow is toward it [11]. There is consequently a region in between where the flow is stationary. The practical implication of this is that only fresh species from the jet can reach the electrode, whereas species from the bulk solution are excluded. This has a number of practical advantages in continuous-flow monitoring applications such as anodic stripping voltammetry (ASV) [17,18]. Figure 2 shows the flow velocity profiles at the WJE.

Albery has obtained analytical solutions for the current distribution at a WJE [19]. Unlike the rotating disk electrode, the WJE is not uniformly accessible. Although this is a disadvantage for kinetic studies, Albery has shown that it is feasible to obtain kinetic data from modified Tafel plots [19]. Application of a ring-disk WJE for collection experiments similar to those previously done at a rotating ring-disk electrode (RRDE) has been described [11]. The collection efficiency was shown to be quite significant.

Gunasingham and Fleet have defined the critical requirements for the design of a flow-through cell that enable true wall-jet behavior [12]: First, although the limiting-current equation does not include a term to express the separation between the inlet nozzle and disk electrode, the

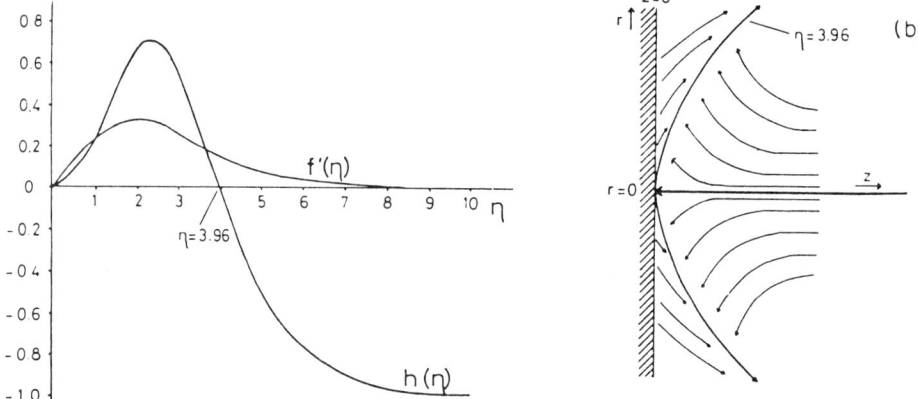

FIG. 2. Velocity flow profiles at the wall-jet electrode: $f'(\eta)$ = radial velocity profile, and $h(\eta)$ = velocity profile normal to electrode surface. Stagnant region occurs at η = 3.96. r = electrode radius, and z is direction normal to electrode. (From Ref. 11, with permission of Elsevier Scientific Publishing.)

assumption is made that the jet of liquid issuing from the inlet does not break up before it impinges on the electrode disk. In general, the jet has been shown to remain intact for up to 10 mm from the inlet for a Reynolds number of 70. However, at a higher Reynolds number, the jet can break up and become turbulent much closer to the inlet. Flow characteristics in the turbulent regime have been studied by Varadi et al. [20,21]. Second, there should be no interference with the development of the boundary layer [12,22]. The practical implication of this is that in the design of the flow-through cell, the nozzle body, back wall, counter electrode, or reference electrode should not be located so close to the working electrode that the boundary layer is disturbed. Third, in the derivation of the limiting-current equation, the assumption is made that the diameter of the free jet issuing from the nozzle is negligible compared to the size of the electrode. In practical terms, this means that the electrode diameter must be at least 10 times larger than the inlet diameter.

Detectors based on the WJ configuration have been widely used in practical applications ranging from HPLC detection [23-29], ASV [29-33], and clinical analysis [34,35]. Figure 3 shows a wall-jet detector (WJD) design that ensures that the boundary layer is not constrained. The reference and counter electrodes are placed close to the working electrode but just outside the boundary layer.

The popularity of the thin-layer detector (TLD) in continuous-flow analysis owes much to the work of Kissinger [36,37]. The configuration

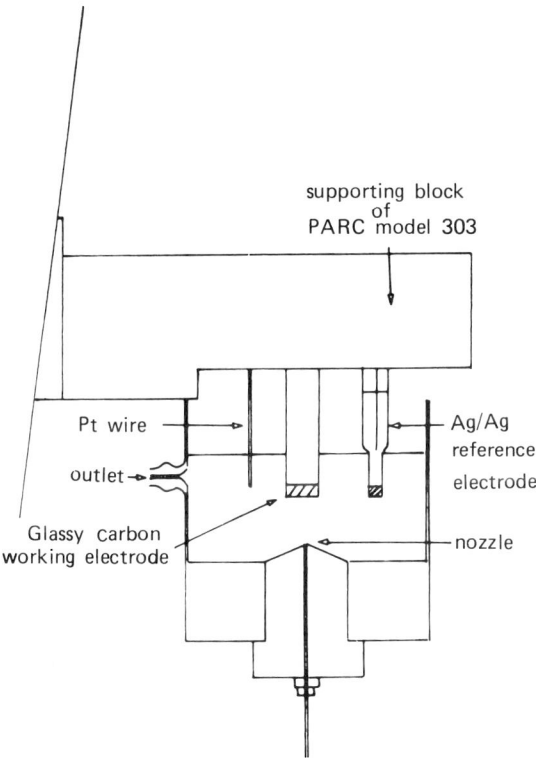

FIG. 3. Large-volume wall-jet detector. The cell is simply a flask with the working electrode centered at the inlet. (From Ref. 27, with permission of the American Chemical Society.)

is perhaps the most widely used for HPLC detection. Recently, the TLD has also been applied to ASV [38].

The derivation of the current response of the TLD is based on a restricted case of the flat-plate electrode with parallel flow where development of the boundary layer is constrained by the wall opposite the electrode. The width of the channel between the working electrode and the opposite wall can be less than the boundary-layer thickness for the usual flow rates employed in flow-injection analysis (FIA) and HPLC (about 1 ml min). Figure 4 shows the velocity profiles in the boundary layer of a flat plate for constrained (channel) and unconstrained (plate) flows.

For the unrestricted case of a flat-plate electrode with parallel flow, where the solution volume is not constrained and the development of the boundary layer is unrestricted, both bulk motion and diffusion are involved in the analyte flux to the electrode. For the restricted case of the thin-layer detector, however, only diffusion is of importance in deriving the limiting-current equation [39].

As Weber has pointed out, some derivations of the current response for the TLD appearing in the literature make erroneous use of correlations derived for the unrestricted case, giving rise to a view that the current response is proportional to the flow rate to the power 1/2 [39].

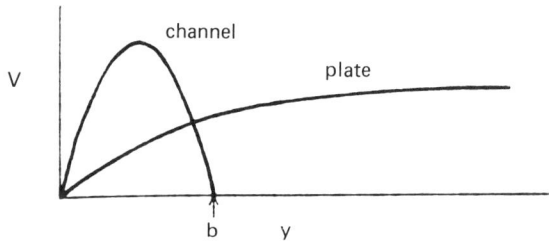

FIG. 4. Velocity profiles in boundary layer for flow in a channel and for unconstrained flow at a plate; b = channel width. (From Ref. 39, with permission of Elsevier Scientific Publishing.)

The correct derivation for the TLD, however, requires consideration of fully developed flow in a channel for which a flow rate dependence of V to the power 1/3 is obtained [40-42]. Anderson and Moldoveanu have presented an analytical solution for the differential convective diffusion equation that substantiates this. The same authors have also done a numerical simulation [43].

The use of channel electrodes in kinetic studies has been described by Matsuda et al. [44]. A two-electrode system can be used to perform collection experiments in a fashion similar to RRDEs [45].

Figure 5 shows a typical design for a TLD that can be used for LC-EC or FI-EC. The channel width is maintained by Teflon spacers. The reference and counter electrodes are placed downstream of the working electrode.

C. Practical Considerations

The specific requirements for a practical electrochemical detector depend largely on the particular application. In general, however, the following apply:

1. Well-defined hydrodynamics
2. Low dead volume
3. High mass-transfer rate
4. High signal/noise ratio
5. Robust, simple, easily maintained design
6. Optimized electrode placement
7. Stable reference electrode

These requirements will now be examined in detail.

Hydrodynamic Characteristics

Often, electrochemical detectors are used with little consideration of the hydrodynamic flow. In many analytical applications, this is acceptable because it is not necessary to have a cell geometry that is hydrodynamically well defined as long as the flow is reproducible and the detector

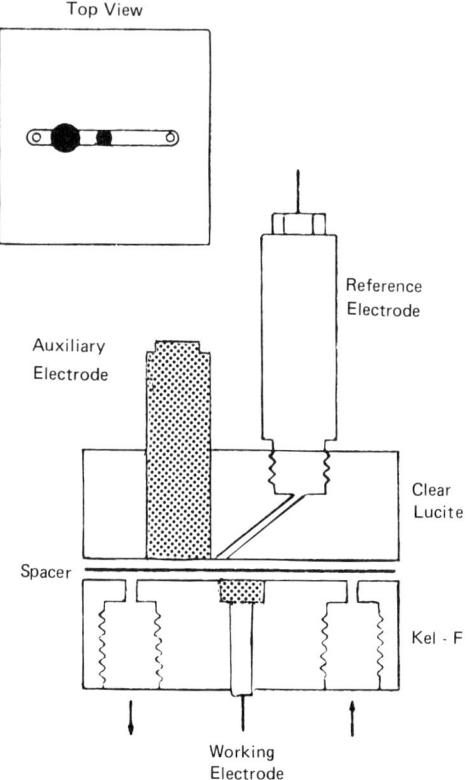

FIG. 5. Design of a thin-layer detector. (From Ref. 38, with permission of Elsevier Scientific Publishing.)

gives reproducible results. For example, although the WJD response can vary between TL and WJ behavior, depending on the inlet-electrode separation [22], it is reproducible at any fixed separation. For the TLD, the current response depends on the position of the inlet with respect to the working electrode and the degree to which the flow is restricted in the vicinity of the working electrode. Knowledge of the hydrodynamic characteristics of the detector (such as the boundary-layer thickness) does, however, enable optimization of the working parameters and maximization of sensitivity. A well-defined response is also necessary in more fundamental work, such as kinetic studies [44,45].

Dead Volume

An important consideration in FI-EC and LC-EC, dead volume, has an immediate bearing on band spreading and, hence, on resolution, sensitivity, and response time. Avoidance of band spreading is particularly critical in LC-EC detection, where there is a need to prevent overlapping of adjacent chromatographic peaks.

The dead volume of flow-through detectors is often identified with their geometric volume. Whereas this is strictly true for spectrometric detectors, the situation is different for electrochemical detectors. This is because, as the solute band passes through the detector, the electrode reaction is confined only to a thin layer adjacent to the electrode.

Comparison of the TLD and WJD is interesting from the standpoint of the description of their dead volume. In the case of the TLD, the geometric volume is purposely constricted to a few microliters so that maximum coverage of the electrode surface and high linear flow rate is achieved [36]. In this case, the dead volume is effectively the geometric cell volume. For the WJD, however, the solution from the inlet impinges perpendicularly on the working electrode and spreads radially over its surface. Here, the dead volume is effectively the boundary layer and is independent of the geometric cell volume [12].

Even in microbore LC-EC applications, where dead volumes of the order of nanoliters are required, either the TLD or WJD can be used. In the former case, a small geometric volume is achieved by reducing the inlet diameter and spacer thickness. For the WJD, equally small dead volumes can be obtained by reducing the inlet diameter and adjusting the inlet-electrode separation.

Estimation of the dead volume of coulometric detectors is complicated by the possibility that the analyte may be totally consumed at the initial, upstream part of the electrode. In this case, the dead volume could be significantly less than the geometric volume of the coulometric cell.

Mass-Transfer Rate

Clearly, a high mass-transfer rate engenders a higher current response. In continuous-flow systems, the mass-transfer rate can be increased by simply increasing the flow rate. A comparison of the response of different detectors has been made by Hanekamp and co-workers [46,47] on the basis of a generalized equation for the limiting current in terms of the dimensionless Reynolds number Re and Schmidt number Sc, given by

$$I = knFDC^0 Sc^\beta wRe^\alpha \tag{8}$$

where $Sc = \nu/D$, α and β are numerical constants that depend on the particular cell geometry, and w is the characteristic electrode width. All other terms have been defined in Table 1.

Although the implication is that the limiting-current response increases with Re, in practice, if Re is too high, it could lead to breakup of the flow into a turbulent regime. For example, the jet emerging from the inlet for the WJD can be characterized as being turbulent if $Re_{Cr} > 1000$ [12].

In FI-EC and LC-EC analysis, although the solute peak current increases with increasing flow rate, the electrolytic efficiency (defined by the total charge measured) decreases in response to the decrease in the residence time as the solute band is swept through the detector [48].

On the basis of the work of Hanekamp and co-workers [46,47], it would appear that the WJD has a greater current sensitivity than the TLD. Equally, however, it is more sensitive to flow fluctuations.

The efficiency of conventional HDV detectors, with small electrode area and operating under amperometric conditions, ranges from 5 to 20% [36]. At very low flow rates (<50 μl min^{-1}) typical for microbore LC applications, efficiency can be greater than 60%. The actual value for a particular application, however, depends as much on the speed of the electrode reaction as on the hydrodynamics.

Large electrode area coulometric cells that approach 100% efficiency have been used in both FI-EC and LC-EC [49–51]. Their effectiveness has been limited, however, primarily because of inherently poor detection limits. At high sensitivity, the signal/noise performance is degraded by high background currents arising from the large electrode surface area. Coulometric cells also present practical difficulties. Electrodes with efficiencies between coulometric and amperometric detectors have also been employed (around 30%) and have been treated by Wang [52].

Noise

Noise may arise from the electronics, the electrode, and the flow system. Electronic noise can be a significant problem, especially at high sensitivity. The noise can arise from several sources, among them, instability in the potentiostat circuit and high-speed transients coupled from digital sources as well as more conventional electronic noise [53].

Weber and co-workers reviewed the sources of noise in continuous-flow electroanalysis [54], and Lankelma and Poppe have described the sources of noise in LC-EC [55]. Noise in electrochemical cells has been modeled using electronic equivalent circuits [56,57]. Three classes of noise were identified; namely, voltage, current, and impedance. Voltage fluctuations in the current-to-voltage convertor that are coupled to the working electrode have been shown to engender higher-frequency noise.

Although increasing the electrode leads to a larger analytical signal, the corresponding increase in noise leads to diminishing returns. Weber has discussed the trade-off between surface area and noise in regard to maximizing the signal/noise [58]. The trade-off also depends on the analyte being detected [59]. High-frequency noise generally becomes a problem at high sensitivities. Lower-frequency noise due to drift can also pose constraints. Drift may arise from gradual change in the working electrode surface resulting either from oxide formation or adsorption of products of the electrode reaction and impurities from the carrier solution. Drift can also be caused by changing characteristics of the carrier solution or by changes in temperature [56,57,60]. Often, use is made of peristaltic or reciprocating pumps, in which case, care must be given to pulse damping. This a particularly important consideration where turbulent flow may be generated. In general, laminar flow is important for low moise [61].

Cell Maintenance

Clearly, a practical electrochemical detector must be easy to handle and maintain. In particular, the working electrode should be readily accessible for periodic mechanical cleaning and polishing. This is one of the practical limitations of the tubular electrode that has restricted its use in routine work.

Compared to the TLD, the WJD is an easier system to maintain. In particular, the cell can be fabricated in such a way that the working, counter, and reference electrodes can be easily removed without dismantling the cell. Because dead volume is defined by the boundary layer, there is no need to maintain a stringent geometric volume. Indeed, as shown in Fig. 3, the cell need be no more than a flask with a hole for the solution inlet [25].

Electrode Placement

An often ignored aspect in the design of electrochemical detectors for flowing stream analysis is the relative placement of the working, counter, and reference electrodes. Although this is often not so crucial for amperometric measurements and for highly conducting aqueous solutions, it is important for potential-scan techniques, especially when fast scan rates or solutions of low conductivity are used. Electrode placement can also have bearing on the performance of pulse techniques. For example, Hanekamp and Nieuwkerk [46] have commented that poor electrode placement may be responsible for the lack of success of differential-pulse techniques in several reported works [62,63]. And Macdonald and Duke obtained improved sensitivity when the electrodes were placed close to each other [64]. Also, the dynamic range can be extended by placing the counter electrode close to the working electrode [37].

In most cell designs, the counter electrode is placed downstream of the working electrode so as to prevent electrolytic products from the former from interfering with detection. This is, however, not a major limitation with the WJD because the radially flowing solution serves to exclude any species from the bulk solution [11,12]. Consequently, both the counter and reference electrode can be placed close to the working electrode surface [25].

Reference Electrode

For a stable and reproducible response, it is crucial to have a stable reference electrode system. This requirement often poses a practical problem in continuous-monitoring applications. In general, silver/silver chloride reference systems have been employed. The advantage of this system lies partly in the simplicity of construction and amenability to miniaturization [65]. Calomel electrodes have been adapted for application to flow-through cells. These electrodes are less sensitive to temperature than the silver/silver chloride system. The disadvantage of both the calomel and silver/silver chloride reference is the need to have a filling solution that can be kept at a constant concentration without cross-contamination from the bulk solution. A salt bridge can, however, be used to minimize this.

Where low-dielectric nonaqueous eluants are used, special consideration must be given to the reference system. Gunasingham and Fleet have described the use of a silver/silver chloride reference in which a salt bridge made of hydroxyethyl cellulose was used to prevent cross-contamination [24]. In a later paper, it was shown that an Ag/Ag^+ reference electrode is preferable because of the reduced junction potential [27]. With the Ag/Ag^+ system, a salt bridge is employed that is made up of the same nonaqueous supporting electrolyte solution as the external cell solution. The reference system could be used for several hours with little change in the reference potential.

Flow System

Flow systems used in continuous-flow analysis can be broadly classified as segmented and nonsegmented systems. In segmented flow systems, the carrier stream is broken up into segments, separated by air or an inert gas (e.g., nitrogen). The sample to be analyzed is mixed with the carrier stream and, for larger sample volumes, can itself be segmented. Usually, the detector gives a steady-state response.

In regard to nonsegmented flow, the most widely applied approach is FI, where the sample is injected into the carrier stream as a discrete plug. In the usual approach, a simple rotary valve injector is employed. By the use of small internal-diameter tubing and low dead-volume fittings, the integrity of the sample plug can be maintained. In general, the sample volume is small (10-1000 µl), and a peak response profile is seen at the output of the detector.

Choice of flow system depends on the particular application. For analyses that require short incubation times, FI is advantageous. However, flexibility is restricted, and this makes FI less suitable for multichannel analyses [66]. Where operations such as on-line dialysis and solvent extraction have to be performed, segmented flow systems enable greater efficiency and throughput.

Stopped flow in combination with FI can be used in applications that need long incubation times although this method requires accurate timing and control. With the availability of inexpensive microcomputers, however, such accuracy is easily accomplished. It is of interest that, even with stopped-flow techniques, high resolution can be maintained. This is because, provided the tubing dimensions are small, little dispersion of the sample takes place when the carrier solution is stationary [66].

A wide variety of high-pressure pumps are commercially available, and most are suitable for use in LC-EC. The most important point of concern is the extent of pulsation. In critical applications and at high sensitivity, use of a pulse damping system may be necessary. Gas-driven solvent systems, which permit completely pulse-free operation, have been employed in low-pressure FI applications. However, for LC-EC, where high pressures are required, such systems are difficult to use, and maintenance is a problem [67].

Often, a simple setup consisting of a peristaltic pump with a glass bulb for pulse damping, as shown in Fig. 6, is suitable. Similar systems have been used for both fundamental studies [68] and analytical applications [69]. A wide operating flow range between 0.5 and 8 ml min is feasible. In order to prevent band spreading, care must be taken in the choice of tubing and connectors so that the dead volume is not unnecessarily increased. In general, tubing connection lengths should be minimized, and a 0.5-mm inside diameter is preferred. All connectors should be flanged and screwed flat to eliminate excess void volume.

Whereas the usual practice is to force the solution into the detector by having the peristaltic pump connected between the carrier solution reservoir and the inlet of the detector, it may be preferable to place the pump at the outlet of the detector and draw solution through it. One of the advantages in regard to the latter is that a single pump can be used for multiple solution lines. Otherwise, a separate pump would

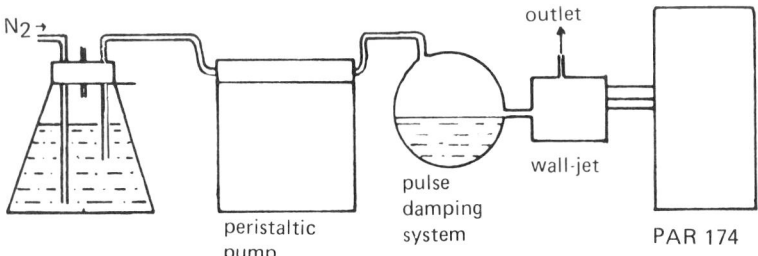

FIG. 6. Simple flow system for HDV studies. (From Ref. 12, with permission of the American Chemical Society.)

be necessary for each line. The flow characteristics can also be better matched in the multiple lines [33].

Valve assemblies for FI can be easily made. Ruzicka and co-workers have described the construction of a simple rotary valve made of Perspex [68]. FI valves are also commercially available at reasonable cost. Use of pneumatic actuation of the valves under microcomputer control has been exploited for fully automated systems that require minimal user intervention [70]. A number of injection schemes have been compared [71–73]. Harrow and Janata have described an FI system that makes use of a number of solenoid-driven pinch valves [71]. The system enabled high reproducibility (0.45% RSD) and is well suited to automation. Kapauan has also described a reliable and inexpensive FI system suitable for automated and continuous use that employs a novel injection mechanism [74]. Good precision was reported to be feasible.

III. ELECTRODE MATERIALS
A. Mercury Electrodes

Mercury electrodes either in the form of the dropping mercury electrode (DME) or hanging mercury drop electrode (HMDE) have a number of shortcomings that do not make them readily amenable to flowing stream analysis. These include high double-layer charging, small anodic potential range, toxicity, handling difficulties, and poor detection limits [75]. Despite this, mercury electrodes have found use in certain applications that benefit from their high hydrogen overpotential and renewable surface [76]. There are also several cases in which the mercury electrode finds unique application because of the participation of mercury in the electrode reaction. Examples of this include detection of tetraalkyl compounds [77–79], amino acids [80], sulphide [81], and steroids [82] and stripping voltammetry of metals [83]. Bond et al. have extensively reviewed the applications of mercury electrodes in flowing stream analysis [54].

Whereas the early designs of flow-through mercury detectors were based on large volume cells, most recent work has been concerned with the development of microcell designs for LC-EC and FI-EC applications [54,84-89]. Figure 7 shows a typical DME detector designed for FI-EC or LC-EC applications.

The DME can be configured with the mercury thread in the capillary either parallel to the solution flow [90] or perpendicular to it [91]. The latter is preferable on grounds of the greater mass-transfer rate it affords [91]. According to Hanekamp and van Nieuwkerk, optimal signal/noise ratios can be obtained with short drop times [46].

One of the difficulties in using mercury electrodes is the need for deoxygenation of the solution [92]. In continuous-flow applications, this can be particularly difficult because of the large solution volumes

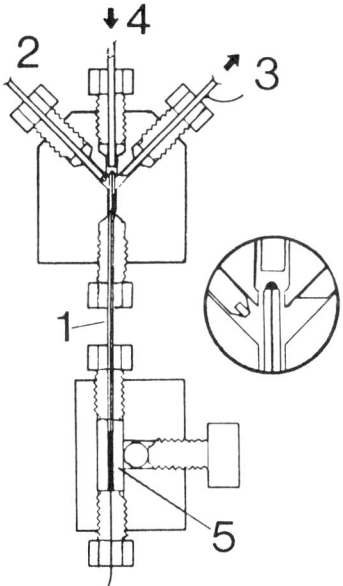

FIG. 7. Design of an SMDE-based microflow cell for LC-EC and FI-EC. 1 = mercury capillary, 2 = reference electrode, 3 = counter electrode, 4 = inlet, 5 = mercury control. (From Ref. 162, with permission of Elsevier Scientific Publishing.)

involved. Although far from ideal in terms of practicality, a number of methods for oxygen removal have been developed, including the use of electrochemical pretreatment [93] and degassing [94].

A hybrid detection system in which a DME and glassy carbon electrode are employed in the same cell configuration for HPLC detection has been described [95]. The DME is held at a negative potential, and the glassy carbon electrode is held at a positive potential; thus, both the reductive and oxidative responses can be simultaneously obtained.

Tenygl and Fleet have outlined current work being carried out for improving the performance of mercury electrodes [76]. This work includes reducing the influence of double-layer charging, improving mechanical stability in continuous-flow analysis, and improving long-term stability and reproducibility. A novel mercury electrode design was reported that has all these attributes. In this design, the mercury electrode is kept in an inverted position, and the mercury drop is mechanically forced to expand and contract alternately in a reproducible manner. The resulting charging current is a symmetric bipolar waveform that can be subtracted. Also, in the process of contracting and expanding, a fresh surface is formed. The mercury electrode is suitable for continuous-monitoring applications.

Transient techniques have been used to obviate the deleterious effects of the charging current [96]. Kowalski and Kubiak have done linear-sweep voltammetry at the DME in flowing solutions [97]. At certain sweep rates, the current is found to be independent of the flow rate. Differential-pulse techniques have been employed effectively [98,99]. Elferink et al. have employed differential pulse at an SMDE detector following HPLC separation [99]. A short pulse width increased sensitivity for the determination of carboplatin.

The application of mercury-film electrodes (MFEs) to anodic stripping voltammetry (ASV) was introduced by Florence [100]. Use of MFEs has since been extended to continuous-flow analysis [30,32-34,38] and also to the detection of organic species [101]. The MFE combines some of the advantages of the conventional mercury electrode (e.g., renewable surface and high hydrogen overvoltage) with the stability and lower charging current of solid electrodes. Computer control has been used to develop an automated MFE that can present a fresh and reproducible surface for each analysis [102].

Amalgam electrodes have also been used in both metal and organic detection [103,104]. For example, Alexander and Akapongkul employed a mercury-copper amalgam electrode for the determination of cadmium [103].

B. Metal Electrodes

The principal requirements for a good electrode material for continuous-flow detection are long-term stability, low residual current, and good electrocatalytic properties. These, however, appear to be contradictory qualities. In general, electrodes with good electrocatalytic activity are more unstable and exhibit high residual currents.

The analytical response, as well as the noise characteristics of electrochemical detectors, is considerably affected by the state of the electrode surface. In general, the active electrode area is quite different from the geometric area, a difference caused by variability in the surface microstructure, which results from physical roughness, adsorbed impurities, and surface oxides, among other factors. As a consequence, it is usually difficult to predict the response of the detector without running a reference standard. In the last few years, considerable effort has been given to developing rigorous methods for conditioning solid electrodes prior to analytical use so as to obtain a consistent response, as well as to improve the electrocatalytic properties.

Noble Metal Electrodes

Noble metal electrodes made of platinum or gold have been widely employed for fundamental electrochemical studies but have found limited usage in practical analytical applications. The limitations of these electrode materials arise from their susceptibility to passivation through surface oxide formation and the tendency toward electrode fouling and poisoning [105-107]. In continuous-flow applications, where it is necessary to use the electrode continuously over several hours, these effects result in diminishing electrode response [108].

For the most part, noble metal electrodes have not been used in continuous-flow analysis except when specific electrocatalytic properties are needed. One such application is the detection of hydrogen peroxide, for which platinum has a much lower overpotential compared to carbon electrodes.

Conditioning of Platinum and Gold Electrodes

Platinum and gold electrodes are notorious for problems associated with surface conditioning. Although considerable work has been done in this regard (e.g., potential cycling and chemical cleaning), it is still very much an art, and it is difficult to achieve any degree of reproducibility in the results.

Use of potential cycling between anodic and cathodic potentials has been advocated as a method of electrode reactivation [109]. Johnson et al. have reported the use of a potential cycle that enables the regular cleaning of the electrode surface [110,111]. The cycle consisted of three repetitive pulses, which enable oxidative cleaning of the electrode, followed by reduction of the electrode and then the actual detection.

Other Metal Electrodes

Apart from the noble metals, only a few other metals have found any practical use, and these have been limited to selected applications that seek to exploit specific electrode reactions involving the electrode material [112-116]. Nickel and copper electrodes have been used to determine amino acids [112]. Nickel electrodes have also been utilized for the determination of ethanol, using FI-EC. Detection was based on the oxidation of ethanol by oxide species formed anodically on the nickel electrode surface [115].

C. Carbon Electrodes

Carbon, in its various forms, has found widespread application in continuous-flow detection. Compared with other materials, carbon electrodes are characterized by a large anodic, and a moderately large cathodic, potential range. At the same time, they display high stability and low residual currents.

Carbon is no exception in regard to the formation of surface oxides. Unlike the case of the metal electrodes, however, oxidation of the electrode surface does not lead to passivation; and, in several cases, it actually has a beneficial effect in making electrode processes more facile. One of the problems, however, is a high background current. In amperometric measurements, there is a need to allow the background current to decay to a steady-state value. The nature and occurrence of carbon-oxygen functionalities have been the subject of numerous papers [117–122], and there is now general consensus that these groups play an active role in electrode processes. Although considerable progress has been made, the mechanism by which this occurs is still a matter for conjecture.

Pyrolytic Graphite

The structure of graphite can be described as a system of infinite layers of fused hexagons held together by weak Van der Waals forces. Pyrolytic graphite has a highly ordered graphitic structure. The material has found limited application compared to glassy carbon, primarily because of its poor electrocatalytic properties [123,124]. Pyrolytic graphite, however, has been shown to have a better cathodic range and has been used for the detection of aromatic nitro compounds [124]. The electrocatalytic activity of pyrolytic graphite electrodes can be enhanced by anodic oxidation [125].

Glassy Carbon Electrodes

Of the various types of carbon materials, by far the most widely used is glassy carbon. One of the earliest uses of glassy carbon electrodes in continuous-flow detection was reported by Fleet and Little [23] in their early work on LC-EC.

In contrast to pyrolytic carbon, glassy carbon has been described as consisting of packets of randomly oriented graphitelike structures [125]. The extreme hardness and strength is attributed to tetrahedral cross linkages between the hexagonal layers. One of the greatest advantages of glassy carbon is its inertness despite high electrocatalytic activity. The electrode can be used over extended periods of time with a reasonably constant response under normal conditions of use.

Carbon Paste

The use of carbon paste electrodes in flow systems was introduced by Kissinger and co-workers [126]. Carbon paste is made from a mixture of carbon powder and an organic solvent such as Nujol. Compared with glassy carbon, carbon paste has the advantage of a lower residual current leading to higher signal/noise responses and correspondingly higher detection limits [127,128]. For example, the detection limits for catecholamines determined by LC-EC has been shown to be 100 fg.

The response of carbon paste electrodes depends to a great extent on the composition of the paste. The larger the percentage of the organic binder in the paste, the lower the residual current. Correspondingly, however, a larger percentage of organic binder leads to lower electrocatalytic activity [129]. Selection of the carbon/organic binder ratio is determined largely by trial and error, and considerable variability exists in the performance of similar formulations prepared by different workers, making it difficult to standardize procedures.

The solubility of the paste in organic solvent precludes the use of carbon paste electrodes in applications where the carrier solution has a significant proportion of organic solvent. Also, carbon paste electrodes are not amenable to reductive detection owing to the high residual current arising from the presence of oxygen in the paste [130].

Composite Carbon Electrodes

As an alternative to carbon paste electrodes, when the aim is to obviate the problem of solubility in organic media, composite carbon electrodes have been made that have shown useful results. In these types of electrodes, the carbon (graphite or carbon black) is mixed with a solid binder in various ratios [131–134]. Among the binders that have been used are polyvinyl chloride, chloroprene rubber, polythene, and Kel-F. A Kel-F-graphite composite electrode has been used in LC-EC with good results [135–137]. The electrode is prepared by compressing Kel-F and graphite into a pellet. A detection limit of 6 pg has been quoted for the determination of phenol [137]. Despite its attractive operating characteristics and machinability, the electrode surface is difficult to condition, and it may require 1 to 4 hr to achieve a stable background for high-sensitivity measurements [136].

Other carbon electrode materials that have been used in continuous-flow analysis include carbon fiber and reticulated vitreous carbon (RVC). Carbon fibers have a diameter of the order of 10 μm. They have been used singly [138] and in arrays [139,140] in LC-EC. Reticulated vitreous carbon is a highly porous carbon that has been applied to coulometric detection [141,142]. One of its inherent advantages is its rigidity, which makes it easy to handle.

Conditioning of Carbon Electrodes

In the case of carbon electrodes, attempts to condition the electrode surface by electrochemical treatment have yielded interesting results [143–151]. Engstrom has shown that treatment can be specific to certain analyte species [145,146]. Kuwana and co-workers have done an extensive study on the effect of electrode pretreatment on the response of glassy carbon electrodes [152–154]. Procedures include vacuum heating [152] and polishing, using a well-defined protocol in which different polishing agents are employed [153]. Kazee et al. have shown evidence that freshly polished glassy carbon has a thin carbon-particle layer on the surface that has a significant influence on electrocatalytic properties [154]. One of the effects of electrochemical pretreatment is thus presumed to be the removal of this layer. Electrochemical and electron spectroscopic studies of highly polished glassy carbon electrodes have been reported by Kamau et al. [155].

Electrochemical pretreatment in continuous-flow analysis has been widely applied. Ravichandran and Baldwin have applied it to both carbon paste [149] and glassy carbon [150]. In the case of the carbon paste, however, the effect of the pretreatment was short-lived because of the deterioration of the electrode. Wang and Peng have shown that pretreatment can be used not only to enhance detection limits but, more significantly, to improve reproducibility in LC-EC using a glassy carbon electrode [151]. The use of a cleaning pulse has been described by Fleet and Little [23] and by Berger [156]. The usefulness of this method, however, is restricted to reversibly adsorbed species. In many cases, such as the phenols where polymer film formation occurs, cleaning would be of little use.

IV. ANALYTICAL TECHNIQUES AND APPLICATIONS
A. Flow-Injection Electrochemcial Detection

Since the early work of Ruzicka, the practical applications of FI analysis have increased considerably. In a recent review, for example, Ruzicka and Hansen have listed over 800 papers dedicated to FI applications [157]. The review shows that the practical use of FI-EC has increased along with the general trend.

FI-EC lends itself to high precision and low detection limits while sample throughput can be high. In general, amperometric measurements are done. As an example of the performance levels that can be achieved, Dieker and Van der Linden have described the determination of Fe(II)/Fe(III), with a detection limit of 10^{-10} M and high sample throughput

[158]. FI-EC has also been widely used for organic determinations. Perhaps the largest area of application is the direct determination of drugs [159,160].

For "real" samples, on-line sample conditioning is necessary both to obviate electrode contamination and to remove interferences. Techniques such as dialysis, derivatization, solvent extraction, and column separation have been applied. Considerable versatility can be achieved. For example, a novel method for the continuous determination of cyanide that includes an on-line distillation system has been described by Pihalar and Kosta [161]. The use of a separation column in conjunction with FI can be considered to be a special case and is separately treated under LC-EC.

Ruzicka and co-workers have introduced the use of potential-scan techniques [162]. Three-dimensional surfaces having axes of current, potential, and time enable greater selectivity although the sensitivity is reduced. The subject is treated in more detail in subsection D.

Another area application of FI-EC is to on-line titrations. These have been described by Janata and Ruzicka [162]. The principle of detection is based on the shift in detection potential resulting from conversion of the analyte at the endpoint.

B. Stripping Voltammetry

Anodic Stripping Voltammetry

Anodic stripping voltammetry (ASV) is a powerful technique for trace metal determination having detection limits in the 10^{-10} to 10^{-12} M range [164,165]. The high detection limit arises from the ability to preconcentrate the metal ions of interest. The early development of ASV in continuous-flow analysis was limited by the inherent difficulties associated with the use of mercury electrodes. However, the development of mercury-film electrodes (MFEs) for ASV by Florence [100] has radically changed this state of affairs. Hydrodynamic ASV (HASV) at an MFE is now a well-characterized technique that lends itself to routine analysis in a wide range of applications.

Mercury-film electrodes used in ASV are usually based on glassy carbon electrodes. Mercury has been described as being deposited on glassy carbon electrodes in the form of small droplets (about 1 μm in diameter) rather than a uniform film [166]. Bately and Florence [167] have, however, found that a thin-film approximation is valid for very thin films (less than 0.5 μm). The stability of mercury films formed on glassy carbon electrodes has been questioned by Yoshida [168]. However, Gunasingham et al. have shown that the film is stable for a wide range of film thicknesses, even at high flow rates [17].

Anodic stripping voltammetry has been employed in both segmented [169] and nonsegmented flow systems [170]. Wang has reviewed these applications [171]. More recently, FI techniques have been employed that enable fine control of timing and sample throughput as high as 24 hr^{-1}. A number of electrode geometries have been employed in ASV that include the tubular [172], thin-layer [38], and wall-jet systems [29-33].

The usual practice of ASV at an MFE involves the deposition of the metal ions of interest onto the mercury film under convective diffusion limited conditions. The solution is then brought to rest, and the deposit is stripped purely under diffusion control. This is the approach taken by Wang and co-workers [30,31]. Recent work has, however, demonstrated the efficacy of performing the stripping step in a flowing carrier stream. Among its other advantages, this measure greatly simplifies and improves the experimental procedure, especially in regard to medium exchange. Also, when stripping is done in a stationary solution, concentration effects resulting from the high concentration of oxidized metal in the reaction layer may adversely affect the stripping peak [173].

There has been some question as to whether a preplated or in situ-plated mercury film is preferable [79]. Generally, an in situ-plated film is employed. Part of the prejudice against preplated films stems from early work with static solutions, where change of solutions in the plating and stripping steps had to be carried out manually. Often, this resulted in exposure of the preplated mercury film to air, causing poor performance. However, Gunasingham et al. have shown that, with a flow system, a preplated film can easily be prepared and may even be preferable because it diminishes contamination of the electrode [33,174]. A preplated film can be used continuously for several analyses, or it can be freshly formed between analyses.

Glassy carbon is the most commonly used electrode material for ASV as it enables formation of a stable mercury film without memory effects [175]. In special cases, however, where stripping analysis is carried out on the bare electrode surface, gold electrodes have been used; Se [176], As [177], and Hg [178] have been determined in this way.

For the case in which the stripping step is done in a quiet solution, the stripping peak current in dc ASV can be described by the theory of deVries and van Dalen [179]. Gunasingham et al. have also applied this theory to describe the stripping current response in flowing solutions at the MF-WJE [174]. The theoretical stripping profile was found to approximate the experimental profile for thin mercury films and low scan rates. Interestingly, the stripping peak current is relatively insensitive to the flow rate as shown in Fig. 8.

Differential-pulse (DP) ASV is being used increasingly for trace metal analysis in preference to dc ASV, primarily because of its inherently greater sensitivity [180]. Figure 8 also shows that the DP stripping peak current is significantly affected by the flow rate. This is chiefly because of the diminishing of the replating effect with increasing flow rate. With proper optimization of the instrument settings and flow conditions, however, it is feasible to obtain a reproducible current response [33].

An advantage of using flow systems in ASV is the convenience it affords in medium-exchange procedures. In medium exchange, the plating and stripping solutions are selected to enable an optimal response. The method can be finely tuned with continuous-flow systems. For example, the

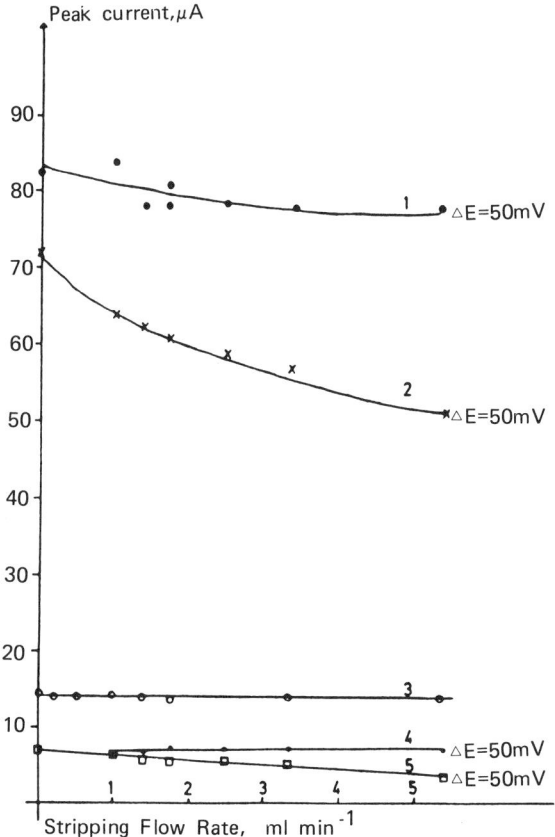

FIG. 8. Effect of flow rate during the stripping step on stripping peak current: 1,4 = differential-pulse scans in sample solutions, 2,5 = differential-pulse scans in blank electrolyte, and 3 = dc scan in blank electrolyte. (From Ref. 174, with permisison of Elsevier Scientific Publishing.)

stripping solution can be altered several times during the stripping scan, so that the best complexing medium for each metal ion of interest can be employed. The approach and its practical applications have been described by Wang and co-workers [30,31] and Gunasingham et al. [33].

One of the practical difficulties in ASV is the removal of dissolved oxygen. Although dissolved oxygen has been shown to have little effect on the plating process, it has a major influence on the stripping scan [174]. In particular, it can lead to instability of the metal amalgam as well as cause irreproducible background and stripping currents; it can also result in loss of sensitivity. In continuous-monitoring applications, the removal of dissolved oxygen becomes more difficult as the procedure must deal with large solution volumes. In segmented flow systems, oxygen removal can be enhanced by the use of an inert gas such as argon or nitrogen for the purpose of flow segmentation [169,181]. In nonsegmented flow systems such as FIA, continuous purging of the carrier solution reservoir is usually carried out with an inert gas. Wang has described the use of a large-volume WJD where purging with the inert gas is restricted to the solution in the detector compartment rather than the entire carrier solution in the reservoir [30]. In the stripping step, however, the solution is brought to rest. Detection limits of 10^{-8} M were reported. Use of an electrochemical scrubber [182] and a large-area glass frit [183] have also been described.

Use of a pulsed stripping waveform such as normal-pulse [184] and differential-pulse [174] has been advocated in the stripping step as a means of obviating the deleterious effects of dissolved oxygen. Although there is a change in the peak potential, the peak current remains unaffected.

Continuous-flow ASV is well suited to computer automation. One of the earliest approaches to the subject was by Kryger and Jagner [185]. In this work, a time-shared minicomputer was used. More recent work by Brown [186], Bond [187], and Christie [188] employed dedicated minicomputer systems. A completely automated ASV system for continuous-monitoring applications has been described by Gunasingham et al. [18]. A dedicated microcomputer was used in conjunction with a pneumatically actuated FI system. Figure 9 is a schematic diagram of the automated hydrodynamic adsorption stripping voltammetry (HASV) system. The computer enables subtle approaches to medium exchange and data handling. Figure 10 shows a typical experimental sequence under computer control. By accurate timing and control of the experimental conditions, each step in the sequence is carried out under optimum conditions. As shown in Fig. 11, background subtraction can be employed to improve the analytical response.

Adsorptive Stripping Voltammetry

The use of adsorption as a means of preconcentrating organic species onto electrode surfaces has been well known for some time [189]. However, it is only recently that its practical applications have been explored, largely because of the efforts of Wang [164]. As a logical step,

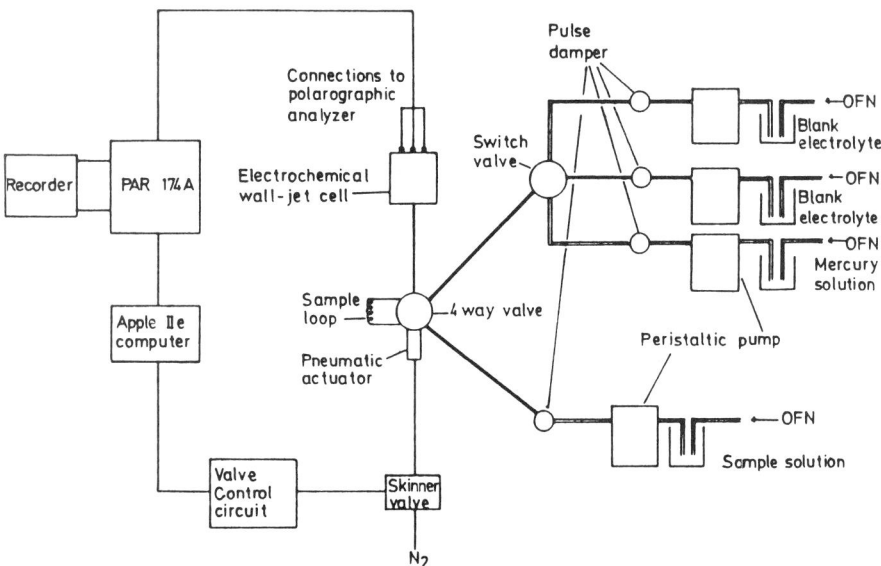

FIG. 9. Schematic diagram of experimental setup for continuous-flow ASV. (From Ref. 174, with permission of Elsevier Scientific Publishing.)

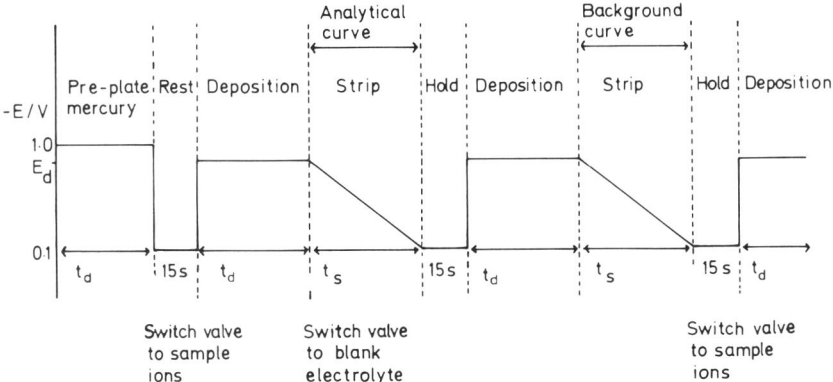

FIG. 10. Potential-time sequence for hydrodynamic ASV. (From Ref. 33, with permission of Elsevier Scientific Publishing.)

Wang has described combining adsorptive stripping with FI. The practical advantages of using FI techniques are similar to those discussed for ASV at an MFE [190]. Perhaps the most interesting is the possibility of using medium exchange to optimize the preconcentration and stripping steps.

C. Liquid Chromatography-Electrochemical Detection

As a routine analytical technique, LC-EC is gaining increasing acceptance. Although it is not as widely applicable as UV-coupled systems, there are a significant number of compounds for which LC-EC is more suitable because of sensitivity or selectivity. For example, it is the most suitable method for the catecholamines, which can be selectively detected at the subpicogram level. A number of reviews have recently appeared on the applications of LC-EC [36,37,54,191,192].

The practical development of LC-EC owes much to the work of Kissinger and co-workers based on the thin-layer detector (TLD). Kissinger and co-workers reported one of the first practical applications in 1973 [126,193]. Another pioneering contribution, contemporary with

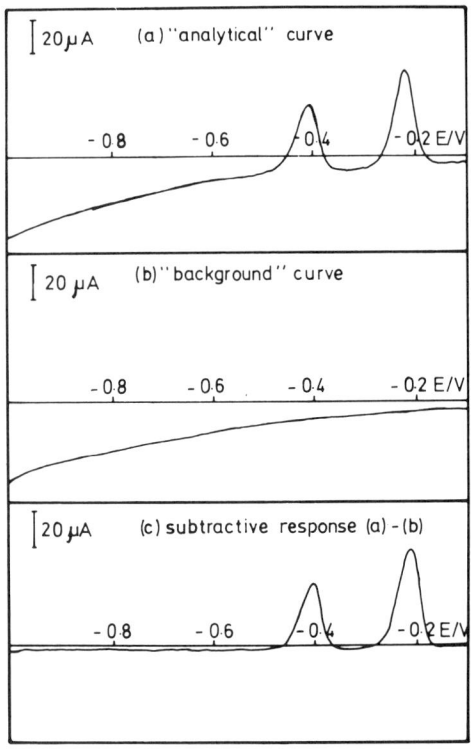

FIG. 11. Use of background subtraction in staircase hydrodynamic ASV. (From Ref. 33, with permission of Elsevier Scientific Publishing.)

the first work of Kissinger, was the development of the wall-jet detector (WJD) by Fleet and Little [23]. Unlike Kissinger's work, where the electrode was a carbon paste electrode, a glassy carbon working electrode was employed. The relative characteristics of the TLD and WJD for LC-EC have been discussed by Gunasingham [25]. Commercial detectors based on the designs of Kissinger and Fleet are now routinely used in laboratories around the world.

One of the essential requirements in LC-EC is the need for a conducting eluant. This can be accomplished by adding a neutral electrolyte or buffer solution to the eluant. In the former case, a 0.01 M KNO_3

solution provides a suitably conducting eluant for all but the most stringent applications, where the nonaqueous fraction of the eluant is high and where extremely high sensitivity may be required. For the case in which the addition of the electrolyte has an adverse effect on the chromatographic separation, postcolumn addition of electrolyte is feasible [194,195]; this, however, can result in a loss of sensitivity resulting from dilution effects.

Oxidative Detection

Most applications of LC-EC have employed oxidative detections. There is a wide range of chemical species that are easily oxidized. Oxidative detections are limited by the positive potential range available at the electrode. Most electrodes are unstable at highly positive potentials because of the formation of surface oxides and breakdown of the solvent/electrolyte.

Most oxidative work is done with the glassy carbon working electrode because of its comparatively large positive potential range and inherent stability, even at very high potentials. With aqueous eluants, electrochemical detection has been employed at potentials as far positive as +1.2 V vs. Ag/AgCl without noticeable deterioration in the electrode response after several hours of continuous use [26]. Potentials even up to +1.3 V are feasible but, in this case, some baseline drift and greater noise are evident at high sensitivity. Also, there is a need to recalibrate the detector response constantly, using an internal standard. Nonaqueous eluants enable a larger anodic oxidative range, as described in the section on normal-phase LC-EC.

Some compounds, such as the polynuclear aromatic hydrocarbons, are electroactive but can be oxidized only at extremely positive potentials. For example, in a work by Gunasingham et al., the detection of carcinogenic PAHs required a working potential of +1.3 V vs. Ag/AgCl.

The Kelgraf electrode [135-137, 196] is well suited to oxidative detection, combining as it does the electrocatalytic properties of graphite and the working characteristics of a microarray electrode. This electrode can be used at highly positive potentials, making it suitable for compounds such as the carbamate pesticides [197]. This is because the current density for oxygen evolution is proportional to the geometric area, whereas the analytical current is relatively enhanced by lateral diffusion, which increases with decrease in electrode area.

Table 2 lists compounds that may be determined by oxidative detection. Some of the most easily determined compounds are those possessing phenolic or catechol moieties. Included in the group are the catecholamines, which represent perhaps the most important applications of LC-EC. The estrogen steroids also possess a phenolic moiety and can be determined at low levels. Table 2 shows that determination of compounds of biological importance is the biggest area of application for LC-EC. This arises from the close correspondence between electrochemical activity and biological activity.

TABLE 2

Selected Oxidative LC-EC Detections

Compound class	Electrode	Reference
1. Polynuclear aromatic hydrocarbons	Glassy carbon	26
2. Estrogens	Glassy carbon	27, 210
3. Carbamate pesticides	Kel-Graff	197
4. Catecholamines	Glassy carbon	37, 196
5. Penicillamine	Gold	198
6. Aceteaminophen metabolites	Glassy carbon	199
7. Aromatic amines	Glassy carbon	200
8. Morphine and derivatives	Glassy carbon	201
9. Phenols	Glassy carbon	202, 24
10. Sulfonamides	Glassy carbon	203

Reductive Detection

Although the potential applications of reductive LC-EC are considerable, in practice, its more widespread use has been limited by the need for scrupulous oxygen removal from the eluant. In recent years, however, a number of methods have been devised that can largely overcome the problem. These include the use of a zinc scrubber [204], inert gas purging at elevated temperature [104], and postcolumn deoxygenation [205]. In order to prevent oxygen from getting into the system after deoxygenation, stainless steel tubing is generally used since plastic tubing, especially Teflon, is highly permeable to oxygen and should be avoided completely. With these precautions, however, high sensitivity and low detection limits are feasible.

A novel approach that gets around the need for stringent oxygen removal involves the use of dual electrodes placed in series [54]. The downstream electrode is kept at a positive potential to monitor the products of the reductive electrode reaction at the upstream electrode. The approach only works, however, if the reduction process generates oxidizable products. There is also a need to select the working conditions carefully to prevent interferences from hydrogen peroxide formed by the reduction of oxygen at the upstream electrode. Further increase in selectivity can be achieved by modulating the reductive potential of the upstream electrode [206].

In most reductive LC-EC work, an SMDE is employed. The applications have been well reviewed by Bond et al. [54]. Mercury amalgam electrodes have been used for specific applications. For example, Bratin et al. have described the determination of aromatic nitro compounds with a Hg-Cu amalgam electrode [104].

Mercury-film electrodes (MFEs) have also gained limited use. Gunasingham et al. have described the use of an automated MFE for LC-EC [102]. Computer control was used to prepare automatically a fresh but reproducible mercury film for each analysis. And, because a fresh film is used for each analysis, electrode contamination is reduced. The increased hydrogen overpotential, as well as better electrocatalytic activity of the mercury, increases the range of applications. Figure 12 shows the recorded traces for successive injections of a mixture of nitro and quinone compounds at the automated MFE. The sequence for the plating and stripping steps is also shown.

Bare glassy carbon has also been used for reductive detection where extreme negative potentials are not required. It is, for example, well suited for the detection of nitro and quinone derivatives.

Some noteworthy examples of practical applications of reductive LC-EC are summarized in Table 3.

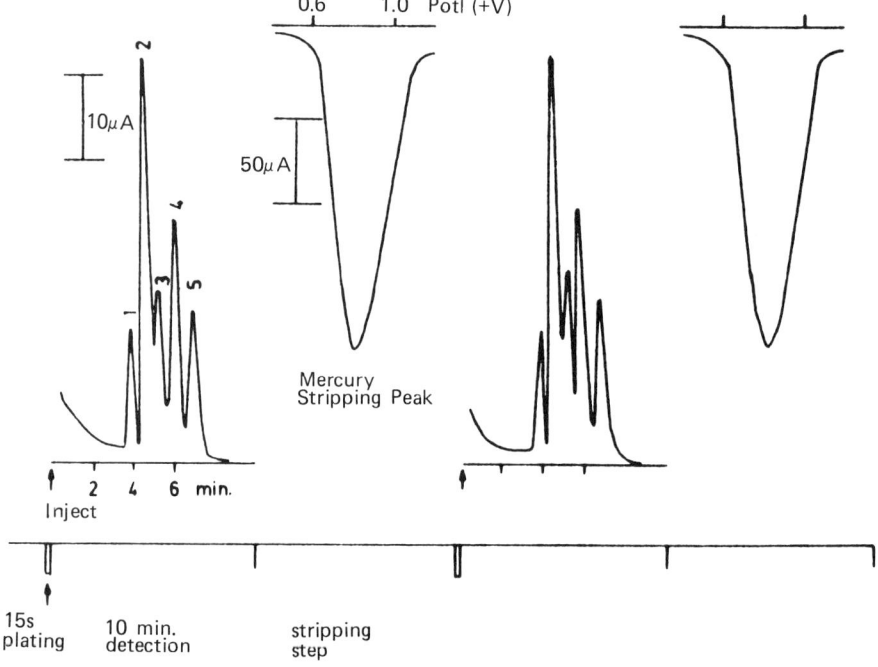

FIG. 12. Detection of successive injections of nitro and quinone compounds using a mercury-film electrode. The mercury film is prepared freshly for each analysis under computer control. (From Ref. 102, with permission of the American Chemical Society.)

TABLE 3

Selected Reductive LC-EC Detections

Compound	Electrode	Reference
1. Benzodiazepines	Glassy carbon	207
2. Pesticides	DME	208
3. Nitrosamines	Glassy carbon	209
4. Nitroaromatics	Hg-Cu	104
	Glassy carbon	204
	Mercury film	102
5. Organometallics	Static mercury drop	54, 78

Normal-Phase Applications

LC-EC has been, in the main, confined to reverse-phase and ion-exchange LC where the eluant is aqueous. Contrary to general belief, however, EC has wide application in normal-phase HPLC, even if the eluant has a low dielectric constant.

Normal-phase LC is frequently used for the separation of polar organic species such as phenols and estrogen steroids. The technique is particularly suited to compounds that have a tendency to lipophilicity by virtue of possessing a large hydrocarbon skeleton.

Normal-phase LC-EC has a number of advantages over reverse-phase systems. Because of the greater solubility of organic compounds in nonaqueous solvents, electrode fouling through adsorption and polymerization of electrolysis products are diminished. A good example is the phenols, which are known to adsorb strongly and polymerize on the electrode surface in aqueous media. It has been shown for these compounds that electrode fouling can be diminished by the use of nonaqueous eluants [23].

Another important benefit of nonaqueous eluants is that a larger working potential range (defined by the solvent-electrolyte breakdown) and lower background currents can be obtained. One of the major problems in using carbon electrodes in aqueous systems is the high background and charging current that arise from redox reactions of surface functionalities. In aqueous solutions, it may take several minutes before the background current reaches a steady state. For nonaqueous systems, however, the background current rapidly reaches a steady state. Also, in aqueous systems, acidic surface oxides can lead to a decrease in the hydrogen overpotential, thus limiting the negative potential range. For example, in neutral aqueous solutions, glassy carbon electrodes have been reported to have a negative potential range of −0.75 V vs. the saturated calomel electrode. Nonaqueous eluants can be used to extend the negative potential range beyond −2 V [27].

Figure 13 is a diagram of a WJD specifically designed for normal-phase use. The cell body is made of PTFE and includes a glass window that allows observation of the internal compartment.

The sensitivity achievable in normal-phase LC-EC has been shown to be comparable to that obtained with reverse-phase systems for several compounds [24, 27, 210]. Figure 14 shows chromatograms obtained with a

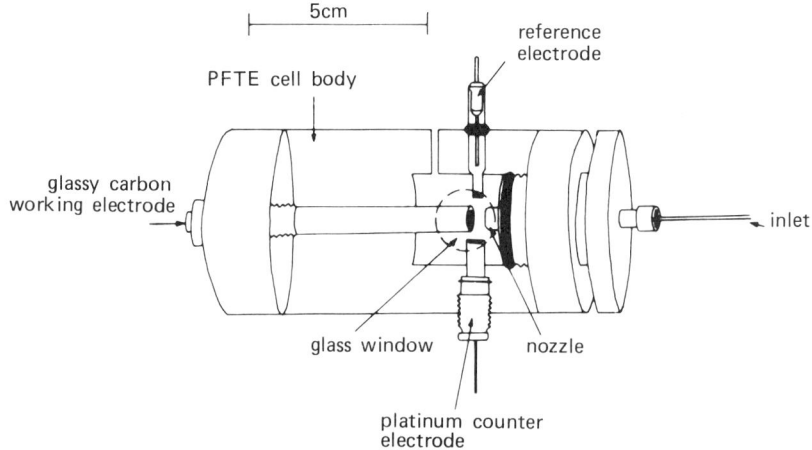

FIG. 13. Design of WJD for normal-phase LC-EC. (From Ref. 210, with permission of Elsevier Scientific Publishing.)

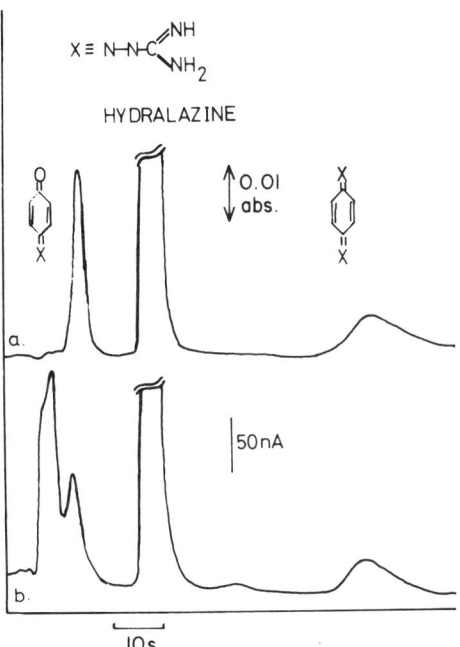

FIG. 14. Comparison of UV and EC detection for the determination of hydralazine and its metabolites. Eluant: hexane/ethanol (4:1) 0.05 M in $(C_4H_9)_4NBF_4$. (a) UV detector 260 nm; (b) WJD at +1.0 V vs. modified Ag/AgCl. (From Ref. 25, with permission of Elsevier Scientific Publishing.)

UV and WJD detector placed in series. In this example, the WJD was found to be more sensitive than the UV detector.

Normal-phase LC-EC does, however, have some drawbacks. One is the need for a special reference system [27]. Another difficulty is the need for a suitable supporting electrolyte. Usually, the electrolyte is added to the eluant, and this can have an adverse effect on the chromatography. Tetrabutyl ammonium fluoroborate has some use for very low dielectric eluants for which there is no suitable electrolyte.

A way of getting around this problem is postcolumn addition, where the usual approach is to add the electrolyte in a more polar solvent at the detector inlet. This, however, leads to band spreading and dilution of the analyte. Gunasingham et al. have proposed the use of a large-volume WJD [27], where the cell conductance is maintained by the presence of electrolyte in the bulk solution of a large-volume WJD rather than at the detector inlet. The advantage of this approach is that band spreading and dilution effects can be minimized. Figure 15 shows the experimental setup, which includes a fluorescence detector placed in series with the WJD for the purpose of comparison. Although, with postcolumn addition, detection limits were reported to have been reduced by a factor of 10, useful results in the nanogram range were reported for easily oxidizable compounds such as the estrogen steroids.

Schieffer has described the use of a Nafion (TM) ion-exchange membrane in conjunction with a coulometric detector that was used in highly resistive eluants without any added electrolyte [211]. Providing a medium for ion conduction was presumed to be the role of the ion-exchange membrane. However, one reason the detector was able to function without

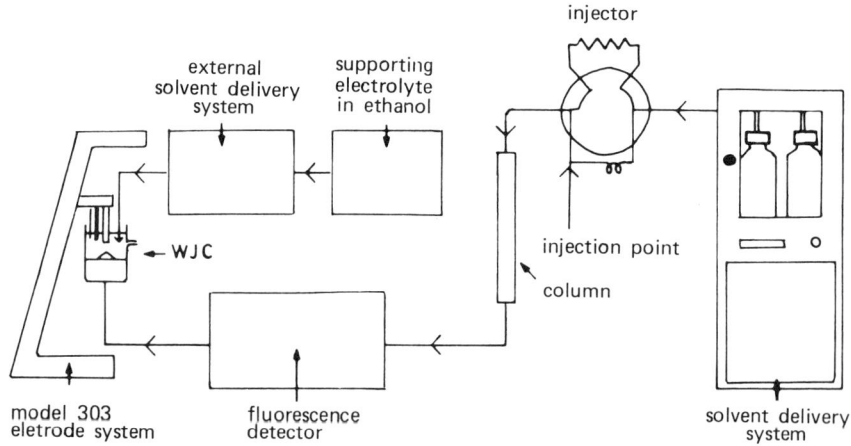

FIG. 15. Experimental setup for postcolumn addition of electrolyte in normal-phase LC-EC. (From Ref. 27, with permission of the American Chemical Society.)

supporting electrolyte may well be that a coulometric electrode was used. Mikkelson and Purdy [212] have shown that coulometric electrodes are generally less susceptible to solution resistance.

Microvoltammetric electrodes having diameters less than 10 μm have been used effectively in normal-phase separations [213,214]. The ability to work without an electrolyte arises from the small faradaic currents (in the picoamp to nanoamp range) generated at these electrodes. Detailed discussion of these electrodes is given in Section VI.A.

A further important drawback of normal-phase LC-EC is the need to purify organic solvents. Impurities in the solvent can lead to electrode fouling and high background currents.

Dual-Electrode Detection

Two or more electrodes in tandem have been applied to enhance the information content of LC-EC. Different schemes have been employed. The most useful approach is to use two electrodes in very close proximity, either in a parallel or a series configuration, as shown in Fig. 16. Parallel electrodes may be positioned adjacent to (same plane), or opposite (facing plane), each other. A common detection strategy is to keep the electrodes at different potentials and monitor the current ratio. The ratio can be used as a quantitative and qualitative measure [215]. The current ratio has been found to be quite precise; interestingly, it

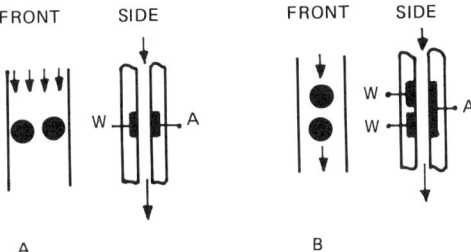

FIG. 16. Configurations for dual-electrode detection: A = parallel-adjacent configuration; B = series configuration. (From Ref. 215.)

is sensitive to differences in molecular structure. The peak ratio method has been used to identify a number of compounds separated from complex mixtures [215–219]. Figure 17 shows the simultaneous chromatograms obtained at parallel-adjacent electrodes, where the electrodes are kept at different potentials for oxidation. The two-electrode mode increases the selectivity of the analysis.

Parallel electrodes can also be used such that one is held positive and the other negative. In this way, both oxidative and reductive plots can be obtained simultaneously [54].

When placed in the opposing configuration the use at a parallel dual-electrode configuration enables the continuous recycling of the redox forms of the analyte if one electrode is kept at the potential for oxidation and the other at the potential for reduction [220,221]. Efficiencies greater than 100% are feasible. In order to achieve reasonable results, however, the spacing between the electrodes should be small and the transit time long, making the configuration well suited to microbore LC applications. Also, the electrode reactions should be reversible or, at least, quasireversible.

Hutchins-Kumar et al. have used a dual-electrode system comprising one bare and one cellulose-acetate-coated electrode placed in parallel as shown in Fig. 18 [222]. By monitoring the response ratios of the two electrodes, additional information regarding solute size can be obtained, thus improving selectivity. The cellulose-acetate film also serves as a protective barrier to exclude larger molecules that may interfere with the electrode reaction. Figure 19 shows typical chromatograms obtained simultaneously at bare and coated electrodes.

In the series electrode configuration, the downstream electrode can be used to monitor the products of the upstream electrode reaction [223, 207]. This can enhance the selectivity and, at the same time, improve the detection limit. The approach is, however, generally suited only for species yielding reversible or quasi-reversible reactions.

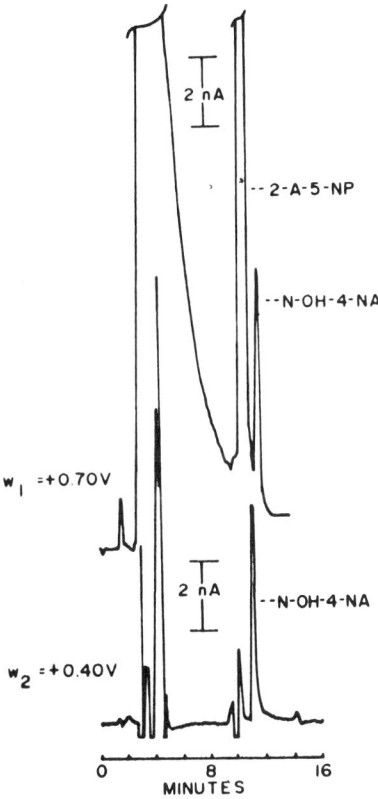

FIG. 17. Use of parallel-adjacent dual electrodes for selective detections. Chromatogram is of mouse liver microsomal incubation with 4-nitroanaline and NADPH. N-OH-4NA: N-hydroxy-4-nitroanaline; 2-A-5-NP: 2-amino-5-nitrophenol. (From Ref. 219, with permission of the American Chemical Society.)

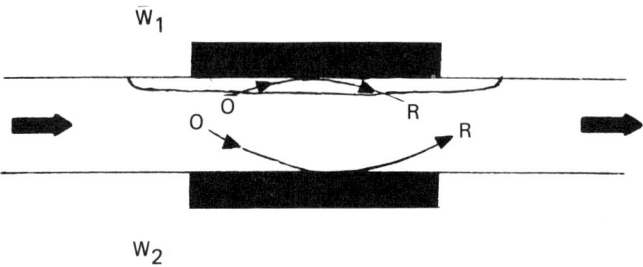

FIG. 18. Dual-parallel electrode detection in opposing configuration. Upper electrode is coated with a cellulosic membrane. (From Ref. 222, with permission of the American Chemical Society.)

Apart from detection of the analyte, a dual-electrode system can be used to compensate for fluctuations such as in flow rate, temperature, changes in the composition of the mobile phase, and static discharges [224]. The difference chromatogram can, however, effectively discriminate against such common-mode noise only if the electrodes are properly matched; otherwise, the electrode responses would have to be normalized [224]. For this reason also, it is necessary that the two electrodes are operated at potentials that are not too far apart.

The concept of dual-electrode detection has been extended to an array of electrodes by Matson et al. [225]. The array electrode system was shown to be useful for resolving coeluting species.

D. Pulse and Potential-Scan Techniques

Whereas dc amperometry at a fixed potential is the most common measurement mode, there is increasing interest in the use of pulse and potential-scan techniques in order to enhance selectivity and sensitivity. Figure 20 shows the various potential-time waveforms that have been employed in continuous-monitoring HDV. Pulse amperometric techniques (e.g., Fig. 20b and 20c) involve measurement at a fixed potential. Often, a differential current is recorded. In theory, pulse techniques should be

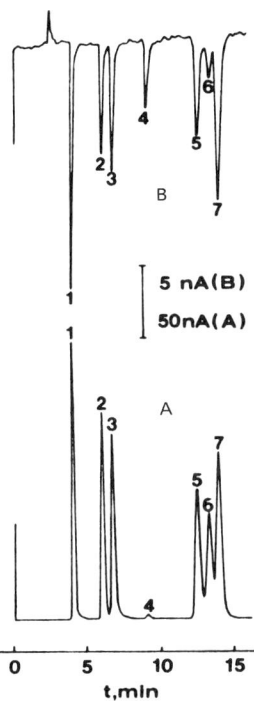

FIG. 19. Dual-electrode detection with electrodes placed in parallel, opposing configuration. (A) Bare electrode. (B) Cellulose-acetate-coated electrode. Operating potential for both electrodes = +900 mV. Mobile phase is 0.15 M chloroacetic acid adjusted to pH 3.5 with phosphoric acid. 1) Norepinephrine, 2) L-dopa, 3) epinephrine, 4) tyrosine, 5) dopamine, 6) a-methyldopa, 7) homogentisic acid. (From Ref. 222, with permission of the American Chemical Society.)

more sensitive compared to dc amperometry. In practice, however, this is rarely achieved because of the increased noise.

Rapid potential-scanning techniques have been used in conjunction with FI or LC to identify compounds on the basis of their characteristic potentials for oxidation or reduction. As shown in Fig. 21, the voltammogram can range between sigmoidal and peak shapes depending on the flow rate and potential-scan rate. Higher flow rates and lower scan rates give rise to steady-state conditions, and a sigmoidal current voltage

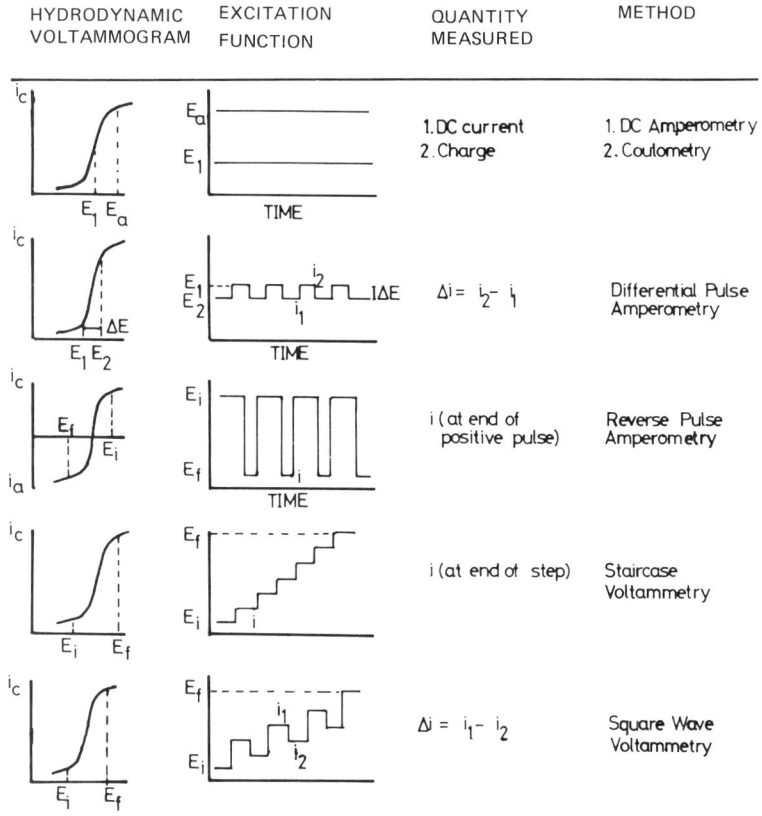

FIG. 20. Potential-time waveforms used in continuous-monitoring HDV. (Adapted from Ref. 215.)

response curve is obtained. For faster scan rates (>0.5 V sec^{-1}) and lower flow rates (example), however, the voltammogram then takes on the characteristic peak profile caused by the decline in the mass-transfer rate to the electrode surface as the working potential passes the formal potential E° because of the depletion effect.

Thorgeson has defined the ratio of the difference between the cathodic peak current and the current at the cathodic switching potential to the cathodic peak current as a means of classifying voltammograms

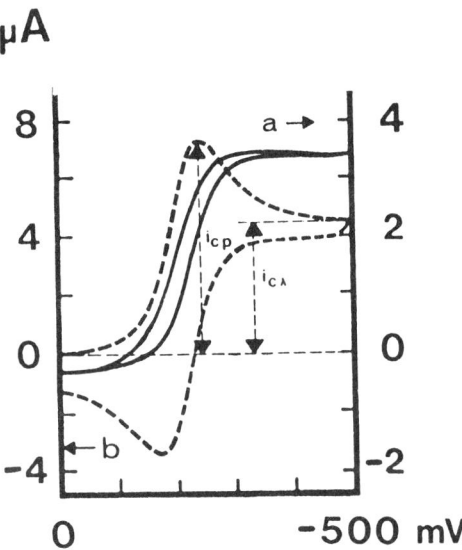

FIG. 21. Rapid-scan voltammetry under hydrodynamic conditions can yield a cyclic voltammogram (– – –) or a hydrodynamic voltammogram (------), depending on the scan rate and flow rate. (From Ref. 162, with permission of the American Chemical Society.)

[162]. In general, as a qualitative and quantitative measure, a cyclic voltammetry (CV) plot is preferred.

In FI or HPLC applications, as the solute band is swept through the detector, the potential is scanned rapidly across the range in which the solute is electrochemically active. The detector response may then be represented as a three-dimensional surface of the current against potential and time.

Several works have been reported where rapid-scan voltammetry (RSV) is used to improve the selectivity of LC-EC [226–229]. For example, RSV has been used as the basis for the identification of a number of neurotransmitters following reverse-phase HPLC separation [226]. Ploegmakers et al. have reported the use of HPLC-RSV for the identification of the anticancer agents eptoside and teniposide [227].

And Gunasingham et al. have described an automated MFE for performing cyclic voltammetry [102] in conjunction with FI or LC. Use of the mercury film enabled well-defined cyclic voltammograms, as shown in Fig. 22, and could be used to identify nitro and quinone compounds.

White et al. have described a scanning microvoltammetric detector for open-tubular HPLC [228]. The detector consisted of a 9-μm carbon fiber inserted into the end of a 15-μm-i.d. column. Scan rates of 1 V $\sec^{-1}$ in the range 0 to +1.5 V vs. Ag/AgCl were feasible. In a later work, a scan and step waveform was used to improve current stability [229]. Significantly reduced distortion due to IR drop was obtained by shortening the electrode length. Also, with the shorter carbon fiber electrode, improved sigmoidal voltammograms were obtained. Improvement in the visual resolution was obtained by converting the sigmoidal voltammograms to peaks by taking a mathematical derivative. Figure 23 shows typical three-dimensional nonderivative and derivative plots [229].

Apart from dc scans, a wide variety of potential waveforms have been reported. Caudill et al. have described the use of a normal-pulse potential waveform that was applied to a channel electrochemical detector [226]. Trojanek and deJong have described fast-scan ac voltammetry, which has made possible better resolution of overlapping peaks [230]. Sammuelsson et al. have reported the application of square wave RSV [231]. Also, Wang and Dewald described the applications of differential-pulse voltammetry to FIA [232].

Gunasingham et al. have reported the use of RSV in normal-phase LC, where use is made of low-dielectric nonaqueous solvents [233]. A staircase waveform was employed, and background subtraction was used to improve peak resolution. The use of RSV in nonaqueous systems has a number of advantages: Better-defined voltammograms can be obtained. Less distortion at high scan rates occurs. And, finally, the useful working range can be extended to more positive and negative

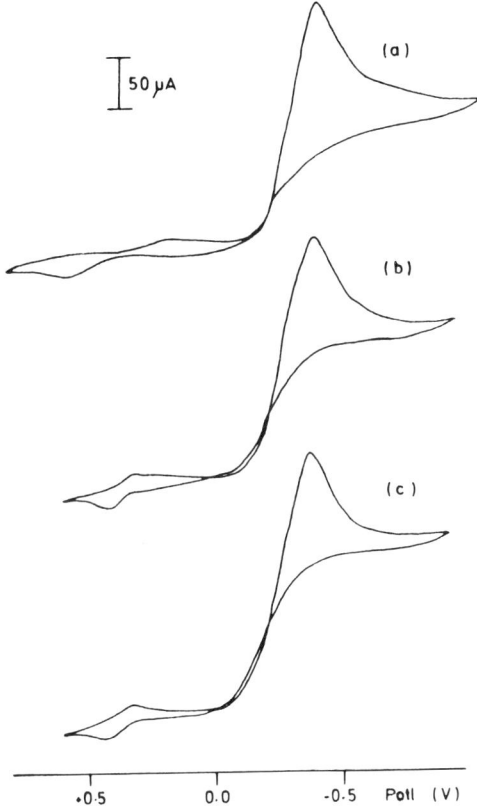

FIG. 22. CVs of 2 mM 4-nitroacetophenone in 0.1 M sodium acetate buffer, pH 4.0 with 30% ethanol obtained with a WJD. (a) Glassy carbon electrode and mercury-film electrodes, (b) 10 Å, (c) 145 Å. Flow rate = 0.7 ml min. (From Ref. 102, with permission of the American Chemical Society.)

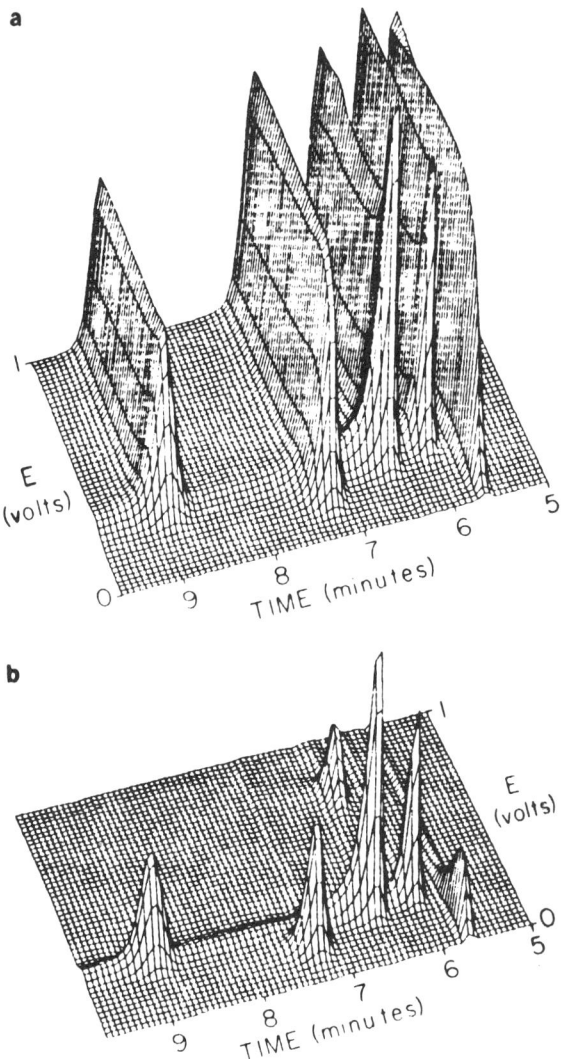

FIG. 23. i-E-t plots following HPLC separation of 0.1 mM ascorbic acid, DL-epinephrine, tyrosine, dopamine, hydroquinone, and catechol: (a) underivatized data, and (b) derivatized and smoothed data. Flow rate = 0.55 cm sec; scan rate = 1 V/sec. (From Ref. 229, with permission of the American Chemical Society.)

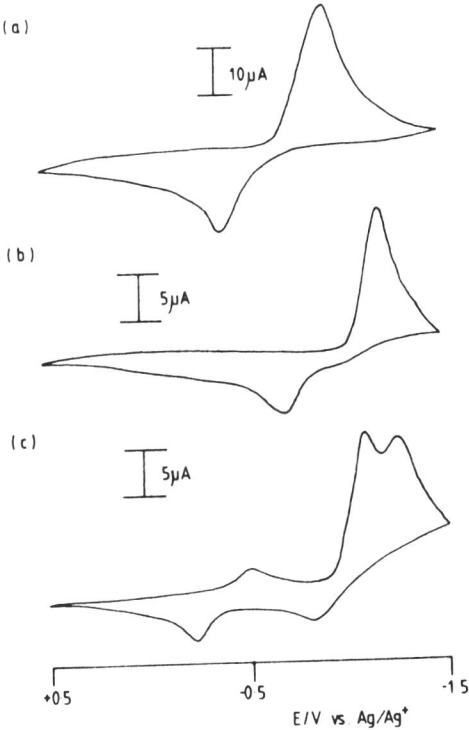

FIG. 24. RSCVs obtained following normal-phase separation: (a) 9.6 μg of 9,10-phenanthrenequinone, (b) anthroquinone, and (c) 1.2 μg of 1-nitroanthroquinone. Eluant = 60/40 ethanol hexane, 0.1 M tetrabutyl-amonium fluoroborate, and 0.1 M acetic acid. Scan rate = 1.25 V/sec and flow rate = 1.0 ml min. (From Ref. 233, with permission of the American Chemical Society.)

potentials. Figure 24 shows typical CVs obtained for three nitro and quinone compounds.

Although RSV is inherently less sensitive than amperometric measurements at a fixed potential, good precision can be obtained for both the peak potentials and peak currents [233]. This is because, in the time scale of the potential scan, the concentration of the solute band in the detector is relatively constant. Computer control is essential, however, to synchronize the CV with the peak elution, so that the CV is taken at the same point on the solute peak profile.

V. INDIRECT METHODS
A. Photoexcitation

Combining photoexcitation and electrochemical detection has proved a fruitful analytical method, which has a number of practical applications, especially in LC-EC. Two reviews have appeared on the subject [234, 235]. The practical benefit of postcolumn photoexcitation includes the ability to generate electroactive species from electroinactive species. Several approaches have been described. One approach involves the irradiation of the eluant by placing the light source between the column and the detector [236]. Photolysis then causes an irreversible chemical conversion of the analyte to an electroactive species, which can be detected amperometrically.

Another approach is to irradiate the working electrode directly as the analyte band passes over it [236]. Here the role of photoexcitation is to promote the analyte to an excited state that has a more facile redox chemistry. The exact nature of this photoexcitation process has not been clearly understood. However, it is known that responsive compounds are restricted to carbonyl compounds and their derivatives that undergo $n \rightarrow p^*$ excitation. With this technique, photogenerated excited states, intermediates, and products with short lifetimes can be detected. Figure 25 shows a typical experimental setup, and Fig. 26 is a schematic diagram of a modified TLD for photoelectrochemical detection. The modification enables direct irradiation of the working electrode surface with an intense light source.

Instead of the analyte being photoexcited, a photoexcitable intermediate can also be used. Weber and co-workers have used a ruthenium-bipyridyl complex for this purpose [237,238]. The excited intermediate can then oxidize or reduce the analyte while it, in turn, is electrochemically detected at the working electrode. The disadvantage of this

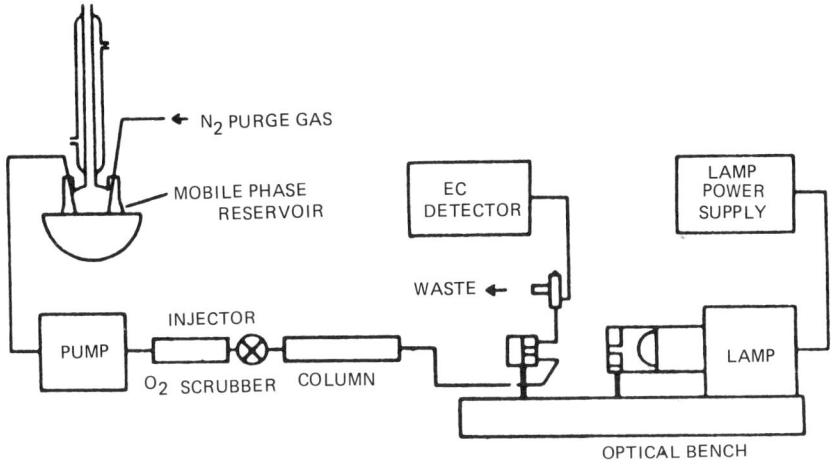

FIG. 25. Experimental setup for photoelectrochemical detection. (From Ref. 237, with permission of the American Chemical Society.)

approach is the possibility that the ruthenium-bipyridyl complex can interfere with the separation in LC applications. Also, from the standpoint of cost, it may be unsuitable for routine analyses unless the reagent is injected in discrete amounts in parallel with the sample.

Finally, it is possible to derivatize the analyte with a tag that is amenable to photolysis. The use of nitroaromatic reagents for amines, amino alcohols, and amino acids has been described [54].

Although the use of photoelectrochemical detection systems in practical analysis is still in its infancy, potential applications have been well recognized. In particular, the combination of fiberoptic systems with electrochemical detection holds considerable promise.

B. Use of an Electroactive Intermediary

The general idea in this indirect method is to employ an electroactive species that undergoes an interaction with the analyte of interest and causes a change in its characteristics that can be monitored. The

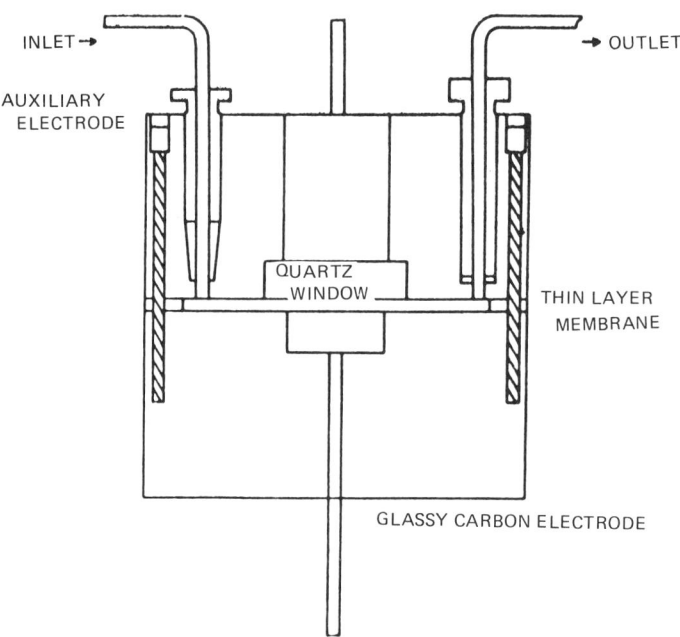

FIG. 26. Thin-layer detector used for photoelectrochemical detection. (From Ref. 237, with permission of the American Chemical Society.)

interaction need not, however, be chemical. Indeed, as the first example shows, it can be simply a physical displacement of the intermediary by the analyte.

Ye et al. have reported an indirect technique that allows monitoring of electroinactive species following HPLC separation [239]. In this approach, the displacement of an electroactive addition (hydroquinone) in the mobile phase by the inactive solute band is monitored by a negative peak. The technique, although not very sensitive, enables monitoring of alcohols and aliphatic carboxylic acids that are electrochemically inactive. A typical chromatogram for the indirect detection of alcohols is shown in Fig. 27.

Horjabri et al. have described an indirect amperometric technique for the determination of metal ions following ion chromatography [240].

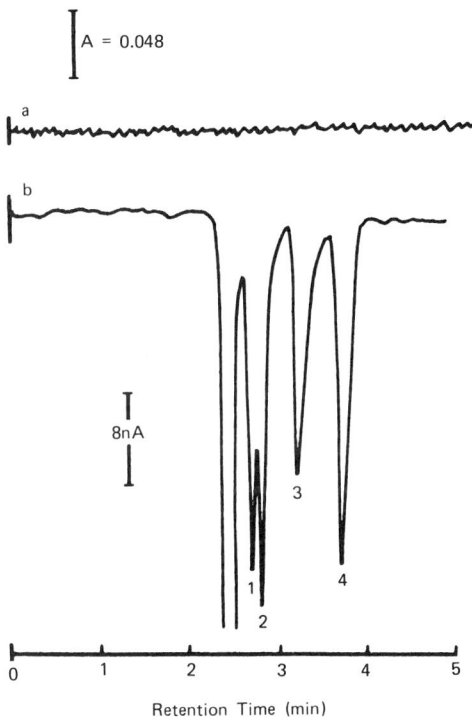

FIG. 27. Indirect detection of mixture of alcohols. Lower trace obtained at working potential of +0.27 V vs. Ag/AgCl with a mobile-phase hydroquinone concentration of 1.0×10^{-3} M. Upper trace obtained by UV absorbance at 210 nm with no hydroquinone added; (1) isopropyl alcohol, (2) 1-propyl alcohol, (3) 1-butyl alcohol, and (4) isobutyl alcohol. (From Ref. 239, with permission from the American Chemical Society.)

excess dithiocarbamate ligand is added postcolumn, and the decrease in the current as the ligand complexes with the metal ion is monitored. Detection limits at the sub-ppm level were obtained for a number of metal ions. An advantage of this method is that oxygen interference is obviated.

Another example uses the dithiocarbamate complex in the opposite sense. Bond and Wallace have used in situ-generated dithiocarbamate complexes for the determination of copper and nickel [241]. The detection is based on the fact that the metal complex can be determined anodically, whereas the free metal ions are electroinactive at positive potentials.

The use of electrogenerated reagents is another indirect method that has gained practical usage. Kok et al. [242] have described the chromatographic determination of phenols using electrogenerated bromine. A two-electrode system is used in which the bromine is generated on the upstream electrode. After reaction with the phenol, the excess bromine is detected at the downstream electrode. A similar approach has been proposed by Albery et al. for the determination of proteins. In this work, a ring-disk WJD was used [243].

C. Indirect Electrode Reactions

Pulsed amperometric detection at platinum or gold electrodes of compounds that are electroinactive under dc amperometric conditions have been reported by Johnson and co-workers [244-249]. Initially, a three-step potential waveform, as shown in Fig. 28A, was employed with alternate anodic and cathodic polarization of a platinum electrode followed by amperometric detection of the adsorbed analyte at an intermediate potential. Applications include the determination of amino acids [244], aminoglycosides [245], carbohydrates [246, 247], and sulfur-containing compounds [248].

More recently, a two-step potential waveform has been proposed (Fig. 28B), as well as the use of a gold electrode in place of platinum [249]. The advantage of gold is that it can be operated at potentials at which oxygen reduction does not occur and surface oxide formation is insignificant. The two-step method at the gold electrode results in increased sensitivities and lower detection limits. Sensitivity is apparently enhanced in alkaline medium.

The use of chemically modified electrodes, where the modifier has the ability to act as a mediator for the electroinactive analyte, is an

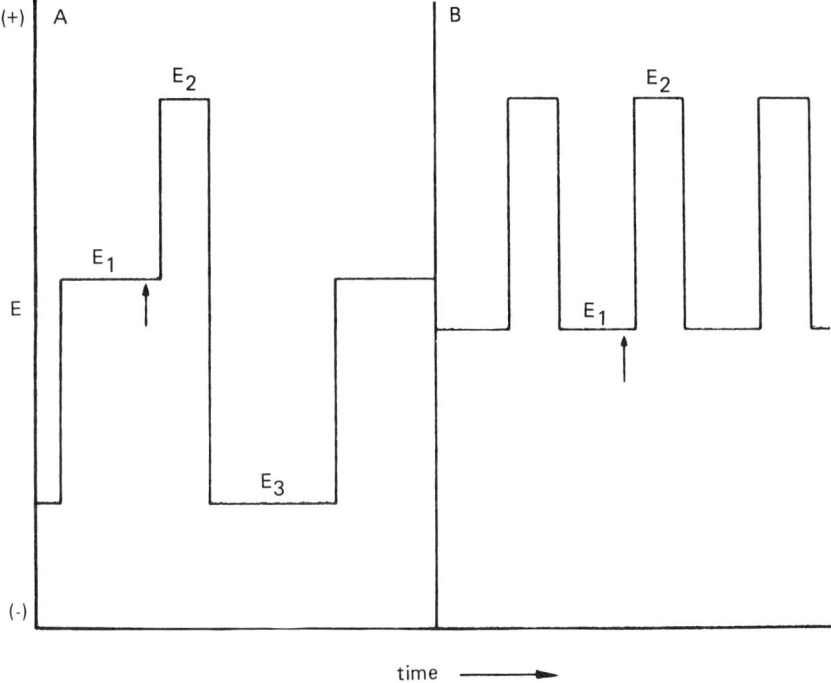

FIG. 28. Three-pulse (A) and two-pulse (B) waveforms for indirect detection of species at platinum and gold electrode. (From Ref. 249, with permission of the American Chemical Society.)

important and developing area. This subject is treated separately in the section on chemically modified electrodes.

D. Chemical Derivatization

The use of chemical derivatization to convert an electroinactive species to an electroactive one has been well applied to LC-EC. Various derivatization strategies have been recently reviewed [54]. Nitroaromatic reagents are perhaps the most widely used ones; they are particularly suitable for carbonyl compounds, amines, and amino acids.

One approach to chemical derivatization is to use a packed column reactor. Enzyme reactors constitute perhaps the most important category that has been used to derivatize compounds selectively. In LC-EC, postcolumn enzyme reactors have been employed for the determination of bile acids, acetyl choline, urea, and amino acids. The area has been reviewed by Dalgard [250].

VI. NEW ELECTRODE SYSTEMS

A. Ultramicroelectrodes in Continuous-Flow Analysis

Ultramicroelectrodes have high mass-transfer rates due to hemispherical rather than linear diffusion [251, 252, 252a,]. A steady-state response is obtained under linear sweep conditions unless scan rates are extremely high (e.g., 100–1000 V sec^{-1}). Figure 29 shows flow-injection hydrodynamic voltammograms at a micro platinum disk electrode.

The theory of microelectrodes in flowing stream analysis has been described for both single-electrode systems and microelectrode arrays [153, 154]. Moldoveanu and Anderson have described the numerical

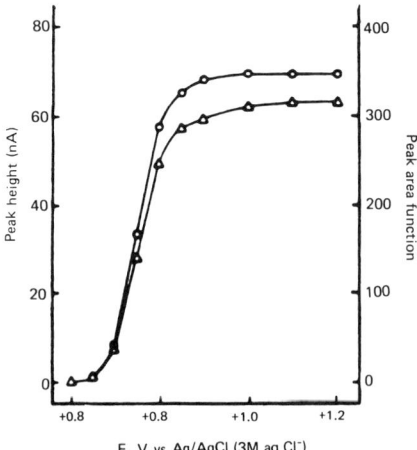

FIG. 29. Flow-injection hydrodynamic voltammograms at a 50 μm platinum microdisk electrode. Sample = 20 μl 1 mM ferrocene in acetonitrile (0.01 M TEAP). ○ = peak height; △ = peak area. (From Ref. 256, with permission of the American Chemical Society.)

simulation for an array of microelectrodes [253]. The optimum geometry for a microarray in a rectangular flow channel detector was investigated by Fosdick and Anderson [254]. Concentration profiles were calculated by using the backward implicit finite difference numerical procedure for electrode arrays. The optimum response was found for electrodes having active sites of constant length, separated by uniform gaps.

Cope and Tallman have shown that the current response for an array of microelectrodes in a thin-layer configuration results in an improved signal-to-noise ratio compared to a single macroelectrode of similar area [255]. There are two possible reasons for this: First, the contribution of lateral diffusion to the analyte flux becomes increasingly significant as the electrode perimeter-to-area ratio increases. Second, as the solute band passes over the array, alternate depletion of the diffusion layer (owing to reaction of the analyte at an upstream electrode) and replenishing of the diffusion layer (as it passes a downstream insulating area) occur. These effects, however, become significant only for small electrode diameters (<25 μm) and low flow rates (a few μl min).

A number of practical advantages arise from the use of microelectrodes in continuous-flow analysis. First, because the faradaic current is in the picoamp to nanoamp range, problems of ohmic potential drop are obviated. The usefulness of this lies in the possibility of eliminating the need for a supporting electrolyte, something that is especially useful in normal-phase LC, where organic solvents having extremely low dielectric constants are commonly used [214,256]. Eliminating the supporting electrolyte can also enhance the working potential range. Also, because of the low current levels, a two-electrode configuration can be employed, thus simplifying the electrochemical control circuit. Second, the dead volume is much smaller than can ever be achieved by macroelectrodes. For example, Knecht et al. have employed a 9-μm carbon fiber for open tubular LC where the column diameter is only 15 μm [257]. Finally, microelectrodes have a much lower dependency on flow rate.

Microelectrode arrays have been used for a number of interesting applications to LC and FI [258,259,197]. Anderson et al. have shown the use of microarray flow detectors at high positive potential for the detection of carbamate pesticides [197]. A novel feature of this work is the capacity of the electrode to be used beyond the potential limit for solvent breakdown.

Khoo et al. have described a micro platinum ring electrode that was employed for HPLC detection. Detection limits of 100 pg for catecholamines were reported [213]. The advantage of the ring electrode is that it affords a comparatively large surface area while ensuring microelectrode properties. The ring electrode is also easy to fabricate.

Although microelectrodes offer novel possibilities in continuous-flow analysis, there are a number of practical difficulties that limit their use in routine analysis with the present state of technology. For example, the measurement of faradaic currents of the order of picoamps can pose difficulties because, at that level, noise becomes comparable. Indeed, at high sensitivity, elaborate procedures, such as the use of a Faraday cage, have to be devised for noise reduction. Present methods of fabricating microelectrodes leave much to be desired. For example, noise associated with carbon-fiber microelectrodes can arise from impurities in the epoxy resin used in their fabrication. One of the important developments, therefore, is the use of microfabrication techniques such as vapor deposition and screen printing using ceramic and glass substrates [260,261].

B. Chemically Modified Electrodes

The ability to synthesize microstructures on electrode surfaces has given rise to prospects for novel electrochemical sensors with tailor-made properties to suit specific applications. From an analytical viewpoint, the goals of chemical modification include increased electrode stability, enhanced electrocatalytic performance, and exploitation of specific properties of the chemical modifier to develop new sensing mechanisms as well as to restrict conventional mechanisms that may occur at the bare electrode [262].

The practical use of chemically modified electrodes to FI-EC and LC-EC is still at its early stages, and only a few applications have been reported. The prospects are extremely bright, however, and the next few years should see a surge in activity in this area.

The simplest approach to chemical modification is to adsorb a species that can reduce the overpotential of the analyte electrode reaction. For example, Wang and Freiha adsorbed α-alumina on glassy carbon for LC-EC [263]. The use of a graphite electrode sputtered with palladium and gold has been described by Appelqvist et al. for the determination of hydrogen peroxide in flowing solutions [264]. The electrode system is interesting, combining as it does the electrocatalytic activity of palladium and gold with the stability of the graphite electrode.

Another example of the use of adsorbed modifier is due to Castner and Hawkridge, who employed a methyl viologen-modified gold foil electrode in a channel flow electrode for mediating the electron transfer of metalloproteins [265].

Adsorption on graphitic carbon electrodes is facilitated by the high pi-electron density of the basal plane [262]. Carbon electrodes have been widely used for the adsorption of organic mediators. Persson et al. [266] have described a phenoxine mediator adsorbed on graphite that facilitates the oxidation of NADH. The electrode has been used in conjunction with FI using the WJD. Use of ferrocene mediators adsorbed on graphite has been described by Cass et al. for the determination of glucose [267]. These electrodes have recently been applied to continuous-flow monitoring [268].

Carbon paste electrodes are a convenient approach to chemical modification. The chemical modifier is simply mixed together with the graphite/binder paste. Baldwin and co-workers have described the use of chemically modified carbon paste electrodes as catalytic voltammetric sensors [269-271]. Use of cobalt phthalocyanine as the modifier enables reduction in the overpotential by several hundred millivolts for the determination of oxalic and several α-keto acids [269]. Because of the reduction in overpotential, increased selectivity is engendered. For example, with the cobalt phthalocyanine carbon paste electrode, quantification of keto acids in urine was done with little sample pretreatment.

Figure 30 shows cyclic voltammograms obtained at a cobalt phthalocyanine carbon paste electrode, and Fig. 31 shows the applications of this electrode to LC-EC.

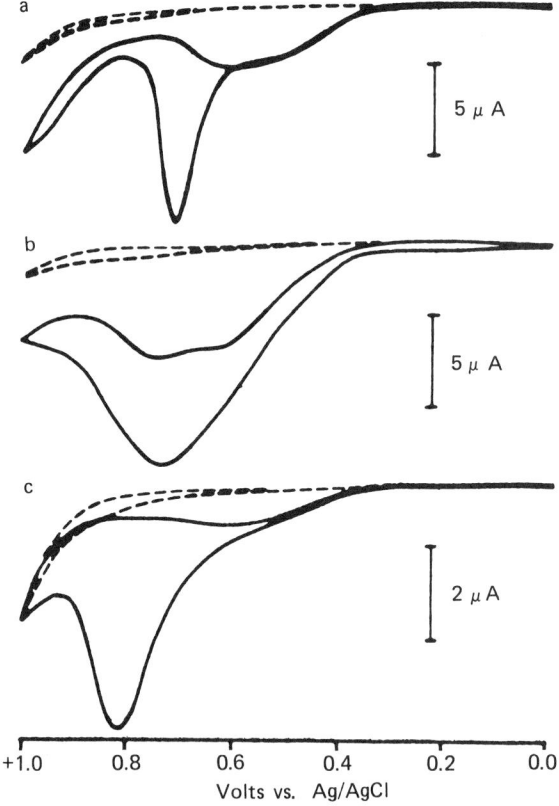

FIG. 30. Cyclic voltammograms obtained at a cobalt phthalocyanine carbon paste electrode. (From Ref. 269, with permission of the American Chemical Society.)

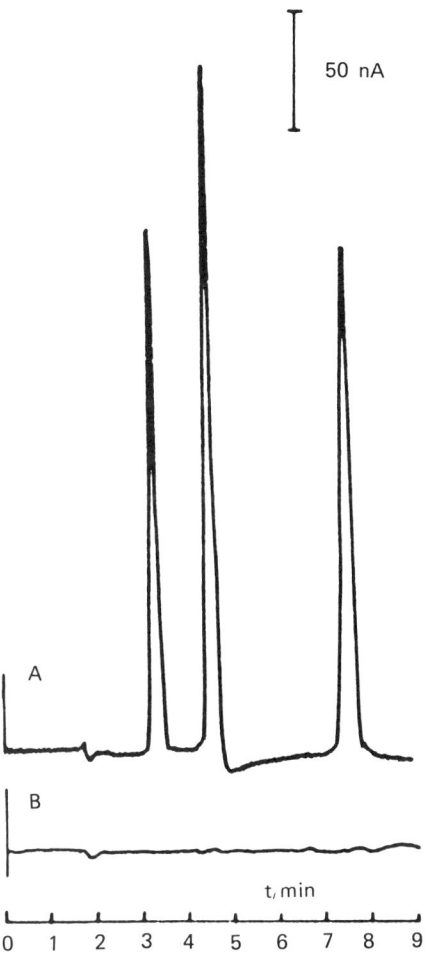

FIG. 31. LC-EC traces at a cobalt phthalocyanine-modified carbon paste electrode: A = modified electrode; B = unmodified electrode. (From Ref. 269, with permission of the American Chemical Society.)

Polymer-coated electrodes constitute another major class of chemically modified electrodes. The polymer is either electropolymerized or physically deposited on the surface from solution in a variety of ways [262]. With care, a well-defined surface can be achieved. Although characterization and use of polymer-modified electrodes is still in its infancy, experimental results that have been obtained indicate enormous potential for the development of practical devices.

Cox and Kulkarni have used a poly(4-vinyl pyridine) polymer coating impregnated with $IrCl_6^{3-}/IrCl_6^{4-}$ for the determination of nitrite [272]. In this system, the $IrCl_6^{3-}/IrCl_6^{4-}$ acts as a mediator, reducing the overpotential for the nitrite electrode reaction. The polymer coating serves the dual function of a binder for the mediator as well as a physical protective coating. For example, it attenuates interference from Pb(II), Mn(II), and Fe(II) and obviates poisoning by SCN^-.

Polypyrrole-coated electrodes have been used in the detection of electroinactive anions. The principle of operation is based on the doping of the anion into the polypyrrole electrode, which results in an anodic analytical response. The electrode is regenerated by changing the potential to a more negative one [273]. The electrode is responsive to phosphate, carbonate, and acetate and has been used to determine these species in conjunction with FI.

C. Enzyme Electrodes

An enzyme electrode consists of an enzyme layer immobilized onto an electrode, over which a semipermeable membrane is fixed. In the amperometric mode, enzyme electrodes respond to changes in substrate concentration, measuring the current produced by the oxidation or reduction of one of the species involved in the enzymatic reaction that takes place in the trapped enzyme/membrane layer. For example, an amperometric electrode for the determination of glucose can use the enzyme glucose oxidase to catalyze the reaction given by

$$\beta\text{-D Glucose} + O_2 \xrightarrow{\text{Glucose oxidase}} \text{Gluconolatone} + H_2O_2$$

Either the decrease in O_2 levels or the H_2O_2 produced can be monitored. The measured current is proportional to the substrate concentration in the bulk solution. The greatest sensitivity occurs when both the permeability and amount of enzyme on the electrode is as large as possible. However, too great a permeability can limit the dynamic range.

Various approaches have been taken for the immobilization of the enzyme layer, including adsorption, gel entrapment, and covalent coupling. The details of the various methods have been reviewed by Carr and Bowers [274] and Guibault [275]. Covalent coupling is generally the preferred approach because of the greater lifetime and stability it affords. Choice of the membrane system is an often neglected area. In actual fact, the membrane serves two critical functions. First, it serves to exclude compounds in the sample, on the basis of size, from interfering with the electrode reaction, which is a particularly vital function in the analysis of biological fluids. Second, it serves as a diffusion barrier, dependent on the membrane thickness, and thereby controls the response time and also the dynamic range of the enzyme electrode. There is a trade-off between membrane thickness and pore size and, in general, it is preferable to employ thin membranes with small pore size.

For usual applications of HDV, the assumption is made that the reaction at the electrode is instantaneous, for enzyme electrodes, however, consideration must be given to the diffusion-controlled reaction within the enzyme/membrane system. The relationship between peak shape and hydrodynamic characteristics of enzyme electrodes operating in hydrodynamic systems has been described theoretically [276–282]. Mell and Maloy have used digital simulation to model the steady-state response of amperometric electrodes [281]. In general, for thick membrane systems, the response of enzyme electrodes is relatively unaffected by convective diffusion from the bulk solution, and transport within the membrane/

enzyme layer takes place mainly by diffusion. For thin membranes, however, some influence is expected.

Amperometric enzyme electrodes have recently gained much interest for their continuous-monitoring applications, particularly in combination with FI [282–287]. Olsson and co-workers have given a theoretical description of amperometric enzyme/membrane electrodes in FI [282]. The response curve was derived assuming a rectangular (plug) injection, Gaussian dispersion, and a diffusion-limited electrode response. The interaction of the sample profile with the enzyme/membrane system is treated as a series of independent processes. The conclusion was that a membrane electrode with small characteristic diffusion time and high enzyme activity should be used to ensure high sample throughput and high sensitivity. Sample throughput of 300 samples/hr was stated, and good precision of less than 1% is feasible. The sample volume can have an important bearing on the dynamic range. In general, the smaller the volume, the higher the upper limit, and vice versa.

Practical applications of enzyme electrodes in continuous monitoring include the monitoring of biochemical processes (e.g., fermentation), on-line patient monitoring, and clinical analysis.

VII. ON-LINE MONITORING APPLICATIONS

A. Process Monitoring

Despite the progress that has been made in the development of chemical process-control systems, the sensor element remains the weak link. Although a wide range of sensors are available for the measurement of physical parameters such as conductivity, flow, temperature, pressure, and volume, apart from the pH electrode, only a few applications of gas sensors and ion-selective electrodes have appeared.

Chemical sensors, including voltammetric sensors, fail primarily because they are unable to meet the stringent requirements of process control. Most significantly, they often cannot be used in harsh environments over long periods. There is also the problem that interfering compounds, usually present in complex chemical processes, can mask the analytical signal and lead to deterioration of the sensor.

In general, there are three approaches to process monitoring: off-line, in-line, and on-line. In off-line methods, a discrete sample is obtained manually from the process, and the analysis is carried out in some remote laboratory site. The main drawback with this approach is that real-time control is difficult to achieve. Also, it is manpower-intensive and prone to sampling errors.

With in-line monitoring, the detector is simply placed in the process stream. Here the problem is that sample conditioning is not possible and, consequently, the analyte could be masked by interfering compounds present in the process. There is also a good chance that the working electrode performance will eventually deteriorate because of electrode fouling and poisoning.

On-line monitoring seeks to combine sample conditioning with real-time control. The process stream is automatically sampled and, by the use of a continuous-flow system, sample conditioning can be done on-line. Whereas both segmented and nonsegmented flows can be employed, the latter—in particular—FI, is preferable.

FI techniques can be used to sample automatically the process stream. The ability to handle small sample volumes permits rapid sample throughput. This is an important advantage as it enables real-time control. Sample conditioning can be effected through a number of well-developed on-line techniques, such as dialysis, solvent extraction, and derivazation. Advances in the FI technique in the last few years have produced some elegant demonstrations of its utility. Van der Linden [288] has given a short review on the development of FI techniques for on-line process control. Fleet and co-workers have summarized the important requirements for HDV systems to be suitable for use in process control [289]. The basic limitation is the changing electrode characteristics over long periods of use.

One of the few reported works of process monitoring by FI-EC was provided to Lundback [290], who described the use of amperometric detection for the determination of hydrogen peroxide in pickling baths for copper and copper alloys. A sample throughput of 75 hr^{-1} was quoted.

Biochemical Processes

Use of voltammetric detectors in the control of biochemical processes is attracting increasing interest. Enzyme electrodes have shown particular value for the selective monitoring of specific species [274]. For example, the monitoring of alcohol levels in fermentation systems can be done with an alcohol oxidase-based amperometric electrode. Although enzyme-modified electrodes have been shown to work well, they do suffer from limited operational lifetime. An alternative approach using an immobilized enzyme reactor unit prior to the measurement step is showing considerable promise [274].

The use of electrodes with immobilized biologically active cells for the monitoring of biochemical processes is an interesting area of development. Whole cells, as well as plant and animal tissues, have been used [291]. As an example of the practical application of these detectors, Hikume et al. have described the use of a microbial sensor for the on-line measurement of acetic acid in fermentation processes [292].

B. Patient Monitoring

With the persistent demand for improved health care, there is increasing interest in on-line patient monitoring. The applications of HDV sensors in this regard have great potential. The applications of enzyme electrodes, referred to previously, is one major area of activity.

In a typical on-line patient-monitoring system, discrete samples are obtained from the patient and passed through various sample conditioning units before injection into the detector. One simple approach is to use an artery-vein shunt similar to the one used in kidney dialysis [293].

Perhaps the most effective demonstration of HDV systems in the area of patient monitoring has been in the monitoring of glucose levels in blood [293–295]. These systems can be used to predict insulin requirements for a portable insulin pump. Figure 32 shows a typical system for on-line glucose monitoring.

The ultimate goal of on-line patient monitoring is the development of fully implantable sensors that can continuously monitor physiologically important species, such as blood glucose. Despite significant progress, really useful devices are still some way off. The main problem is the limited lifetime of the enzyme electrode and the deterioration of the materials used in the construction of the sensors [296].

Corrosion Monitoring

Electrochemical measurements are commonly used for the study of corrosion [297]. The advantage of continuous-flow methods in most cases is that they enable convenient study of the effect of hydrodynamic parameters on the corrosion rate [298]. General information often sought

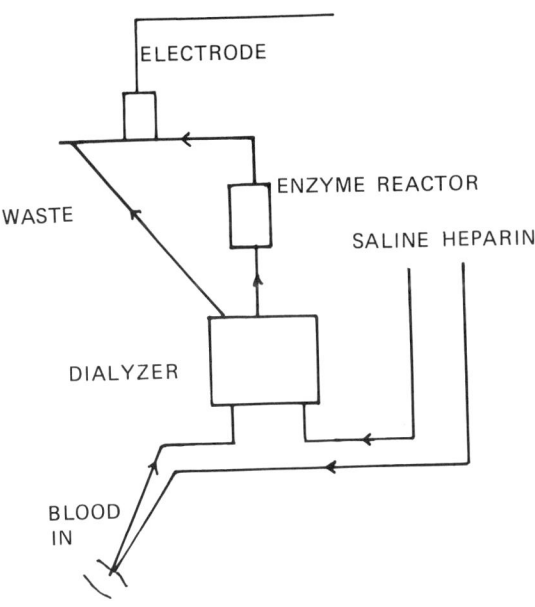

FIG. 32. Glucose analyzer for on-line patient monitoring.

in corrosion studies includes the rate of corrosion, the corrosion mechanism, knowledge of the effect of hydrodynamic parameters, and the composition of the solution surrounding the corroding element of interest. Among various possible configurations, the tubular and WJ electrodes have been used for on-line corrosion studies.

C. Computer Control and Automation

The computer is gaining increasing usage for experimental control in FI-EC and LC-EC. Apart from the well-developed use of computers for generating the applied potential waveform and acquiring and processing experimental data, a less recognized use is the synchronization, timing, and control of the actual physical operations, such as sample injection and carrier stream selection. In FI-EC techniques in which high sample throughput is desirable, computer control is essential for synchronizing operations, which may be carried out simultaneously or in sequence. Examples of the use of computer control in various FI-EC and LC-EC techniques have been described in previous sections.

One of the general advantages of using a computer is the flexibility afforded by software control. A change in any control operation simply involves changing the program without the need to alter the hardware. Apart from flexibility, software control can also provide a user-friendly interface that enables easier handling of the instrumental control by inexperienced persons. The increasing availability of improved software and the continual decrease in price should see significant improvements in the development of user interfaces.

The present need in on-line process monitoring is for more intelligent systems that can function independently without human intervention. This is particularly important in on-line process-monitoring applications where there is a requirement for use over extended periods of time. For example, where there is deterioration in the electrode performance because of surface fouling, a computer can be used to track the electrode response and compensate for the change. Several techniques along this line have been developed. These include background subtraction as applied to ASV and RSV and common mode noise rejection in dual-electrode systems [33, 233].

The concept of the smart detector, in which a microprocessor is dedicated to a single detector, has been proposed [289]. The smart detector would be capable of routinely performing self-diagnostic and self-calibration functions. The ultimate concept of the smart detector is the integration of all the functions of the detector, including the associated electronics, into a single system.

GLOSSARY OF ABBREVIATIONS

HDV	Hydrodynamic voltammetry
TLD	Thin-layer detector
WJD	Wall-jet detector
LC	Liquid chromatography
LC-EC	Liquid chromatography-electrochemical detection
FIA	Flow-injection analysis
FI-EC	Flow injection-electrochemical detection
CV	Cyclic voltammetry
RSV	Rapid-scan voltammetry
DME	Dropping-mercury electrode
MFE	Mercury-film electrode
ASV	Anodic stripping voltammetry
RRDE	Rotating ring-disk electrode
RDE	Rotating disk electrode
Re	Reynolds number

REFERENCES

1. O. H. Muller, J. Am. Chem. Soc. 69:2992 (1947).
2. W. Kemula, Rocz. Chem. 26:281 (1952).
3. W. Nernst, Z. Phys. Chem. 47:52 (1904).

4. V. G. Levich, *Physicochemical Hydrodynamics*, Prentice-Hall, Englewood Cliffs, NJ, 1962.
5. H. Matsuda, J. Electroanal. Chem. *16*:153 (1968).
6. J. Jordan, R. A. Javick, and W. E. Ranz, J. Am. Chem. Soc. *80*:3846 (1958).
7. H. Matsuda and J. Yamada, J. Electroanal. Chem. *30*:261 (1971).
8. W. J. Blaedel and L. N. Klatt, Anal. Chem. *38*:879 (1966).
9. H. Matsuda, J. Electroanal. Chem. *15*:109 (1967).
10. J. Yamada and H. Matsuda, J. Electroanal. Chem. *44*:189 (1973).
11. W. J. Albery and C. M. A. Brett, J. Electroanal. Chem. *148*:201 (1983).
12. H. Gunasingham and B. Fleet, Anal. Chem. *55*:1409 (1983).
13. D. T. Chin and C. H. Tsang, J. Electrochem. Soc. *125*:1461 (1978).
14. P. L. Meschi, D. C. Johnson, and G. R. Luecke, Anal. Chim. Acta *124*:315 (1981).
15. S. Prabhu and J. L. Anderson, Anal. Chem. *59*:157 (1987).
16. M. B. Glauert, J. Fluid Mech. *1*:625 (1956).
17. H. Gunasingham, K. P. Ang, and C. C. Ngo, Anal. Chem. *57*:505 (1985).
18. H. Gunasingham, K. P. Ang, C. C. Ngo, P. C. Thiak, and B. Fleet, J. Electroanal. Chem. *186*:51 (1985).
19. W. J. Albery, J. Electroanal. Chem. *191*:1 (1985).
20. M. Varadi, Anal. Chim. Acta *80*:31 (1975).
21. M. Varadi, M. Gratzl, and E. Pungor, Anal. Chim. Acta *83*:1 (1976).
22. J. M. Elbicki, D. M. Morgan, and S. G. Weber, Anal. Chem. *56*:978 (1984).
23. B. Fleet and C. J. Little, J. Chromatogr. Sci. *12*:747 (1974).
24. H. Gunasingham and B. Fleet, J. Chromatogr. *43*:261 (1983).
25. H. Gunasingham, Anal. Chim. Acta *159*:139 (1984).
26. H. Gunasingham, B. T. Tay, K. P. Ang, and L. L. Koh, J. Chromatogr. *285*:103 (1984).
27. H. Gunasingham, B. T. Tay, and K. P. Ang, Anal. Chem. *56*:2422 (1984).
28. W. E. van der Linden and J. W. Dieker, Anal. Chim. Acta *119*:1 (1980).

29. B. Fleet and H. Gunasingham, *Proceedings of the International Meeting on Chemical Sensors*, Kodansha, Fukuoka, Japan, 1983, p. 558.
30. J. Wang and H. D. Dewald, Anal. Chem. *56*:156 (1984).
31. J. Wang and B. A. Freiha, Anal. Chem. *57*:1776 (1985).
32. W. J. Albery and C. M. A. Brett, J. Electroanal. Chem. *148*:211 (1983).
33. H. Gunasingham, K. P. Ang, C. C. Ngo, and P. C. Thiak, J. Electroanal. Chem. *198*:27 (1986).
34. C. Wechter, N. Sleszynski, J. J. O'Dea, and J. Osteryoung, Anal. Chim. Acta *175*:45 (1985).
35. G. J. Moody, G. S. Sanghera, and J. D. R. Thomas, Analyst *111*(6):605 (1986).
36. P. T. Kissinger, J. Chem. Educ. *60*:308 (1983).
37. P. T. Kissinger, Anal. Chem. *49*:447A (1977).
38. J. A. Wise, W. R. Heineman, and P. T. Kissinger, Anal. Chim. Acta *172*:1 (1985).
39. S. G. Weber, J. Electroanal. Chem. *145*:1 (1983).
40. K. Brunt and C. H. P. Bruins, J. Chromatogr. *172*:39 (1979).
41. Y. Hirata, P. T. Lin, M. Novotny, and R. M. Wightman, J. Chromatogr. *181*:287 (1980).
42. J. L. Anderson and S. Moldoveanu, J. Electroanal. Chem. *175*:67 (1984).
43. J. L. Anderson and S. Moldoveanu, J. Electroanal. Chem. *179*:107 (1984).
44. K. Aoki, K. Tokuda, and H. Matsuda, J. Electroanal. Chem. *79*:49 (1977).
45. W. J. Albery, *Electrode Kinetics*, Clarendon Press, Oxford, England, 1975.
46. H. B. Hanekamp and H. J. van Nieuwkerk, Anal. Chim. Acta *121*:13 (1980).
47. H. B. Hanekamp and H. G. deJong, Anal. Chim. Acta *135*:351 (1982).
48. H. Gunasingham, B. T. Tay, and K. P. Ang, Anal. Chim. Acta *176*:143 (1985).
49. D. J. Curran and T. P. Tougas, Anal. Chem. *56*:672 (1984).
50. W. J. Blaedel and J. Wang, Anal. Chem. *51*:799 (1979).

51. J. Wang and H. D. Dewald, J. Chromatogr. 285:281 (1984).
52. J. Wang, Anal. Chim. Acta 127:157 (1981).
53. P. Jayaweera and L. Ramaley, Anal. Instrum. 15:259 (1986).
54. D. C. Johnson, S. G. Weber, A. M. Bond, R. M. Wightman, R. E. Shoup, and I. S. Krull, Anal. Chim. Acta 180:193 (1986).
55. J. Lankelma and H. Poppe, J. Chromatogr. 125:375 (1976).
56. J. M. Davis and J. C. Giddings, Anal. Chem. 55:418 (1983).
57. D. M. Morgan and S. G. Weber, Anal. Chem. 56:2560 (1984).
58. S. G. Weber, Electroanalytical Symposium, BAS Press, West Lafayette, IN, 1985, p. 100.
59. D. J. Miner, Anal. Chim. Acta 134:101 (1982).
60. H. W. van Rooijen and H. Poppe, J. Liq. Chromatogr. 6:2157 (1983).
61. W. M. Peterson, Am. Lab. 11:69 (1979).
62. D. G. Swatzfager, Anal. Chem. 48:2189 (1976).
63. W. A. McCrehan and R. A. Durst, Anal. Chem. 50:2108 (1978).
64. A. McDonald and P. D. Duke, J. Chromatogr. 83:331 (1973).
65. D. T. Sawyer and J. L. Roberts Jr., Experimental Electrochemistry for Chemists, Wiley, Toronto, 1974.
66. L. R. Snyder, Anal. Chim. Acta 114:3 (1980).
67. C. J. Little, Ph.D. Thesis, London University, London, England, 1975.
68. J. Ruzicka, E. H. Hansen, and A. U. Ramsing, Anal. Chim. Acta 134:55 (1982).
69. E. H. Hansen and J. Ruzicka, J. Chem. Educ. 56:677 (1979).
70. H. Gunasingham, K. P. Ang, and B. Fleet, Electroanalytical Symposium, BAS Press, W. Lafayette, IN, 1985.
71. J. F. Harrow and J. Janata, Anal. Chem. 55:2461 (1983).
72. B. Rocks and C. Riley, Clin. Chem. 28:409 (1982).
73. C. Riley, L. H. Assiett, B. F. Rocks, R. A. Sherwood, J. D. Watson, and J. Morgan, Clin. Chem. 29:332 (1983).
74. A. F. Kapauan and M. C. Magno, Anal. Chem. 58:509 (1986).
75. P. T. Kissinger, in P. T. Kissinger and W. R. Heineman (Eds.), Laboratory Techniques in Electroanalytical Chemistry, Marcel Dekker, New York, 1984, pp. 611–635.

76. J. Tenygl and B. Fleet, *Electroanalytical Symposium*, BAS Press, West Lafayette, IN, 1985, p. 84.
77. A. M. Bond and N. N. McLachlan, J. Electroanal. Chem. *182*:367 (1985).
78. A. M. Bond and N. M. McLachlan, Anal. Chem. *58*:756 (1986).
79. A. M. Bond, *Modern Polarographic Techniques in Analytical Chemistry*, Marcel Dekker, New York, 1984, p. 611.
80. S. B. Thomson, D. J. Tucker, and M. H. Briggs, J. Chromatogr. *338*:205 (19XX).
81. A. M. Bond, I. D. Heritage, G. G. Wallace, and M. J. McCormick, Anal. Chem. *54*:582 (1982).
82. J. E. Anderson, A. M. Bond, I. D. Heritage, R. D. Jones, and G. G. Wallace, Anal. Chem. *54*:1702 (1982).
83. A. M. Bond, H. A. Hudson, and P. A. van den Boash, Anal. Chim. Acta *127*:121 (1981).
84. A. G. Fogg and A. M. Summan, Analyst *109*:1029 (1984).
85. A. Kimla and F. Strafelda, Collect. Czech. Chem. Commun. *29*:2913 (1964).
86. Y. Okinaka and I. M. Kolthoff, J. Electroanal. Chem. *73*:3326 (1957).
87. W. J. Blaedel and J. H. Strohl, Anal. Chem. *33*:1631 (1961).
88. H. B. Hanekamp, W. H. Voogt, P. Bos, and R. W. Frei, Anal. Chim. Acta *118*:81 (1980).
89. K. Toth, G. Nagy, Z. Feher, G. Horvoi, and E. Pungor, Anal. Chim. Acta *114*:45 (1980).
90. W. Kutner, J. Debowski, and W. Kemula, J. Chromatogr. *191*:47 (1980).
91. L. Michel and A. Zatka, Anal. Chim. Acta *105*:109 (1979).
92. H. B. Hanekamp, P. Bos, and R. W. Frei, J. Chromatogr. *186*:489 (1979).
93. H. B. Hanekamp, W. H. Voogt, P. Bos, and R. W. Frei, Anal. Lett. *12*:175 (1979).
94. S. J. Lyle and M. I. Saleh, Talanta *28*:251 (1981).
95. W. Kutner and W. Kemula, Chromatographia *17*:322 (1983).
96. J. Wang, E. Ouziel, C. Yarnitzky, and M. Ariel, Anal. Chim. Acta *102*:99 (1978).
97. Z. Kowalski and W. Kubiak, Anal. Chim. Acta *159*:129 (1984).

98. A. M. Bond, H. A. Hudson, and P. A. van den Bosch, Anal. Chim. Acta 127:121 (1981).
99. F. Elferink, W. J. F. Van der Vijgh, and H. M. Pinedo, Anal. Chem. 58:2293 (1986).
100. T. M. Florence, J. Electroanal. Chem. 27:273 (1970).
101. S. Stojek, B. Stepnik, and Z. Kublik, J. Electroanal. Chem. 74:277 (1976).
102. H. Gunasingham, B. T. Tay, and K. P. Ang, Anal. Chem. 58:1578 (1986).
103. P. W. Alexander and U. Akapongkul, Anal. Chim. Acta 148:103 (1983).
104. K. Bratin, P. T. Kissinger, and C. S. Bruntlet, J. Liq. Chromatogr. 4:1777 (1981).
105. R. Woods in A. J. Bard, Ed., Electroanalytical Chemistry, Vol. 9, Marcel Dekker, New York, 1976, p. 20.
106. J. Giner, Electrochim. Acta 9:63 (1964).
107. R. N. Adams, Electrochemistry at Solid Electrodes, Marcel Dekker, New York, 1969.
108. C. Bollett, P. Oliva, and M. Caude, J. Chromatogr. 149:625 (1977).
109. A. MacDonald and P. D. Duke, J. Chromatogr. 83:331 (1981).
110. D. S. Austin, J. A. Polta, P. C. Tang, T. D. Cabelka, and D. C. Johnson, J. Electroanal. Chem. 168:227 (1984).
111. J. A. Polta and D. C. Johnson, J. Liq. Chromatogr. 6:1727 (1983).
112. B. S. Hui and C. O. Huber, Anal. Chim. Acta 134:211 (1982).
113. W. T. Kok, U. A. T. Brinkman, and R. W. Frei, J. Chromatogr. 265:17 (1983).
114. R. J. Davenport and D. C. Johnson, Anal. Chem. 46:1971 (1974).
115. T. N. Morrison, K. G. Schick, and C. O. Huber, Anal. Chim. Acta 120:75 (1980).
116. W. Buchberger, K. Winsaer, and C. Breitwieser, Fresenius Z. Anal. Chem. 311:517 (1982).
117. R. E. Panzer and P. J. Elving, Electrochim. Acta 20:635 (1975).
118. D. Laser and M. Ariel, J. Electroanal. Chem. 52:291 (1974).
119. T. Nagaoka and T. Yoshino, Anal. Chem. 58:1037 (1986).
120. H. Gunasingham and B. Fleet, Analyst 107:896 (1982).

121. L. Falat and H. Y. Cheng, Anal. Chem. 54:2108 (1982).

122. D. C. S. Tse and T. Kuwana, Anal. Chem. 20:1315 (1978).

123. B. R. Hepler, S. G. Weberand, and W. C. Purdy, Anal. Chim. Acta 102:41 (1978).

124. W. L. Caudill, G. P. Houck, and R. M. Wightman, J. Chromatogr. 227:331 (1982).

125. F. C. Cowlard and J. C. Lewis, J. Material Sci. 2:507 (1967).

126. P. T. Kissinger, C. Refshauge, R. Dreiling, and R. N. Adams, Anal. Lett. 6:465 (1973).

127. R. Keller, A. Oke, I. Mefford, and R. N. Adams, Life Sci. 19:995 (1976).

128. E. J. Caliguri and I. N. Mefford, Brain Res. 296:156 (1984).

129. M. E. Rice, Z. Galus, and R. N. Adams, J. Electroanal. Chem. 143:89 (1983).

130. J. Lindquist, Anal. Chem. 45:1006 (1973).

131. K. Stulik, V. Pacakova, and B. Starkova, J. Chromatogr. 213:41 (1981).

132. K. Stulik and V. Pacakova, J. Chromatogr. 203:269 (1981).

133. R. J. Fenn, S. Siggia, and D. J. Curran, Anal. Chem. 50:1067 (1979).

134. D. N. Armentrout, J. D. McLean, and M. W. Long, Anal. Chem. 51:1039 (1979).

135. J. E. Anderson, D. E. Tallman, D. J. Chesney, and J. L. Anderson, Anal. Chem. 50:1051 (1978).

136. D. J. Chesney, J. L. Anderson, D. E. Weisshaar, and D. E. Tallman, Anal. Chim. Acta 124:321 (1981).

137. D. E. Weisshar, D. E. Tallman, and J. L. Anderson, Anal. Chem. 53:1809 (1981).

138. L. A. Knecht, E. J. Guthrie, and J. W. Jorgenson, Anal. Chem. 56:479 (1984).

139. W. L. Caudill, J. O. Howell, and R. M. Wightman, Anal. Chem. 54:2532 (1982).

140. K. Stulik, V. Pacakova, and M. Podolak, J. Chromatogr. 298:225 (1984).

141. J. Zak and T. Kuwana, J. Am. Chem. Soc. 104:5514 (1982).

142. J. Zak and T. Kuwana, J. Electroanal. Chem. 150:645 (1983).

143. A. N. Strohl and D. J. Curran, Anal. Chem. 51:1045 (1979).
144. D. J. Curran and T. P. Tougas, Anal. Chem. 56:672 (1984).
145. R. C. Engstrom, Anal. Chem. 56:2310 (1982).
146. R. C. Engstrom and V. A. Strasser, Anal. Chem. 56:136 (1984).
147. H. W. Rooijen and H. Poppe, Anal. Chim. Acta 130:9 (1981).
148. J. Wang and L. D. Hutchins, Anal. Chim. Acta 167:325 (1985).
149. K. Ravichandran and R. P. Baldwin, Anal. Chem. 56:1744 (1984).
150. K. Ravichandran and R. P. Baldwin, Anal. Chem. 55:1782 (1983).
151. J. Wang and T. Peng, Anal. Chem. 58:1787 (1986).
152. D. T. Fagan, I. F. Hu, and T. Kuawana, Anal. Chem. 57:2759 (1985).
153. D. E. Seisshaar and T. Kuwana, Anal. Chem. 57:378 (1985).
154. B. Kazee, D. E. Weisshaar, and T. Kuwana, Anal. Chem. 57:2736 (1985).
155. G. N. Kamau, W. S. Willis, and J. F. Rusling, Anal. Chem. 57:545 (1985).
156. T. A. Berger, Hewlett-Packard Corp., Avondale, PA, U.S. Patent No. 4,496,454, Jan. 29, 1985.
157. J. Ruzicka and E. H. Hansen, Anal. Chim. Acta 114:19 (1985).
158. J. H. Dieker and W. E. Van der Linden, Anal. Chim. Acta 114:267 (1980).
159. H. K. Chan and A. G. Fogg, Anal. Chim. Acta 111:281 (1979).
160. A. Ivaska and F. Smyth, Anal. Chim. Acta 114:283 (1980).
161. B. Pihalar and L. Kosta, Anal. Chim. Acta 114:275 (1980).
162. N. Thørgerson, J. Janata, and J. Ruzicka, Anal. Chem. 55:1488 (1983).
163. T. P. Tougas, J. M. Jannetti, and W. G. Collier, Anal. Chem. 57:1377 (1985).
164. J. Wang, *Stripping Analysis*, VCH Publishers, Deerfield Beach, FL, 1985.
165. T. R. Copeland and R. K. Skogerboe, Anal. Chem. 46:1257A (1974).
166. M. Stulikova, J. Electroanal. Chem. 48:33 (1973).
167. G. E. Bately and T. M. Florence, J. Electroanal. Chem. 55:23 (1974).

168. Z. Yoshida and S. Kihara, J. Electroanal. Chem. 95:159 (1979).
169. J. Wang and M. Ariel, Anal. Chim. Acta 101:1 (1978).
170. J. Wang and H. D. Dewald, Anal. Chem. 56:156 (1984).
171. J. Wang, Am. Lab. 7:14 (1983).
172. S. H. Lieberman and A. Zirino, Anal. Chem. 46:20 (1974).
173. A. C. M. Almeida Motu, J. Buffle, S. P. Kounaves, and M. Goncalves, Anal. Chim. Acta 172:13 (1985).
174. H. Gunasingham, K. P. Ang, and C. C. Ngo, J. Electroanal. Chem. 215:123 (1986).
175. H. Gunasingham, K. P. Ang, and C. C. Ngo, J. Electroanal. Chem., in press.
176. R. W. Andrews and D. C. Johnson, Anal. Chem. 48:1056 (1976).
177. J. Wang and G. Greene, J. Electroanal. Chem. 154:261 (1983).
178. D. Jagner, M. Josefson, and K. Aren, Anal. Chim. Acta 141:147 (1982).
179. W. T. deVries and E. van Dalen, J. Electroanal. Chem. 14:315 (1967).
180. P. Valenta, L. Mart, and H. Rutzel, J. Electroanal. Chem. 82:327 (1977).
181. T. A. Berger, Ph.D. Thesis, London University, London, England, 1975.
182. J. Wang and H. D. Dewald, Anal. Chem. 55:933 (1983).
183. J. Wang and M. Ariel, Anal. Chim. Acta 99:89 (1978).
184. M. Wojciechowski, W. Go, and J. Osteryoung, Anal. Chem. 57:155 (1985).
185. L. Kryger and D. Jagner, Anal. Chim. Acta 80:255 (1975).
186. S. D. Brown and B. R. Kowalski, Anal. Chim. Acta 107:227 (1979).
187. A. M. Bond and B. S. Grabaric, Anal. Chim. Acta 88:227 (1977).
188. J. H. Christie and R. A. Osteryoung, Anal. Chem. 48:869 (1976).
189. C. A. H. Chambers and J. K. Lee, J. Electroanal. Chem. 14:309 (1967).
190. J. Wang and B. A. Freiha, Anal. Chem. 55:1285 (1983).
191. A. M. Krstulovic, H. Colin, and G. A. Guiochon, in *Advances in Chromatography*, Vol. 0, Marcel Dekker, New York, 198.

192. K. Brunt, *Trace Analysis*, Vol. 1, Academic Press, Orlando, FL, 1981.
193. P. T. Kissinger and W. R. Heineman, Eds., *Laboratory Techniques in Electroanalytical Chemistry*, Marcel Dekker, NY, 1984.
194. M. Lemar and M. Porthault, J. Chromatogr. *130*:372 (1977).
195. O. Hiroshima, S. Ikenoya, M. Ohmae, and K. Kawabe, Chem. Pharm. Bull. *29*:451 (1981).
196. I. N. Mefford, J. Neurosci, Methods *3*:207 (1981).
197. J. L. Anderson, K. K. Whiten, J. D. Brewster, T. Y. Ou, and W. K. Nonidez, Anal. Chem. *57*:1366 (1985).
198. F. Kreuzig and J. Frank, J. Chromatogr. *218*:615 (1981).
199. J. M. Wilson, J. T. Slattery, A. J. Forte, and S. D. Nelson, J. Chromatogr. *227*:453 (1982).
200. V. Concialini, G. Chiavari, and P. Vitali, J. Chromatogr. *258*:244 (1983).
201. R. V. Smith and D. W. Humphrey, Anal. Lett. *14*:601 (1981).
202. E. M. Lores, T. R. Edgerton, and R. F. Moseman, J. Chromatogr. Sci. *19*:466 (1981).
203. M. A. Alawi and H. A. Reussel, Chromatographia *14*:704 (1981).
204. Z. Jin and S. M. Rappaport, Anal. Chem. *55*:1778 (1983).
205. R. E. Reim, Anal. Chem. *55*:1188 (1983).
206. W. A. MacCrehan and R. A. Durst, Anal. Chem. *53*:1700 (1981).
207. W. A. McCrehan, Anal. Chem. *53*:74 (1981).
208. J. G. Koen and J. F. K. Huber, Anal. Chim. Acta *51*:303 (1970).
209. R. Samuelson and R. A. Osteryoung, Anal. Chim. Acta *123*:97 (1981).
210. H. Gunasingham, B. T. Tay, and K. P. Ang, J. Chromatogr. *159*:139 (1985).
211. G. W. Schieffer, Anal. Chem. *57*:2745 (1985).
212. S. R. Mikkelson and W. C. Purdy, Anal. Chem. *59*:244 (1987).
213. S. B. Khoo, H. Gunasingham, K. P. Ang, and B. T. Tay, J. Electroanal. Chem. *216*:115 (1987).
214. S. B. Khoo, H. Gunasingham, and K. P. Ang, submitted.
215. K. Bratin and P. T. Kissinger, J. Liq. Chromatogr. *4*:321 (1981).
216. D. A. Roston, R. E. Shoup,, and P. T. Kissinger, Anal. Chem. *54*:1417 (1982).

217. R. E. Shoup and G. S. Mayer, Anal. Chem. 54:1164 (1982).
218. D. A. Roston and P. T. Kissinger, Anal. Chem. 53:1695 (1981).
219. D. M. Radzik, J. S. Brodbelt, and P. T. Kissinger, Anal. Chem. 56:2927 (1984).
220. S. G. Weber and W. C. Purdy, Anal. Chem. 54:1757 (1982).
221. S. A. McCintock and W. C. Purdy, Anal. Chim. Acta 148:127 (1983).
222. L. D. Hutchins-Kumar, J. Wang, and P. Tuzhi, Anal. Chem. 58:1019 (1986).
223. D. A. Roston and P. T. Kissinger, Anal. Chem. 54:429 (1982).
224. C. E. Lunte, P. T. Kissinger, and R. E. Shoup, Anal. Chem. 57:1541 (1985).
225. W. R. Matson, P. Langlais, L. Volicer, P. H. Gamache, E. Bird, and K. A. Mark, Clin. Chem. 30:1477 (1984).
226. W. L. Caudill, A. G. Ewing, S. Jones, and R. M. Wightman, Anal. Chem. 53:1877 (1983).
227. H. H. J. L. Ploegmakers, M. J. M. Mertens, and W. J. van Oort, Anal. Chim. Acta 174:71 (1985).
228. J. G. White, R. L. St. Claire, and J. W. Jorgenson, Anal. Chem. 58:293 (1986).
229. J. G. White and J. W. Jorgenson, Anal. Chem. 58:2992 (1986).
230. A. Trojanek and H. De Jong, Anal. Chim. Acta 141:115 (1982).
231. R. Samuelsson, J. O'Dea, and J. Osteryoung, Anal. Chem. 52:2215 (1980).
232. J. Wang and H. D. Dewald, Anal. Chim. Acta 153:325 (1983).
233. H. Gunasingham, B. T. Tay, and K. P. Ang, Anal. Chem. 59:262 (1987).
234. I. S. Krull, C. M. Selavka, X.-D. Ding, K. Bratin, and G. Forcier, J. Forensic Sci. 29:449 (1984).
235. W. R. LaCourse and I. S. Krull, Trends Anal. Chem. 4:118 (1985).
236. W. R. LaCourse, I. S. Krull, and K. Bratin, Anal. Chem. 57:1810 (1985).
237. J. M. Elbicki, D. M. Morgan, and S. G. Weber, Anal. Chem. 57:1746 (1985).
238. S. G. Weber, D. M. Morgan, and J. M. Elbicki, Clin. Chem. 29:1665 (1983).
239. J. Ye, R. P. Baldwin, and K. Ravichandran, Anal. Chem. 58:2337 (1986).

240. H. Hojabri, A. G. Lavin, G. G. Wallace, and J. M. Riviello, Anal. Chem. 59:54 (1987).
241. A. M. Bond and G. G. Wallace, Anal. Chem. 55:718 (1983).
242. U. T. Kok, U. A. Brinkman, and R. W. Frei, Anal. Chim. Acta 162:19 (1984).
243. W. J. Albery, L. R. Svanberg, and P. J. Wood, J. Electroanal. Chem. 162:45 (1984).
244. J. A. Polta, D. C. Johnson, and K. E. Merkei, J. Chromatogr. 324:407 (1985).
245. S. Hughes and D. C. Johnson, Anal. Chim. Acta 132:11 (1981).
246. S. Hughes and D. C. Johnson, Anal. Chim. Acta 149:149 (1983).
247. T. Z. Polta and D. C. Johnson, J. Electroanal. Chem. 209:159 (1986).
248. T. Z. Polta, G. R. Luecke, and D. C. Johnson, J. Electroanal. Chem. 209:171 (1986).
249. G. G. Neuberger and D. C. Johnson, Anal. Chem. 59:150 (1987).
250. L. Dalgard, Trends Anal. Chem. 7:185 (1986).
251. R. M. Wightman, Anal. Chem. 53:1125A (1981).
252. J. O. Howell and R. M. Wightman, Anal. Chem. 56:524 (1984).
252a. R. M. Wightman and D. O. Wipf, this volume, Chap. 3.
253. S. Moldoveanu and J. L. Anderson, J. Electroanal. Chem. 185:239 (1985).
254. L. E. Fosdick and J. L. Anderson, Anal. Chem. 58:2481 (1986).
255. D. K. Cope and D. E. Tallman, J. Electroanal. Chem. 188:21 (1985).
256. J. W. Bixler and A. M. Bond, Anal. Chem. 58:2859 (1986).
257. L. A. Knecht, E. J. Guthrie, and J. W. Jorgenson, Anal. Chem. 56:479 (1984).
258. W. L. Caudill, J. O. Howell, and R. M. Wightman, Anal. Chem. 54:2532 (1982).
259. W. L. Caudill, A. G. Ewing, S. Jones, and R. M. Wightman, Anal. Chem. 55:1877 (1983).
260. W. Thormann, P. Van den Bosch, and A. M. Bond, Anal. Chem. 57:2764 (1985).
261. M. A. de Abreu and W. C. Purdy, Anal. Chem. 59:204 (1987).

262. R. M. Murray in A. J. Bard, Ed., *Electroanalytical Chemistry*, Vol. 13, Marcel Dekker, New York, 198X.

263. J. Wang and B. Freiha, Anal. Chem. 56:2266 (1984).

264. R. Appelqvist, G. Marko-Varga, L. Goton, and G. Johansson, *Proceedings of the 2nd International Conf. on Chemical Sensors*, Elsevier, 1986, p. 6.

265. J. F. Castner and F. M. Hawkridge, J. Electroanal. Chem. 143:217 (1983).

266. B. Persson, L. Gorton, and G. Johansson, *Proceedings of the 2nd International Conference on Chemical Sensors*, Elsevier, 1986, p. 11.

267. A. E. G. Cass, G. Davis, G. D. Francis, H. A. O. Hill, W. J. Ashton, I. J. Higgens, E. V. Plotkin, L. D. Scott, and A. P. F. Turner, Anal. Chem. 56:667 (1984).

268. H. Gunasingham, K. P. Ang, and V. S. Herath, submitted.

269. K. M. Korfhage, K. Ravichandran, and R. P. Baldwin, Anal. Chem. 56:1514 (1984).

270. M. K. Halbert and R. P. Baldwin, Anal. Chem. 57:591 (1985).

271. L. M. Santos and R. P. Baldwin, Anal. Chem. 58:849 (1986).

272. J. A. Cox and K. R. Kulkarni, Analyst 111:1219 (1986).

273. Y. Ikariyama and W. R. Heineman, Anal. Chem. 58:1803 (1986).

274. P. W. Carr and L. D. Bowers, *Immobilized Enzymes in Analytical and Clinical Chemistry*, Wiley, New York, 1980.

275. G. G. Guibault, *Analytical Uses of Immobilized Enzymes*, Marcel Dekker, New York, 1984.

276. W. J. Blaedel, T. R. Kisse, and R. C. Boguslaski, Anal. Chem. 44:2030 (1972).

277. D. A. Gough and J. K. Leypold, Appl. Biochem. Bioeng. 3:175 (1981).

278. H. F. Hameka and G. A. Rechnitz, Anal. Chem. 53:1586 (1981).

279. J. J. Kulys, Enzyme Microb. Technol. 3:344 (1981).

280. J. K. Leypold and D. A. Gough, Anal. Chem. 56:2896 (1984).

281. L. D. Mell and J. T. Maloy, Anal. Chem. 47:299 (1975).

282. B. Olsson, H. Lundbaek, G. Johansson, F. Scheller, and J. Nentwig, Anal. Chem. 58:1046 (1986).

283. B. Watson, D. N. Stifel, and F. E. Semersky, Anal. Chim. Acta 106:233 (1979).

284. K. Matsumoto, M. Naotsuka, Y. Sirasaka, T. Nomura, and Y. Osajima, Agric. Biol. Chem. 46:2749 (1982).
285. D. A. Gough, J. Y. Lucisano, and P. H. S. Tse, Anal. Chem. 57:2351 (1985).
286. H. Gunasingham, in press.
287. J. R. Wilson, Trends Anal. Chem. 3:223 (1984).
288. Van der Linden, Anal. Chim. Acta 179:91 (1986).
289. B. Fleet, T. Caohuu, S. Das Gupta, B. Cardoza, C. E. Small, and R. Khoyetsian, *Proceedings of the International Meeting on Chemical Sensors*, Kodansha, 1983, p. 738.
290. H. Lundback, Anal. Chim. Acta 145:189 (1983).
291. G. A. Rechnitz, Trends Anal. Chem. 5:172 (1986).
292. M. Hikume, T. Kubo, T. Yasuda, I. Karube, and S. Suzuki, Anal. Chim. Acta 109:33 (1979).
293. R. K. Mayfield, F. M. Sullivan, J. A. Colwell, and H. I. Wohtmann, Diabetes 32:908 (1983).
294. A. H. Clemens, D. L. Hough, and P. A. D'Oruzio, Clin. Chem. 28:1899 (1982).
295. A. U. Abel, A. Muller, and U. Fischer, Biomed. Biochim. Acta 43:577 (1984).
296. W. H. Ko, *Implantable Sensors for Closed-loop Prosthetic Systems*, Futura, New York, 1985.
297. F. L. La Que, J. Electrochem. Soc. 116:73 (1969).
298. B. Poulson, Corrosion Sci. 23:391 (1983).

Electrochemical Aspects of
Low-Dimensional Molecular Solids

Michael D. Ward

Central Research and Development Department
E.I. du Pont de Nemours & Company, Inc.
Experimental Station 328, Wilmington, Delaware

I. Introduction   182
II. Description of Low-Dimensional Solids   184
III. Redox Properties of Molecular Solids   196
    A. Electrochemical Properties of Molecular Solid Components   196
    B. Role of Redox Behavior in Solid-State Properties   202
IV. Electrochemical Preparation of Low-Dimensional Solids   220
    A. Basic Principles   220
    B. Examples   223
    C. Mechanistic Aspects   243
    D. Role of Electrochemical Parameters in Crystallization   254
    E. Electrochemical Doping of Low-Dimensional Phthalocyanines   263
V. Low-Dimensional Solids as Electrode Materials   267
    A. Characteristics of Solid Electrodes   267
    B. Potential Applications of "Synthetic Metal" Electrodes   280
    C. Other Electrode Materials   293
VI. Summary   296
    Appendix: Definition of Acronyms   297
    References   299

I. INTRODUCTION

The trend toward electronic devices with very small dimensions and desirable electronic characteristics has led to extensive investigations of molecular materials, that is, those materials designed on the basis of molecular principles. The intense interest in these materials has been aroused mainly by the promise of facile modification of their physical and electronic properties through rational approaches, which would facilitate investigation of structure-function relationships and possibly the design of molecular-scale electronic devices [1,2]. Interest in molecular materials was heightened with the discovery of the organic conductor TTF-TCNQ (TTF = tetrathiafulvalene, TCNQ = tetracyanoquinodimethane) [3,4], followed by the discovery of superconductivity in $(TMTSF)_2X$ (TMTSF = tetramethyltetraselenafulvalene; X = PF_6^-, ClO_4^-, AsF_6^-, ReO_4^-) [5]. Since that time, the area has grown substantially, including the discovery of numerous organic conductors and superconductors [6-8], electrically conductive organometallic solids [9], and a vast number of conducting polymers [10]. Electronic conductivities at room temperature ranging from semiconducting and insulating values (ca $10^{-9}-10^{-1}$ Ω^{-1} cm^{-1}) to actual metallic conductivity (ca $1-10^5$ $\Omega^{-1} cm^{-1}$) have been observed, in addition to the superconducting materials that exhibit zero resistivity at very low temperatures. Optoelectronic properties, such as nonlinear optical phenomena arising from unique lattice structures and intermolecular interactions in these materials [11,12], and cooperative magnetic behavior [13-15], such as ferromagnetism [16], have recently expanded the scope of possibilities for molecular materials.

This review will concern the class of molecular materials designated as <u>molecular solids</u>, which are defined as containing <u>discrete molecular components</u> that have been condensed into an organized framework, generally as a result of intermolecular interactions, crystal forces, or Van der Waals forces. As a corollary, molecular solids can generally be redissolved to their molecular components. They are therefore distinguished from other molecularly designed systems that possess extended covalent networks, such as conducting [10] or redox polymers [17] (e.g., polypyrrole, polyacetylene, or polyvinylferrocene), inorganic chain compounds [e.g., Zintl ions (18)], or two-dimensional layered materials (e.g., metal chalcogenides and molybdates).

Because molecular solids are composed of discrete molecular components, the ability to manipulate and understand the properties of the isolated constituents on the molecular level allows rational modification of the properties of the condensed solid. The properties of the individual molecules play an important role in the unique intermolecular interactions in the crystalline solid state, which ultimately result in the cooperative electronic behavior commonly observed in these materials. To the extent that the electrochemical properties are generally indicative of the electronic structure of molecular constituents, redox behavior plays an important role in the solid-state properties of a low-dimensional material. The electrochemical properties of the individual components can therefore facilitate understanding of the physical properties of the molecular solid as well as the design of new materials.

The purpose of this article is to present an overview of various electrochemical aspects of molecular solids. Although discussion of detailed physical and electronic properties is not intended, some digression will be unavoidable where these factors possess some relevance to the electrochemical properties of the solids or their constituents. Section II briefly introduces nomenclature and describes some basic concepts of molecular solids in order to facilitate discussion in the following sections for those who are unfamiliar with this area. Other sections describe the electrochemical properties of common molecular components, the role of redox potentials in the properties of molecular solids, electrochemical synthetic techniques, and the use of these materials as solid electrodes. An appendix defining acronyms used throughout the text is also included.

II. DESCRIPTION OF LOW-DIMENSIONAL SOLIDS

Molecular solids are composed of molecular species usually arranged in an ordered crystalline lattice. In many cases, planar species with π systems (e.g., napthalene, tetracyanoethylene, hexachlorobenzene) crystallize in low-dimensional or quasi-one-dimensional lattices [19]. Structural low dimensionality in organic solids has been well recognized for decades [20]. Although this allows for efficient close packing of molecular components with minimization of void space in the crystalline lattice [21], many aromatic compounds do not exhibit this feature. Rather, low-dimensional lattices commonly result from favorable Van der Waals or electrostatic interactions. More importantly for this class of materials, low dimensionality is observed when intermolecular π-π interactions lead to face-to-face stacking of molecular species. These interactions result in configuration interaction, and electronic delocalization in cases where extended band structure is involved, resulting in energy lowering of the system compared to other lattices. Anisotropy may be present as a purely structural property, or as an electronic property such as the conductivity anisotropy commonly exhibited by organic conductors. As will be readily apparent in this review, conductivity is especially significant for many open shell radical species that give rise to partially filled bands. When at least one of the components of a molecular solid is capable of intermolecular charge-transfer interactions in the solid state, the compound is commonly referred to as a charge-transfer (CT) solid. Since CT interactions are tantamount to movement of electronic charge, molecular species that are redox-active are generally capable of supporting these interactions.

The alignment of molecules into low-dimensional lattices with their molecular planes parallel is commonly accompanied by donor-donor, acceptor-acceptor, or donor-acceptor interactions involving energetically accessible HOMOs, LUMOs, or POMOs, which are the origin of many of the interesting electronic properties of these materials. A species is frequently described as a donor if it has surrendered charge from its

HOMO and, conversely, as an acceptor if it has acquired charge via additional occupation of its LUMO. Indeed, the formation of many molecular solids is commonly accompanied by an in situ charge transfer from an electron donor (D) to an electron acceptor (A). However, caution should be exercized when using the terms donor and acceptor; they do not necessarily reflect the direction of charge transfer during synthesis since the CT properties of the solid are independent of the route employed to make that material. For example, formation of the generic ionic solid X^+Y^- via a net transfer of unit charge from Y^{2-} to Y^{2+} results in the same material as that realized from neutral X and Y species [Eqs. (1) and (2)]. In these cases, the donor-acceptor nomenclature can be somewhat vague, and assignment is frequently historical, often depending on whether the ionization potential (donor) or the electron affinity (acceptor) of a constituent better describes its role in the properties of the solid.

$$X + Y \longrightarrow X^+Y^- \qquad (1)$$

$$X^{2+} + Y^{2-} \longrightarrow X^+Y^- \qquad (2)$$

Although there are exceptions, most well-known CT solids possess donors and acceptors that are to some degree positively and negatively charged, respectively. That is, an ionic solid will usually be described as D^+A^-. It is also important to note that the net amount of charge transferred, and therefore the charge residing on the molecular constituents, is not necessarily integral. Low-dimensional solids either possess segregated stacking motifs in which donors and acceptors are aligned in separate discrete stacks, or display mixed stack arrangements in which the one-dimensional chains are composed of alternating donor and acceptor molecules (Scheme 1).

Materials that possess open shell radical π donors or π acceptors that crystallize in segregated stacks commonly exhibit electrical conductivity owing to intermolecular π overlap, which results in formation of a conduction band and the delocalization of charge [22]. The extent of overlap

Scheme 1

```
.       .              .
.       .              .
.       .              .
D       A              D
D       A              A
D       A              D
D       A              A
D       A              D
D       A              A
.       .              .
.       .              .
.       .              .
```

Segregated stack   Mixed stack

ultimately defines the bandwidth of the conductor 4t, where t is defined as the charge-transfer integral, which is a measure of the overlap of the molecular wave functions and, accordingly, the delocalization of the electrons into a band that tends to lower the energy of the system. This quantity is analogous to the Huckel resonance or overlap integral β commonly used in molecular systems. For a diatomic or bimolecular system, charge-transfer mixing results in the formation of in-phase bonding and out-of-phase antibonding levels. For an extended solid, the orbital overlap of N molecules results in the formation of band structure with N states formed from each molecular orbital (Scheme 2). It is readily seen

Scheme 2

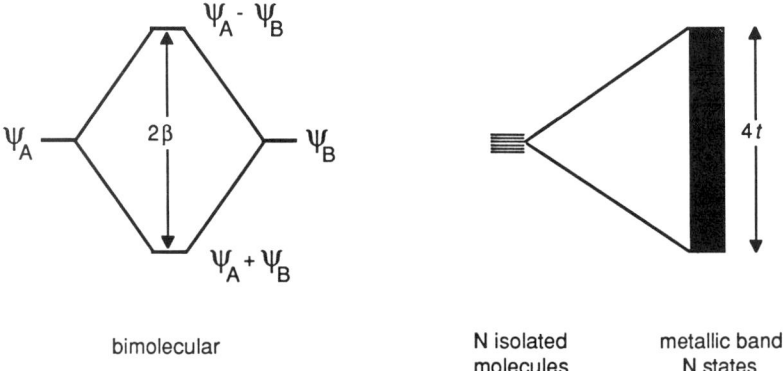

bimolecular   N isolated molecules   metallic band N states

that open shell radicals, with one electron per HOMO, result in a half-filled band. Segregated stack solids have been shown to exhibit metallic, semiconducting, and superconducting behavior [23], the nature of the conductivity depending to a large extent on the degree of molecular overlap, interplanar separation, amount of band filling, degree of interstack interactions, and presence (or absence) of structural distortions. These aspects are discussed in more detail in Sec. III.B. Magnetic behavior, owing primarily to antiferromagnetic coupling and mobile triplet excitons, has also been observed [24].

Two of the most renowned one-dimensional conductors, TTF-TCNQ and $(TMTSF)_2ClO_4$ [25], exemplify the diversity of segregated stack materials (Figs. 1,2). The former possesses separate stacks of TTF and TCNQ molecules, in which the average charge on each species is actually nonintegral; i.e., $TTF^{0.59+}TCNQ^{0.59-}$. Both TTF and TCNQ play a role in the CT ineractions that give rise to electrical conductivity, as expected from the redox properties of these compounds, which indicate facile oxidation and reduction, respectively. In contrast, the metallic conductivity and superconductivity observed in $(TMTSF)_2ClO_4$ is due solely to the TMTSF donor stacks that formally contain $TMTSF^{0.5+}$ species, as the ClO_4^- ions simply maintain electroneutrality in the lattice. The counterion, being electrochemically inert in the potential regime of TMTSF, does not play an active role in charge transfer. In both TTF-TCNQ and $(TMTSF)_2ClO_4$, high electrical conductivity is facilitated by the nonintegral charge ($\rho < 1$) and mixed valent nature of the constituents (see Sec. III).

Mixed stack one-dimensional solids generally do not possess significant electronic conductivity owing to the different orbital energies and symmetries of the molecular components, which results in charge localization and inhibited electron movement. That is, these materials generally do not possess the band structure required for conductivity. However, as a result of intermolecular π-π interactions, these materials exhibit interesting optical properties stemming from Mulliken donor-acceptor interactions

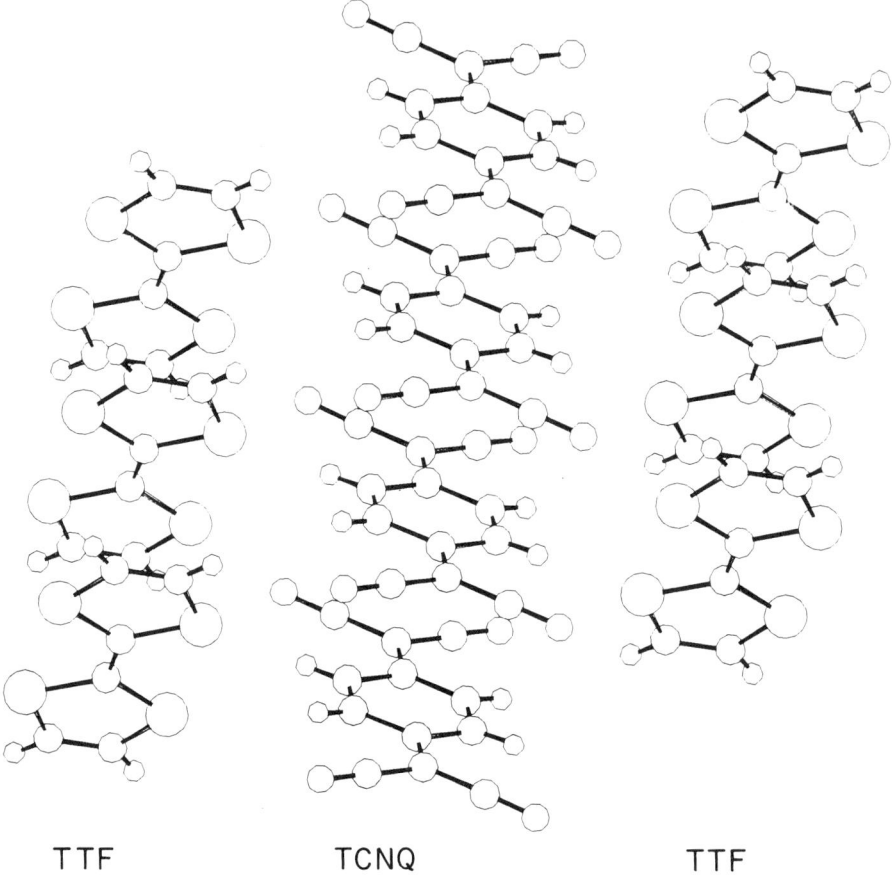

FIG. 1. Segregated stacks of TTF and TCNQ molecules in (TTF)(TCNQ) viewed roughly normal to the stacking axis.

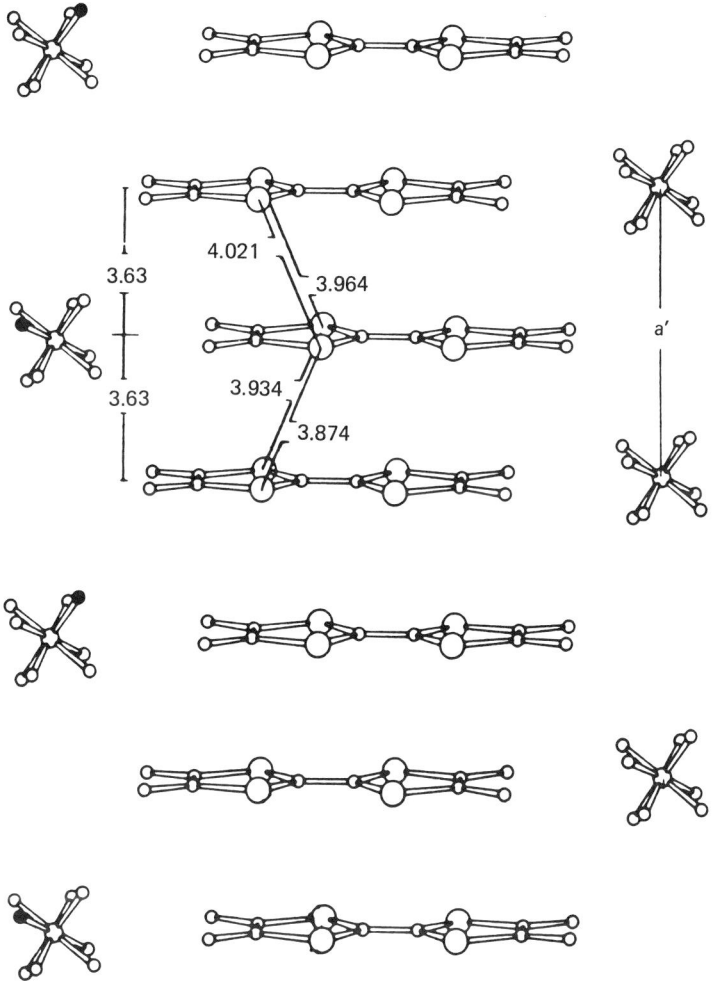

FIG. 2. Side view of the stack in TMTSF$_2$ClO$_4$ clearly illustrating the presence of segregated stacks of donor cations and ClO$_4^-$ anions. Dimerization of TMTSF and disorder of the anions is evident. (From Ref. 25.)

[26], including thermo [27] and piezochromism [28], as well as cooperative magnetic phenomena such as ferromagnetism [20,21]. These interactions commonly result in mixing of ground and excited state wave functions and frequently nonintegral charges on the donors and acceptors. For simplicity, the components are generally described as possessing integral charge, and "neutral," "ionic," and "doubly ionic" CT solids have been reported (Scheme 3). Examples of mixed stack solids include

Scheme 3

Neutral: .....DADADADA.....
Ionic: .....$D^+A^-D^+A^-D^+A^-D^+A^-$.....
Doubly ionic: .....$D^{2+}A^{2-}D^{2+}A^{2-}D^{2+}A^{2-}D^{2+}A^{2-}$.....

organic materials such as TTF-chloranil, a neutral CT solid under ambient conditions (Fig. 3) [29], the ionic $[(C_5Me_5)_2Fe]^+[TCNE]^-$ [20] (Fig. 4) and doubly ionic $[(C_6Me_3H_3)_2M]^{2+}[HCTMP]^{2-}$ [30] (Fig. 5) organometallic complexes.

The compounds that have been reported to form low-dimensional molecular solids are too numerous to list here. Representative examples, including many that are to be discussed in this review, are shown in Fig. 6, where they are classified according to whether they behave as donors or acceptors. For convenience, donors and acceptors are defined as they have generally been used in the literature. The hexacyanotrimethylenemethanediide ($HCTMM^{2-}$) and hexacyanotrimethylenecyclopropanediide ($HCTMCP^{2-}$) dianions are exceptions to the common nomenclature, in which donors as positively charged, as they previously have been assigned as donors based on their properties in solid-state materials [30,31]. However, their monanions could be considered as acceptors. Similarly, the positively charged $[(arene)_2M]^{2+}$ (M = Fe, Ru) [30,31] and $[N,N-dimethylbipyridinium]^{2+}$ (N,N-dimethylbipyridinium = MV) [32,33] cations behave as acceptors in their solid-state doubly ionic complexes.

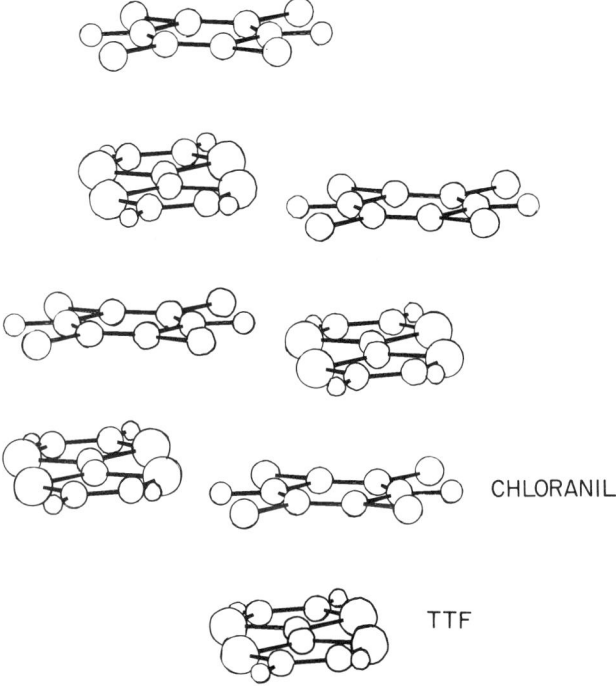

FIG. 3. Two adjacent stacks in the mixed stack solid (TTF)(chloranil) showing the alternating mixed stack motif. This compound has a neutral ground state under ambient conditions.

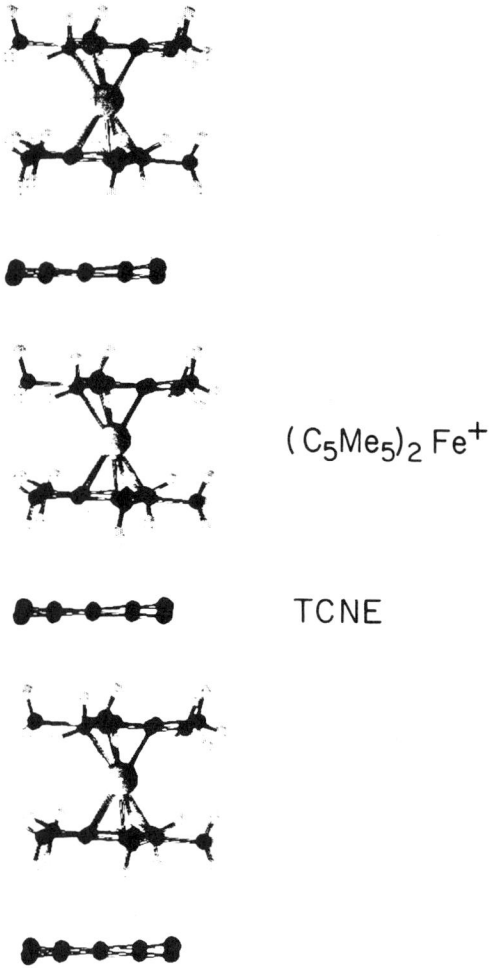

FIG. 4. One-dimensional chain of $(C_5Me_5)_2Fe^+$ and $TCNE^-$ ions in the organometallic ionic comples $[(C_5Me_5)_2Fe]^+[TCNE]^-$.

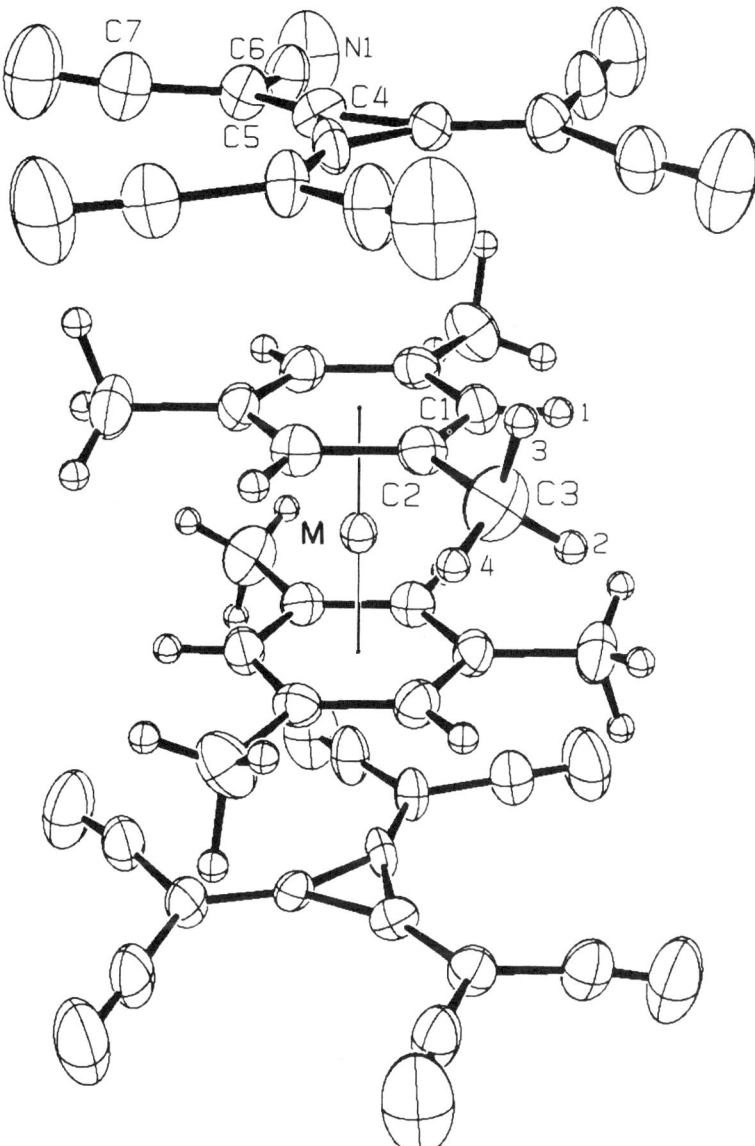

FIG. 5. A segment of the one-dimensional chain in [(C_6Me_3H_3)_2M]-[C_6(CN)_6] (M = Fe, Ru). (From Ref. 30.)

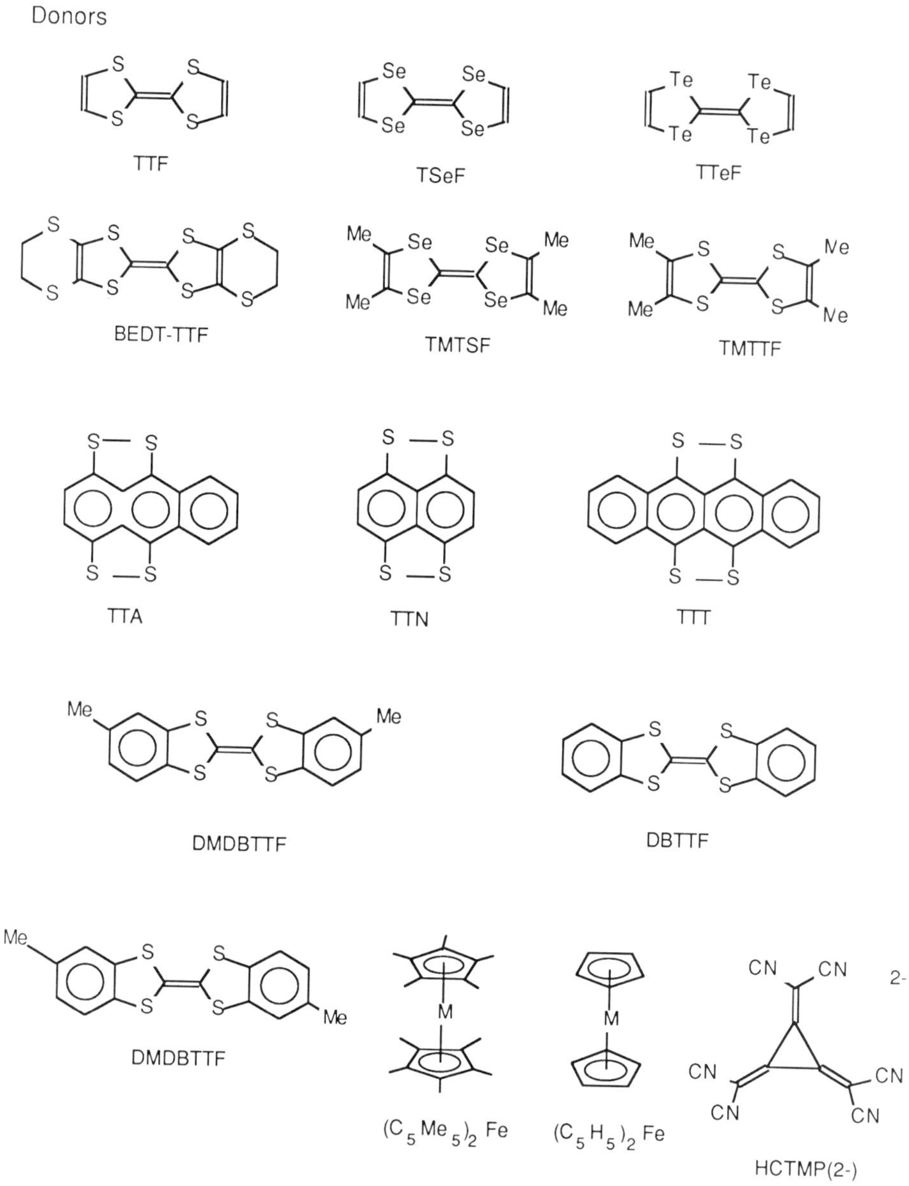

FIG. 6. Examples of molecular components commonly found in one-dimensional solids.

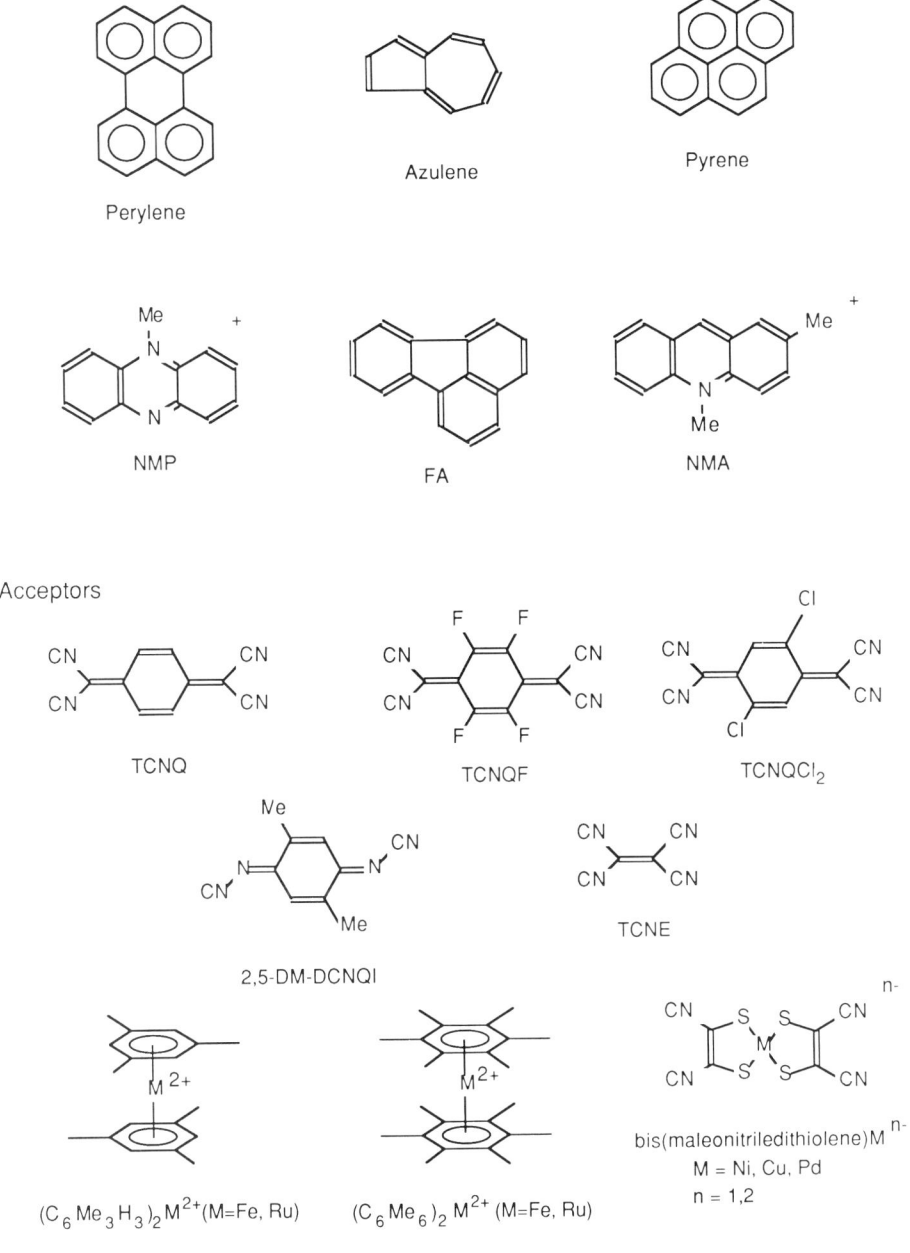

FIG. 6. (continued)

HCTMCP²⁻   HCTMM²⁻   MV²⁺

It is obvious from Fig. 6 that the majority of π donors and π acceptors possess sterically unhindered planar conformations, so that close interplanar approach of molecules and π-π interactions in the solid state are possible. Organometallic derivatives, although not having true molecular planarity, possess planar aromatic π ligands that are capable of intermolecular π-π CT interactions. The molecular structure of organometallic complexes is such that the metal ions are essentially buried inside a hydrocarbon envelope so that the complexes structurally mimic their organic analogs.

III. REDOX PROPERTIES OF MOLECULAR SOLIDS

A. Electrochemical Properties of Molecular Solid Components

The redox potentials of some compounds commonly found in molecular solids are tabulated in Table 1. In general, chemical stability on the voltammetric time scale is a desirable characteristic because this will favor the formation of stable compounds, especially when open shell radical ions are involved. With the exception of some of the organometallic derivatives, most of the compounds have been deliberately chosen because they exhibit reduction potentials in a fairly narrow range and thus are not prone to the deleterious decomposition reactions that may accompany more strongly oxidizing or reducing species. In any case, exclusion of air is generally observed during electrochemical investigations and preparation of CT solids to avoid side reactions with oxygen and water, especially for those compounds with extreme redox potentials.

Donor molecules tend to be electron-rich systems that are rather easily oxidized. In many cases, such as hexamethoxytriphenylene (HMTP),

TABLE 1

Reduction Potentials for Components Commonly Found in One-Dimensional Molecular Solids[a]

	Redox couple				
	$E^0_{(2+/+)}$	$E^0_{(+/0)}$	$E^0_{(0/-)}$	$E^0_{(-/2-)}$	
Compound	Volts vs. SCE				Reference
Donors-organic					
TTF	0.66	0.30			37
TSeF	0.90	0.62			38
TTeF	0.84	0.59			39
DSDSeF	0.63	0.28			37
$(Me_2N)_4C_2$	−0.61	−0.75			40
HCTMM				0.92	41
HCTMCP			1.13	0.34	42
TMPD	0.66	0.29			43
	0.70	0.10			44
HMTP	0.85	0.55			34
HET	0.27	0.02			35
HOC	0.36	−0.11			45
	0.06	−0.44			46
TTT	0.67	0.24			47
TTN	0.93	0.55			48
TTA	0.85	0.41			47
TSeA	0.84	0.42			47
TSeN	0.80	0.40			49
TSeN	0.86	0.47			50
ET (BEDT-TTF)	0.94	0.63			51
DBTTF	1.06	0.72			40
DMDBTTF	1.02	0.68			50
HAB	0.56[b]	−0.11			41
TMTSF	0.81	0.42			52

TABLE 1 (continued)

Compound	Redox couple				Reference
	$E^0_{(2+/+)}$	$E^0_{(+/0)}$	$E^0_{(0/-)}$	$E^0_{(-/2-)}$	
	Volts vs. SCE				
TMTTF	0.73	0.24			52
HMTTeF	0.69	0.40			52
HMTSeF		0.40			53
HMTTF	0.80	0.40			52
Perylene		0.85			54
Azulene		0.71			54
Pyrene		1.16			54
Fluoranthene	1.81[b]	1.50			55
Naphthalene		1.54			54
TAE	−0.47	−1.37			37
TDMAB		+0.05			56

Donors-organometallic

Cyclopentadienyl complexes

Compound	$E^0_{(2+/+)}$	$E^0_{(+/0)}$	$E^0_{(0/-)}$	$E^0_{(-/2-)}$	Reference
$V(C_5H_5)_2$			−0.55	−2.74	57
$Cr(C_5H_5)_2$			−0.55	−2.3	58
$Fe(C_5H_5)_2$			0.41		58
$Co(C_5H_5)_2$			−0.91	−1.88	59
$Ni(C_5H_5)_2$	0.71		−0.09	−1.66	60
$Os(C_5H_5)_2$	1.80		0.76		61, 62
$Ru(C_5H_5)_2$			0.88		61
$Fe(C_5Me_5)_2$	1.58		−0.12		58
$Mn(C_5Me_5)_2$			−0.56	−2.50	63
$Ni(C_5Me_5)_2$	0.31		−0.65		58
$Ru(C_5Me_5)_2$			0.55		64
$Co(C_5Me_5)_2$			−1.47		58
$Cr(C_5Me_5)_2$			−1.04		58

TABLE 1 (continued)

Compound	$E^0_{(2+/+)}$	$E^0_{(+/0)}$	$E^0_{(0/-)}$	$E^0_{(-/2-)}$	Reference
		Volts vs. SCE			
$Fe(C_5H_5)(C_5H_4Me)$		0.33			41
$Fe(C_5H_4Me)_2$		0.29			41
$Fe(C_5H_5)(C_5H_4Ph)$		0.29			65
$Fe(C_5H_4(CH_2)_2C_5H_4)$		0.34			41
$Fe(C_5Me_4H)_2$		0.07			41
$Fe(C_5H_5)(C_5Me_5)$		0.11			41
$Fe(C_5H_4(CH_2)_3C_5H_4)$		0.30			41
$Fe(C_5H_5)(C_6Me_6)$		−1.78			66, 67
$Fe(C_5H_5)(C_6H_6)$		−1.62			66
Arene complexes					
$Mo(C_6H_6)_2$		−0.63			43
$Cr(C_6H_6)_2$	1.06	−0.75			68
$Cr(C_6Me_6)_2$	0.17	−1.30			69
$Cr(C_6Me_3H_3)_2$	0.90	−0.92			68
$Cr(C_6H_5Me)_2$	0.97	−0.81			68
$Cr(Cr_6H_5OMe)_2$		−0.73			68
$Cr(C_6H_5C_6H_5)_2$	0.93	−0.65			68
$Cr(C_6H_5COOEt)_2$		−0.56			68
$Cr(C_6H_5Cl)_2$		−0.34			68
Acceptors-organic					
$n-C_4(CN)_6$			0.60	0.02	70
HCTMCP			1.13	0.34	42
HCTMM				0.92	41
DDQ			0.51	−0.30	71
TCNE			0.15	−0.57	72
TCNQ			0.17	−0.37	41

TABLE 1 (continued)

Compound	Redox couple				Reference
	$E^0_{(2+/+)}$	$E^0_{(+/0)}$	$E^0_{(0/-)}$	$E^0_{(-/2-)}$	
	Volts vs. SCE				
TCNQBr			0.29	−0.22	37
TCNQBrCH$_3$			0.26	−0.22	37
TCNQBr$_2$			0.41	−0.08	37
TCNQClCH$_3$			0.26	−0.23	37
TCNQCl			0.29	−0.22	37
TCNQCl$_2$			0.41	−0.10	37
TCNQEt$_2$			0.11	−0.37	37
TCNQF$_4$			0.53	0.02	37
TCNQICH$_3$			0.25	−0.21	37
TCNQI$_2$			0.35	−0.10	37
TCNQMe$_2$			0.10	−0.38	37
TCNQ(CN)$_2$			0.65	0.09	37
TCNQ(CN)$_4$			1.31	0.51	41
TCNQ(i-Pr)$_2$			0.12	−0.35	37
TCNQ(OEt)(SMe)			0.08	−0.36	37
TCNQ (OMe)			0.07	−0.45	37
TCNQ (OMe)$_2$			−0.01	−0.47	37
TCNQ (OMe) (OCH$_2$OCH$_2$)			0.05	−0.33	37
TCNQ (OMe) (OEt)			−0.02	−0.47	37
TNAP			0.20	−0.17	37
Fluoranil			0.04	−0.05	71
Bromanil			0.00	−0.72	71
Chloranil			0.01	−0.71	71
TCNQF$_2$			0.30	−0.14	41
TCNQF			0.26	−0.29	41
TCNQMe			0.17	−0.34	41

TABLE 1 (continued)

Compound	Redox couple				Reference
	$E^0_{(2+/+)}$	$E^0_{(+/0)}$	$E^0_{(0/-)}$	$E^0_{(-/2-)}$	
	Volts vs. SCE				
OCNAQ			0.26	0.05	73,74
MV^{2+}	−0.69	−1.12			75
2,5-DM-DCNQI			0.08	−0.54	41
Acceptors-organometallic					
$Fe(C_6Me_6)_2^{2+}$	−0.24	−1.11			30
$Fe(C_6Me_3H_3)_2^{2+}$	−0.02				41
$Ru(C_6Me_6)_2^{2+}$	-1.01^c	-1.01^c			76
$Ru(C_6Me_3H_3)_2^{2+}$	−0.75				41
Ru(3,3',5,5'-tetra-methylbiphenyl)$_2^{2+}$	−0.63				77
$Pt(CN)_4^{2-}$				0.72^b	78

[a] See Appendix for definition of acronyms.
[b] Chemically irreversible.
[c] 2e⁻ reduction.

2,3,6,7,10,11-tris(N,N'-diethylethylenediamino)triphenylene (HET) and hexaazaoctadecahydrocoronene (HOC), oxidation to stable dications is also possible. Indeed, reversible electrochemical oxidation of HMTP to higher oxidation states (2+, 3+, 4+) is reversible in trifluoracetic acid [34], and the tri- and tetracations of HET and HOC are stable on the voltammetric time scale [35].

Conversely, acceptor molecules commonly possess electron-withdrawing substituents that stabilize the anions and facilitate reduction at reasonable potentials. This is obvious from the ubiquity of TCNQ in molecular solids. The good electron-accepting properties of polycyano hydrocarbons, combined with their planar conformations, have led to their predominance

HET HTMP HOC

in one-dimensional solids. In some cases, such as the HCTMM and HCTMCP anions, the electron-withdrawing ability of the cyano groups is sufficiently strong that the neutral forms are not stable. Transition metal dithiolates (Fig. 6), which exhibit reversible redox behavior, are not included in Table 1 because they have been rather extensively reviewed elsewhere [36].

B. Role of Redox Behavior in Solid-State Properties

Many features are considered in the design of one-dimensional solids, including structure, polarizability, electron affinity, ionization potential, and redox behavior. As stated in the Introduction, one of the advantages of molecular solids is that their electronic properties are easily and rationally modified because of the flexibility in design of their molecular components. It is obvious from Table 1 that the redox potentials of molecular species within a given family can be rationally modified with substituent groups while the gross structural features of that category of compounds are maintained. In many cases, the electronic properties of a molecular solid can be interpreted to a significant degree on the basis of the redox properties of its components and, conversely, the redox behavior can be used to predict the nature of the CT behavior likely to be observed in a new solid. In this section, the role redox properties play in the extent of charge transfer, the nominal oxidation states of the molecular components, and the conductivity of one-dimensional solids will be described.

The role of redox properties can be illustrated, using as an example the formation of a molecular solid from neutral constituents. Reaction of an electron donor and an electron acceptor will be accompanied by charge transfer if the free energy change for formation of the ionic (or partially ionic) solid is negative. This depends on the difference among donor ionization potential IP_d and acceptor electron affinity EA_a, electrostatic (Madelung) energy associated with formation of the lattice $<e^2/r>$, as well as other terms, defined here as C, which include such factors as electronic delocalization resulting from π-π intermolecular overlap and the formation of band structure [Eq. (3)]. Charge transfer becomes more likely if the donors and acceptors are easily oxidized and reduced, respectively, and the electrostatic term is large. The actual degree of charge transfer ρ may vary from $0 < \rho < 1$, depending on the relative magnitudes of the terms on the right side of Eq. (3). This trend has been clearly demonstrated for several TCNQ complexes in which very high IP_d donors resulted in $\rho = 0$ complexes, and low IP_d donors (e.g., alkali metals) resulted in $\rho = 1$ complexes [79]. Donors with intermediate IP_d values favor solids with nonintegral values, ρ being determined by a variety of methods including infrared spectroscopy. In general, smaller Madelung energies and more positive values of $(EA_a - IP_d)$ tend to favor weak charge transfer and mixed valent salts ($\rho < 1$).

$$\Delta G \text{ (charge transfer)} = (EA_a - IP_d) - <e^2/r> + C \tag{3}$$

The value of ρ is an important consideration in the design of molecular solids since it is manifested as the amount of charge residing on the molecular components which, in turn, significantly affects the solid-state electronic properties. In a segregated stack one-dimensional solid, the value of ρ has a significant effect on its conductivity. A $\rho = 1$ species possesses a full unit charge and a singly occupied HOMO which, in the extended solid, leads to a half-filled band and metallic conduction. However, structural distortions, referred to as Peierls distortions, commonly occur in one-dimensional chains with incompletely filled bands accompanied by splitting of the band into a lower energy filled band and a higher energy empty band. For the originally half-filled band, this results in the onset of semiconducting or insulating behavior, depending on the magnitude of the Peierls gap E_g (Scheme 4). These distortions occur

Scheme 4

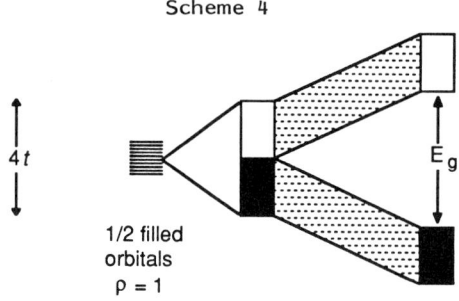

1/2 filled
orbitals
$\rho = 1$

1/2 filled
metallic band

when the energy lowering associated with the splitting of the partially filled band, conceptually analogous to the Jahn-Teller effect in molecular species, exceeds that required for the structural distortion of the lattice. This effect is commonly observed when electronic localization or disorder is present in the lattice.

The value of ρ also plays an important role in the significance of Coulomb correlation effects, which result from the Coulomb repulsion energy U between charge carriers. Conductivity can be described according to a site hopping mechanism by considering the energy requirements for carrier mobility among discrete energy levels within a conduction band, as illustrated in Scheme 5 (the site hopping depiction arises because each of these states originated from an orbital of a single molecular species). The first case illustrates a half-filled ($\rho = 1$) band in which $U < 4t$, that is, the Coulomb repulsion between carriers in one

Scheme 5

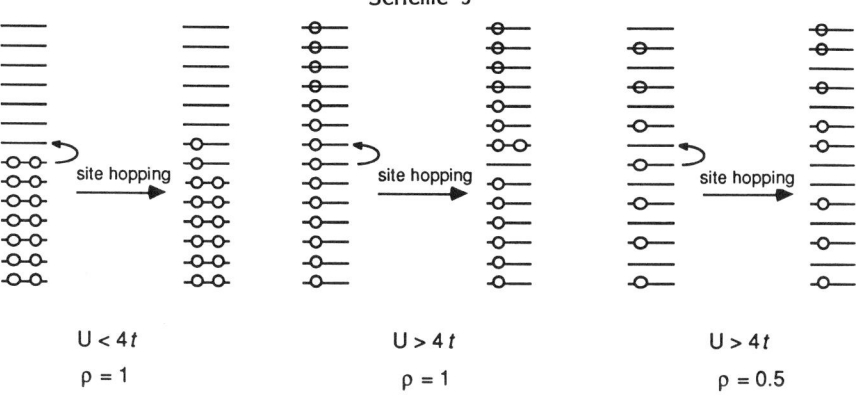

state, is small compared to the bandwidth. Under these conditions, charge hopping requires only the infinitesimal amount of energy needed for promotion of an electron to the next highest unoccupied state, resulting in metallic conductivity. In the second case, the states are half-filled, but the intrasite Coulomb repulsion is large. Charge hopping necessarily results in double occupation of the sites; since Coulomb repulsion is large, this results in activated conductivity. The energy U is referred to as the Mott transition energy, and these materials are commonly referred to as Mott semiconductors or insulators [80]. Conversely, the presence of nonintegral charge alleviates the unfavorable intrasite coulombic interactions within the stack during electron transport.

This behavior can be conceptually described for bands where $\rho = 0.5$, which can be considered to possess equivalent amounts of occupied and unoccupied states. Electron transport can occur via hopping without the formation of energetically unfavorable doubly occupied states.

This behavior can also be depicted in molecular terms as in the lower part of Scheme 3, which shows an A^- acceptor chain. When this approach is used, double occupation of a state is equivalent to the formation of an A^{2-} site (the same scheme applies for donor molecules D^+ and D^{2+}), and U is the difference between the ionization potential and electron affinity of A^- (or D^+). Semiconducting or insulating behavior results when U is large because the cost in energy for placing two electrons on one A^{2-} site during charge transport is too great. Mixed valent species, however, can exhibit enhanced conductivities since the magnitude of U can be smaller in these cases. When $\rho = 0.5$, at short times, the one-dimensional acceptor stack can be considered to possess equal numbers of A^0 and $A-$ sites, and charge hopping can occur without formation of A^{2-} or D^{2+}. Although *intersite* coulombic repulsion is still present, it is significantly less than the *intrasite* repulsion in $\rho = 1$ chains because of the larger separation between charges [81].

The advantage of this molecular view can be readily seen: the energy associated with the formation of A^{2-} or D^{2+} sites can be significant if *intramolecular* coulombic repulsions between charges in these species is large, that is, small EA values for A^- or large IP values for D^+. This suggests that proper design of the molecular species can ameliorate these repulsive interactions. Later, we shall see that the electrochemical redox properties can be correlated with these properties and thus play an important role in determining U.

Since the redox potentials of molecular components of CT solids can be correlated with either EA_a or IP_d, the electrochemical properties of these species play a significant role in the properties of molecular solids. It has been reported that solvation energies do not vary markedly among different systems [82]. As a result, good correlation between $E_{1/2}$ or

$E°$ values and vertical ionization potentials for various aromatic hydrocarbons [83] and alkyl-substituted benzenes has been found [84]. The latter group was reported to exhibit standard redox potentials that expermentally followed Eq. (4). Theoretical considerations suggested that $E°_{Ar}$ can be described by Eq. (5), where $\Delta G_r°$ is the reorganizational energy of an ArH^+ cation, $\Delta G_s°$ the solvation energy term, F the Faraday constant, and C a constant dependent on the working electrode and reference electrode. The same principles hold for the electron affinity of acceptor molecules.

$$E°_{Ar} = 0.71 IP_d - 3.68 \tag{4}$$

$$E°_{Ar} = IP_d + [(\Delta G_r° + \Delta G_s°)/F] + C \tag{5}$$

The relationship between redox potential and IP_d and EA_a can allow relative comparisons between molecules under identical experimental conditions (e.g., solvent, electrode material), particularly if the species have similar molecular structure and the CT solids have similar solid-state structure. This allows prediction of whether electron transfer will accompany formation of the CT solid and, accordingly, the degree of ionic charge on the molecules in the solid state. The mixed stack solids $[(C_5H_5)_2Fe][TCNE]$ [85] and $[(C_5Me_5)_2Fe]^+[TCNE]^-$ [20] are simple examples of this principle. The difference in redox potential between TCNE and these donors, $\Delta E°_{a-d} = E°_a - E°_d$, differs due to the greater electron-releasing character of the C_5Me_5 ligands in $(C_5Me_5)_2Fe$ ($\Delta E°_{a-d}$ = +0.26 V) compared to the C_5H_5 ligands in $(C_5H_5)_2Fe$ ($\Delta E°_{a-d}$ = -0.27 V). As a result, $[(C_5H_5)_2Fe][TCNE]$ is nominally neutral ($\rho = 0$), whereas $[(C_5Me_5)_2Fe]^+[TCNE]^-$ is nominally ionic ($\rho = 1$).

These examples illustrate that, if $\Delta E°_{a-d}$ is very positive, electron transfer will probably occur with formation of an ionic solid. The free energy lowering owing to electrostatic interactions in an ionically charged solid and electronic delocalization resulting from band structure further facilitates electron transfer. However, a negative $\Delta E°_{a-d}$ value does not necessarily rule out formation of an ionic solid if the Madelung term and electronic delocatlization compensate for the lack of redox driving force. This appears to be the case in (TTF)(TCNQ), which is nominally ionic even though $\Delta E^\circ_{a-d} = -0.13$ V. Of course, extremely negative values of ΔE°_{a-d} reduce the chance for electron transfer in formation of the molecular solid.

Redox properties have been shown to correlate with ρ in segregated one-dimensional solids which, in turn, dramatically affects the conductivity of these materials. The presence of very strong donors or acceptors results in full charge transfer and the formation of $\rho = 1$ Mott insulators. When $\rho = 0$, there is not sufficient carrier density to support conductivity. In contrast, high conductivity is commonly observed when $\rho < 1$ resulting from the presence of moderate donors and acceptors. In a rather exhaustive study of organic CT solids derived from TCNQ derivatives in which both components were redox-active, it was reported that the conductivity correlated with the equilibrium constant for electron transfer in solution [Eq. (6)] [86]. For large values of K ($K = 10^{0.2}-10^6$) or very small values ($K < 10^{-4}$), the CT solids

$$\log K = \frac{\log[D^+][A^-]}{[D^\circ][A^\circ]} = \frac{(E_a^\circ - E_d^\circ)}{0.059} = \frac{\Delta E^\circ_{a-d}}{0.059} \qquad (6)$$

were poor conductors, whereas solids with intermediate values ($K = 10^{0.2}-10^{-4}$) gave reasonable conductivities. The explanation for this behavior was that the value of log K, which according to Eq. (6) is directly proportional to ΔE°_{a-d}, correlates with the equilibrium concentration of D^+, A^-, D°, and A°, and therefore ρ, in the solid. At the extreme values of K, $\rho = 0$ or 1, representing the limits of no and complete charge transfer, respectively, both cases result in poor conductivity. Intermediate K values, however, result in nonintegral values of ρ, and reasonable conductivity resulting from reduced intrasite repulsion is observed. The dependence of ρ on ΔE°_{a-d} obeys the behavior illustrated in Fig. 7, as derived from Eq. (6). Ideally, $\rho = 0.5$ when

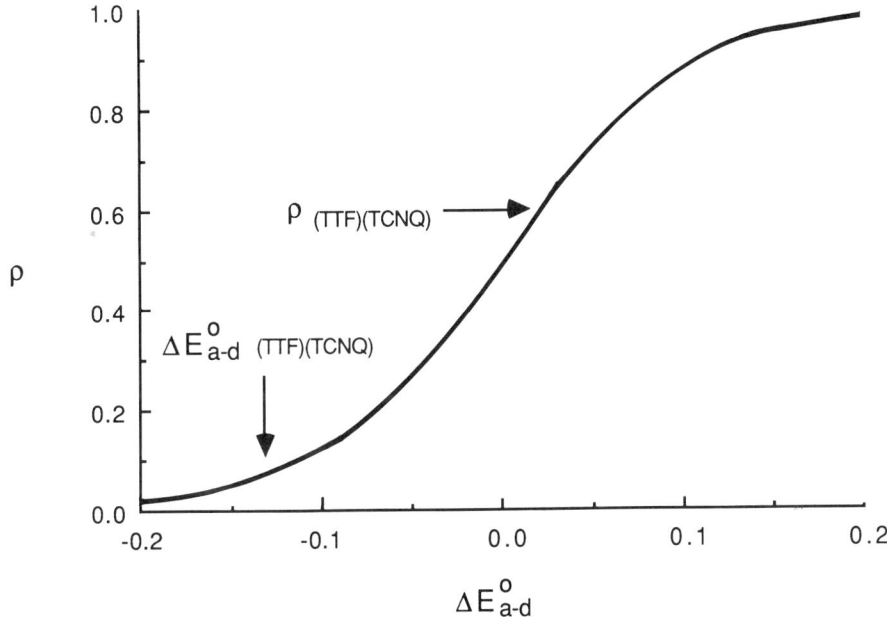

FIG. 7. The ideal dependence of ρ, the degree of charge transfer, on ΔE°_{a-d}, the difference in the standard redox potentials of the acceptor and donor species, as derived from Eq. (6). The actual values of ρ and ΔE°_{a-d} for (TTF)(TCNQ) are indicated in the figure.

$\Delta E^\circ_{a-d} = 0$ V. However, other energetic terms in the solid state will alter the ionicity of the solid so that the curve more accurately describes the expected trends if it is displaced to the left. The displacement of a given compound from the ideal curve in Fig. 7 is a measure of these energetic terms in the solid state, and discrepancies arising from comparisons between ΔE°_{a-d} and $(IP_d - EA_a)$. For (TTF)(TCNQ), it has been determined that $\rho = 0.59$, whereas the ideal curve would predict $\rho \approx 0.05$, reflecting the greater extent of electron transfer driven by Madelung energies and electronic delocalization.

The trend in resistivity for various complexes formed with TTF and TCNQ derivatives is summarized in Fig. 8. A maximum in conductivity

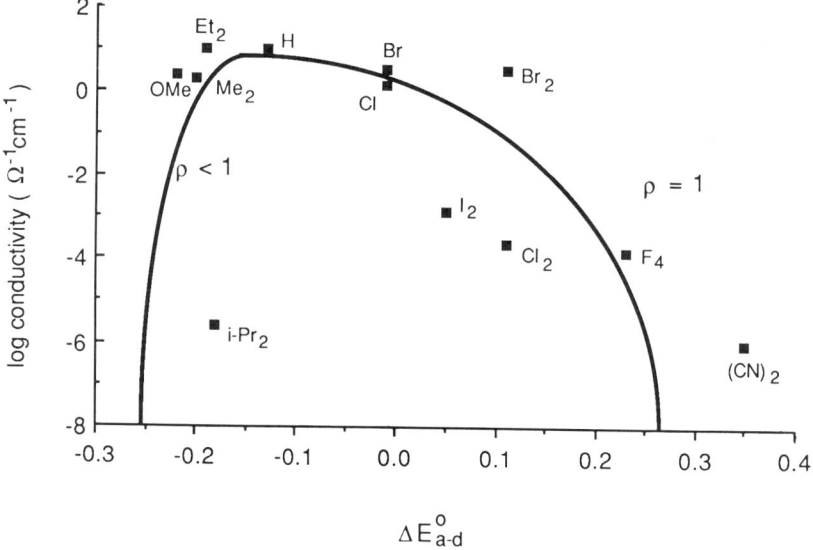

FIG. 8. Observed dependence of conductivity on ΔE°_{a-d} for TTF complexes with differently substituted TCNQ acceptor species. (Adapted from Refs. 37 and 86.)

is observed for (TTF)(TCNQ), for which $K = 10^{-2.2}$. At the extremes are complexes containing halogenated TCNQ species with very positive redox potentials resulting in $\rho = 1$ and complexes with alkyl-TCNQ acceptors that are not readily reduced, resulting in $\rho = 0$. TCNQ complexes with TMSA, which has a very positive redox potential ($E_{ox} = 0.90$ V), were also poor conductors since no charge transfer occurs ($\rho = 0$ and $K < 10^{-4}$). In this case, the conductivity for different TCNQ salts increased with larger values of K owing to small incremental increases in charge transfer.

TMSA

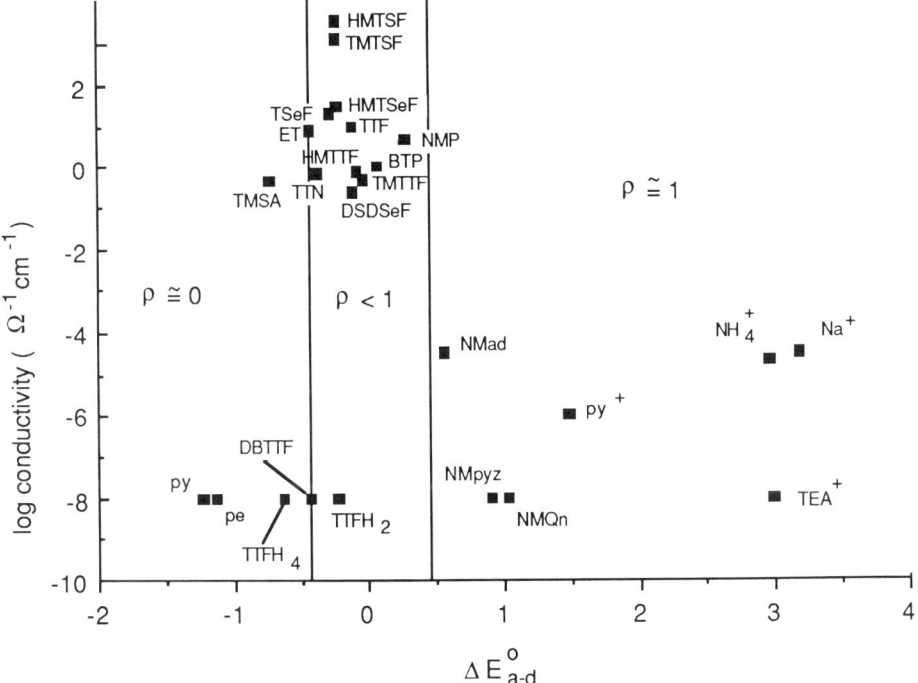

FIG. 9. Observed dependence of conductivity on ΔE°_{a-d} for TCNQ complexes with various donor molecules. The value of ρ is shown for different regions in the figure. It is clear that the most highly conducting complexes are observed when $|\Delta E^\circ_{a-d}| < 450$ mV. (Some of the values were taken from a similar plot in Ref. 87.)

Similar trends have been observed for TCNQ complexes with various donors (Fig. 9) [87]. Alkali metal and tetraalkylammonium salts of $TCNQ^-$ are generally poor conductors since $\rho = 1$ in the $TCNQ^-$ anion stacks as a result of the low ionization potential of the counterions. Conversely, neutral $\rho = 0$ complexes are formed when weak donors such as perylene or pyrene are used. Only those materials with intermediate ΔE°_{a-d} values tend to possess nonintegral values of ρ and correspondingly reasonable conductivity. In summary, although quantitative agreement between ΔE°_{a-d} and ρ is not to be expected, conductivity properties can be predicted with reasonable confidence based on readily measured solution redox potentials.

It has also been suggested that higher conductivity in molecular solids can be achieved by reducing the intrasite coulombic repulsion in D^{2+} or A^{2-} [22]. For example, acceptor molecules with electron-withdrawing groups at remote locations of the molecule would result in smaller coulombic repulsion. This would effectively reduce the energy difference between A^- and A^{2-}, reducing the energy required for electron transport via site-to-site hopping (Scheme 5). For example, the 11,11,12,12-tetracyano-2,6-quinodimethane dianion ($TNAP^{2-}$) would be expected to exhibit less intramolecular coulombic repulsion compared to

TNAP

$TCNQ^{2-}$ since the two electrons, which are primarily localized on the exocyclic methylidene carbons, are more distant in $TNAP^{2-}$. The energy U required for charge transport in the site hopping description is the difference between coulombic repulsion from charge on the same molecule (U_1) and two charges on adjacent molecules (U_2). When $U \gg 4t$ (the bandwidth), the one-dimensional solid will be insulating. For simple $\rho = 1$ salts, conductivity can be enhanced by minimizing U.

$$U = U_1 - U_2 \tag{7}$$

The role of redox behavior in this strategy becomes evident if one considers that site hopping is analogous to the disproportionation processes for the donors and acceptors [Eqs. (8) and (9)]. According to this analogy, U corresponds to $\Delta E°_{1-2}$, the difference between the standard potentials for the first and second reductions [Eq. (10)].

$\rho = 1$

$$D^+ + D^+ \longrightarrow D^0 + D^{2+} \qquad U = U_1 - U_2 \propto \Delta E^\circ_{1-2(D)} \qquad (8)$$

$$A^- + A^- \longrightarrow A^0 + A^{2-} \qquad U = U_1 - U_2 \propto \Delta E^\circ_{1-2(A)} \qquad (9)$$

$$\Delta E^\circ_{1-2} = E_1^\circ - E_2^\circ \qquad (10)$$

$\rho = 0.5$

$$D^+ + D^0 \longrightarrow D^0 + D^+ \qquad U = \Delta E^\circ = 0 \qquad (11)$$

$$A^- + A^0 \longrightarrow A^0 + A^- \qquad U = \Delta E^\circ = 0 \qquad (12)$$

The importance of reducing intrasite repulsion is demonstrated by comparison of ΔE°_{1-2} of different donors and acceptors with the conductivity of their one-dimensional solids. For example, solids derived from TNAP ($\Delta E^\circ_{1-2} = 0.38$ V) [88,89] and TCNQ ($\Delta E^\circ_{1-2} = 0.42$ V) have been reported to be highly conducting, whereas TCNQF$_4$ ($\Delta E^\circ_{1-2} = 0.51$ V) and TCNE ($\Delta E^\circ_{1-2} = 0.72$ V) salts are insulators. Whereas site hopping will be energetically unfavorable for $\rho = 1$ when ΔE°_{1-2} is large ($U_1 > U_2$), for $\rho = 0.5$ the CT process is nominally isoenergetic using the disproportionation analogy, minimizing the barrier to site hopping [Eqs. (11) and (12)]. Small values of ΔE°_{1-2}, however, would still facilitate site hopping in these salts as the contribution of more highly charged sites to the electronic properties of the solid cannot be completely ignored.

Mixed stack CT solids also exhibit properties that depend strongly on redox behavior of their constituents. These solids generally exhibit Mulliken CT interactions, which result in rather intense optical absorptions, commonly in the visible region of the electromagnetic spectrum. The absorption is tantamount to electron transfer from the donor HOMO to the neighboring acceptor LUMO. In a neutral ($\rho = 0$) linear chain, this results in the creation of a nominally charged D^+A^- pair [Eq. (13)].

$$\ldots DADADADA\ldots \longrightarrow \ldots DAD^+A^-DADA\ldots \qquad (13)$$

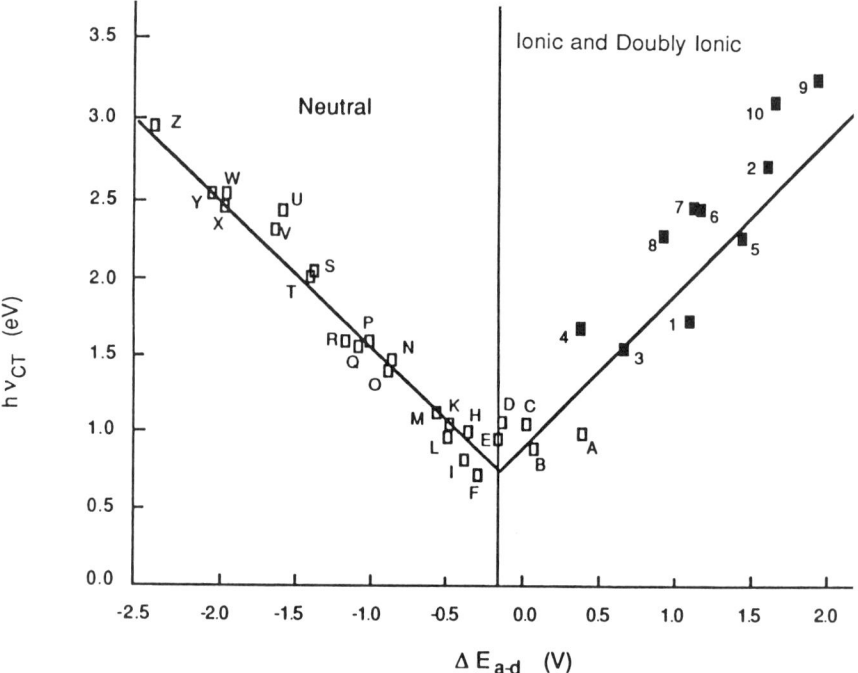

FIG. 10. Dependence of $h\nu_{CT}$, the charge-transfer transition energy, on $\Delta E^\circ_{a\text{-}d}$. The values for the neutral and ionic complexes were taken from Ref. 91, and those for the doubly ionic complexes from Refs. 30, 31, and 33. The dividing line between neutral and ionic/superionic regions is located in the same position as shown in Ref. 91. The slope = 1 lines are drawn arbitrarily, and no best fit is implied.

The energy of the CT absorption $h\nu_{CT}$ depends on the difference between IP_d and EA_a [90] and the change in electrostatic Madelung energy $\langle e^2/r \rangle$, realized on formation of the excited state [Eq. (14)]. Given the relationships between E° and either IP_d or EA_a, it is perhaps not surprising that attempts have been made to correlate $h\nu_{CT}$ with $\Delta E^\circ_{a\text{-}d}$. Mixed stack CT complexes with both nominally neutral and ionic [91] as well as doubly ionic ground states [30-32] exhibit fairly good agreement between these quantities (Fig. 10). The behavior illustrated in

TABLE 2

References for Symbols in Fig. 10

Symbol	Compound
1	$[MV]^{2+}[HCTMCP]^{2-}$
2	$[MV]^{2+}[HCTMM]^{2-}$
3	$[Fe(C_6Me_6)_2]^{2+}[HCTMCP]^{2-}$
4	$[Fe(C_6Me_3H_3)_2]^{2+}[HCTMCP]^{2-}$
5	$[Ru(C_6Me_6)_2]^{2+}[HCTMCP]^{2-}$
6	$[Ru(C_6Me_3H_3)_2]^{2+}[HCTMCP]^{2-}$
7	$[Fe(C_6Me_6)_2]^{2+}[HCTMM]^{2-}$
8	$[Fe(C_6Me_3H_3)_2]^{2+}[HCTMM]^{2-}$
9	$[Ru(C_6Me_6)_2]^{2+}[HCTMM]^{2-}$
10	$[Ru(C_6Me_3H_3)_2]^{2+}[HCTMM]^{2-}$
A	(TMPD)(TCNQF$_4$)
B	(DMP)(TCNQ)
C	(TMPD)(TCNQ)
D	(TMPD)(chloranil)
E	(TMDAP)(TCNQ)
F	(TTF)(chloranil)
G	(TTF)(fluoranil)
H	(DBTTF)(TCNQ)
I	(DEDMTSeF)(TCNQEt$_2$)
J	(TMDAP)(fluoranil)
K	(TTF)(dichlorobenzoquinone)
L	(Perylene)(TCNQF$_4$)
M	(Perylene)(DDQ)
N	(Perylene)(TCNE)
O	(Perylene)(TCNQ)
P	(TTF)(dinitrobenzene)
Q	(Perylene)(chloranil)
R	(Pyrene)(TCNE)
S	(Pyrene)(chloranil)

TABLE 2 (continued)

Symbol	Compound
T	(Anthracene)(chloranil)
U	(Hexamethylbenzene)(chloranil)
V	(Naphthalene)(TCNE)
X	(Anthracene)(PMDA)
Y	(Anthracene)(tetracyanobenzene)
Z	(Phenanthrene)(PMDA)

Fig. 10 can be conceptually understood on the basis of redox potential dependence on the HOMO and LUMO energies of the donors and acceptors. The acceptor level of neutral $\rho = 0$ solids is at higher energy than the donor level and, as ΔE°_{a-d} becomes smaller (i.e., less negative), $h\nu_{CT}$ decreases. The converse is true for the ionic and doubly ionic solids. (Note: the donor/acceptor nomenclature for the doubly ionic compounds in Fig. 10 has been reversed compared to their use in the literature to make the sign convention of their ΔE°_{a-d} values compatible with the previously reported neutral and ionic values). Deviations from the slope = 1 line in Fig. 10 result from differences between donor and acceptor solvation energies and electrostatic terms owing to their different molecular and electronic structure. These deviations appear to be more evident for the doubly ionic solids, as may be expected since the factors contributing to discrepancies may be magnified for highly charged species.

$$h\nu_{CT} = (IP_d - EA_a) - \langle e^2/r \rangle \qquad (14)$$

The molecular level relationship between $h\nu_{CT}$ and ΔE°_{a-d} can be demonstrated by the doubly ionic DA complexes $[(arene)_2M]^{2+}[HCTMCP]^{2-}$ and $[(arene)_2M]^{2+}[HCTMM]^{2-}$ (M = Fe, Ru, arene = $C_6Me_3H_3$, C_6Me_6) [30,31]. The ground and excited states of these complexes were assigned based on the observed electrochemical behavior of their constituents; the dications were reducible at potentials far more negative (for arene = $C_6Me_3H_3$, $E°_{Fe}$ = +0.02, $E°_{Ru}$ = −0.75) of those required for oxidation of the anions ($E°_{HCTMCP}$ = +0.43, $E°_{HCTMM}$ = +0.92). The CT transition in this case was therefore tantamount to electron transfer from the dianion donor to the dication acceptor [Eq. (15)] to form a singly charged excited state.

$$\cdots D^{2-}A^{2+}D^{2-}A^{2+}D^{2-}A^{2+}D^{2-}A^{2+} \longrightarrow \cdots D^{2-}A^{2+}D^{-}A^{+}D^{2-}A^{2+}D^{2-}A^{2+}$$

(15)

These complexes also demonstrated that more precise comparisons between optical CT properties of mixed stack solids and redox potentials can be realized for complexes in which the individual components are structurally and electronically similar. The difference in CT transition energy, $\Delta h\nu_{CT}$, between two complexes with different organometallic acceptors but the same dianion donor species was *exactly* equivalent to the difference in redox potentials, ΔE_{redox} = 0.77 eV, of the $(C_6Me_3H_3)_2$-M^{2+} acceptors (Fig. 11). The good agreement was attributed to the essentially identical structure of the dications in which the metal ions are buried within an organic matrix defined by aromatic ligands. Under these conditions, $\Delta G°_r$, $\Delta G°_s$, and $\langle e^2/r \rangle$ terms [Eqs. (5) and (14)] are independent of the metal ion, and the difference in redox potentials of the cations accurately reflects the change in electron affinity. The CT energy was also found to increase for the $[HCTMM]^{2-}$ complexes compared to the $[HCTMCP]^{2-}$ complexes owing to the more positive redox potential (i.e., large IP_d) of the former. However, precise agreement was not obtained because the two ions differ significantly in structure. Similar behavior was observed for the totally organic doubly ionic mixed stack complexes $[MV]^{2+}[HCTMM]^{2-}$ ($h\nu_{CT}$ = 1.7 eV) and $[MV]^{2+}$-$[HCTMCP]^{2-}$ ($h\nu_{CT}$ = 2.7 eV) in which the MV^{2+} dication also assumes the role of the acceptor [32,33]. These results clealry illustrate that an easily determined experimental parameter such as E° can be used to understand

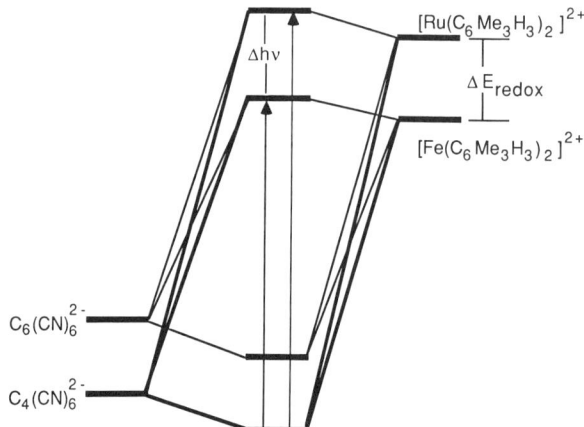

FIG. 11. Schematic representation of the charge-transfer transitions responsible for the visible absorptions in (arene)$_2$M^{2+} complexes. The difference in the redox potentials of the cations (ΔE_{redox}) is nearly equivalent to difference in the charge-transfer energies ($\Delta h\nu_{CT}$) for a given anion. The relationship of the HOMO of $[C_4(CN)_6]^{2-}$ and $[C_6(CN)_6]^{2-}$ is illustrated by the lower energy (higher ionization potential) of the former. The observed CT absorptions are in accord with this scheme, although the difference in ionization potentials of the two anions cannot be determined from ΔE_{redox} of the anions.

and, in many cases, predict the optical properties of mixed stack DA solids, as well as the electronic nature of the ground and excited states.

Redox properties of molecular species may also play a role in the design of ferromagnetic one-dimensional solids. Because of the possible cooperative effects between donors and acceptors, there has been considerable effort to prepare organic [92,93] and organometallic [94] ferromagnetic materials based on mixed stack DA complexes. Much of this effort has involved the design of molecules having energetically accessible triplet states which, on admixing into the ground state, will induce ferromagnetism. Recently it has been shown that a series of decamethylferrocinium-polycyanoanion CT solids display bulk ferromagnetism [95,96].

Scheme 6

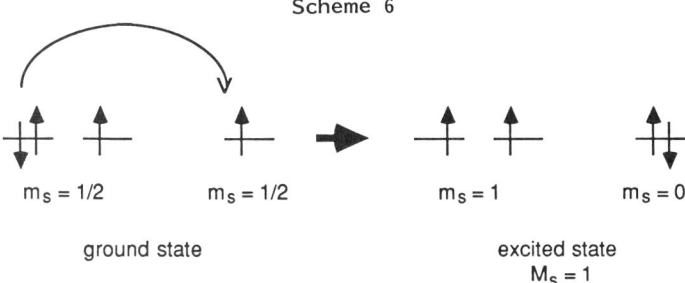

A model has been proposed [97], based on an earlier model by McConnell [98], which considers admixing of triplet states in DA complexes. Briefly, mixing of degenerate, high spin-multiplicity excited states with the ground state wavefunctions (i.e., configuration interaction) provides a mechanism for ferromagnetic stabilization. It was argued that the formation of high spin states can be envisioned to occur via virtual "forward" or "retro" electron transfer from the ground state (Scheme 6). For example, virtual forward charge transfer in a CT solid with a D^+A^- ground state resulting in a $D^{2+}A^{2-}$ excited state is favorable if the doubly charge entities are accessible. If this state possesses high spin multiplicity by virtue of degeneracy and orbital population, as shown in Scheme 4 for a d^3-s^1 system, ferromagnetic stabilization may be achieved. Although, by definition, virtual charge transfer processes will always be uphill from the ground state, mixing becomes more feasible as electron transfer becomes less energetically demanding. Therefore, design of these systems should include consideration of the relative redox potentials of the donors and acceptors, particularly with regard to $D^+ \rightarrow D^{2+}$ and $A^- \rightarrow A^{2-}$ processes. Some compounds listed in Table 1 have been designed for this purpose. For example, dications of HET, HMTP, and HOC are electrochemically accessible, and the last has been reported to exhibit behavior consistent with a ground state triplet [45]. Although redox potentials are not the only consideration due to electrostatic terms and conformational restrictions in the solid state, this readily obtainable

IV. ELECTROCHEMICAL PREPARATION OF LOW-DIMENSIONAL SOLIDS

A. Basic Principles

information can certainly be useful for screening likely candidates for molecular ferromagnets.

A common method for the preparation of molecular solids involves the reaction of a neutral electron donor and a neutral electron acceptor, which very often is accompanied by electron transfer to form an ionic solid [Eq. (16)]. In many cases, growth of high-quality ionic crystals can be achieved by conventional recrystallization techniques in which the components are redissolved either into their original neutral forms if electron transfer occurs only in the solid state or into their ionic forms if the electron transfer has sufficient driving force in solution. Crystal size, morphology, and quality can, to a large extent, be controlled by factors such as solvent, rate of cooling, and concentration. If the DA solid possesses a significant degree of electrostatic Madelung stabilization owing to its ionicity, negligible solubility can result. The extended nature of one-dimensional solids in which significant stabilization of the solid is realized through intermolecular interactions also tends to diminish solubility. Whereas negligible solubility can facilitate isolation of the CT complex, it can also prohibit the use of standard recrystallization techniques. Under these conditions, one commonly must resort to metathesis methods; that is, reaction of the oxidized donor cation and reduced acceptor anion [Eq. (17)].

$$D^\circ + A^\circ \longrightarrow D^+A^- \qquad (16)$$

$$D^+ + A^- \longrightarrow D^+A^- \qquad (17)$$

Metathesis methods generally rely on slow interdiffusion of D^+ and A^-, either through quiet solutions or through fritted glass separators between compartments containing both species. Unfortunately, slow, reproducible growth of high-quality crystals for structural studies and for measurement of electronic properties is not always possible since introduction of D^+ and A^- cannot always be predictably and precisely regulated.

Less frequently used approaches to the synthesis of low-dimensional DA solids have included electrochemical methods, which are possible because of the redox nature of molecular solid constituents. For example, electrochemical preoxidation of neutral donors has been demonstrated for the preparation of TTF and TSeF halide phases [99,100]. In these cases, the donor molecules were oxidized in CH_3CN, and the halide salts precipitated by addition of tetraalkylammonium halides. For the TTF complexes, the ratio of TTF/TTF^+ was controlled by the extent of electrolysis of TTF solutions. The CT solid was subsequently formed by slow crystallization of the electrochemically prepared mixtures and tetraethylammonium halide solutions over a period of months. The value of ρ, and therefore the conductivity, was determined by the TTF/TTF^+ ratio in the solution. The preparation of $(TSeF)Br_{0.8}$ and $(TSeF)Cl$ was similarly accomplished. It was also noted that the composition of these phases could be affected by difference between the standard potentials of the D/D^+ and Z^-/X_3^- redox couples. Accordingly, the higher conductivity of $(TSeF)Br_{0.8}$ was attributed to the mixed valent character ($\rho < 1$) of the donor molecules resulting from a positive $\Delta E°$ for Eq. (18), which made available neutral TSeF for formation of the mixed valent salt. The same is true for the TTF-iodide phases [Eq. (19)]

$$2\ TSeF^+ + 3Br^- \longrightarrow 2\ TSeF + Br_3^- \qquad \Delta E = +0.17 \qquad (18)$$

$$2\ TTF^+ + 3I^- \longrightarrow 2\ TTF + I_3^- \qquad \Delta E = +0.21 \qquad (19)$$

Although the above examples demonstrate that electrochemical techniques are useful for the preparation of CT solids, they do not represent any clear difference between, or advantage over, more conventional metathetical methods (the D/D^+ ratio could also be adjusted simply by preparation of solution directly from D and D^+). An alternative and

TABLE 3

TTF and TSeF Halides Prepared by Electrochemical Preoxidation of Donors

Chlorides	Bromides	Iodides
$(TTF)Cl_2$	$(TTF)Br_2$	
$(TTF)Cl$	$(TTF)Br$	
$(TTF)Cl_{0.77}$	$(TTF)Br_{0.76}$	$(TTF)I_{0.72}$
$(TTF)Cl_{0.68\pm0.02}$	$(TTF)Br_{0.59\pm0.02}$	$(TTF)I_{0.69\pm0.02}$
$(TSeF)Cl$	$(TSeF)Br_{0.8}$	

widely exploited electrochemical method for the preparation of CT solids (although not as frequently used as conventional methods) is electrocrystallization of the solid directly at the working electrode, in which one of the reagents is introduced through its electrochemical formation at the electrode. This method is most viable when the desired complex is insoluble under the crystal growth conditions and only one of the molecular species is electrochemically active at potentials either applied or incurred under potentiostatic or galvanostatic conditions, respectively, at the working electrode. Oxidation of a donor molecule in the presence of an acceptor anion or, conversely, reduction of an acceptor in the presence of a donor cation results in crystal growth at the electrode if the concentration of the DA complex exceeds its solubility limit at the electrode [Eqs. (20) and (21)].

$$D - e^- \xrightarrow{A^-} D^+A^- \qquad (20)$$

$$A + e^- \xrightarrow{D^+} D^+A^- \qquad (21)$$

The apparatus reported for electrocrystallization generally comprises a two-compartment cell to minimize contamination from products of the counterelectrode reaction (Fig. 12). Generally, platinum electrodes have been used, although there is no intrinsic reason for avoiding other

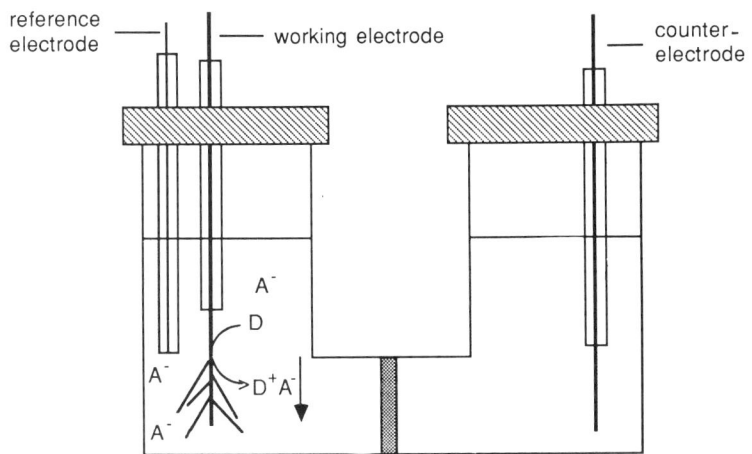

FIG. 12. Schematic representation of a conventional H cell used for electrocrystallization. In this example, donor molecules are oxidized at the anode in the presence of A^-, resulting in crystal growth at the electrode. A glass frit is commonly used to prevent contamination of the desired compound with products of the counterelectrode reaction.

electrode materials. If potential control of the working electrode is desired, an appropriate reference electrode is required. Electrolysis is simply performed for an appropriate length of time, and the crystals are harvested from the electrode on completion.

B. Examples

The first reported electrocrystallization of one-dimensional materials was the preparation in 1971 of perchlorate complexes of pyrene, perylene, and azulene with the general formula $(D)_2ClO_4$ [101]. The work did not focus on electrocrystallization but rather was motivated by interest in the magnetic properties of paramagnetic ionic crystals [102]. Electrochemical preparation was performed by electrolysis with an applied voltage of approximately 2V across the cell; the cell was constructed so that the actual current density was approximately 1 mA cm^{-2} using wire electrodes. The result was the growth of black crystals on the anode,

FIG. 13. Consecutive photographs of growth of $(Py)_2ClO_4$ on a platinum anode. (From Ref. 101.)

as shown for $(Py)_2ClO_4$ in Fig. 13. The rapid crystal growth was attributed to the conductivity of these crystals. Interestingly, crystal formation at the electrode was accompanied by a sharp increase in the conductance of the cell. This was attributed to the fairly high energy required for nucleation, of a crystal, which is followed by less energetically demanding crystal growth on those nuclei. The importance of nucleation was demonstrated by the observation that crystals did not form if fire-polished platinum was used as the anode. Additionally, since the crystals are conductive, it is also plausible that the effective electrode area increases as the crystals form on the electrode, resulting in higher conductance. The rate of crystal growth was surmised to depend on the diffusion of the hydrocarbon to the electrode. The stoichiometry of these phases indicated one neutral molecule for every cation, resulting in $\rho = 0.5$. Although conductivity was not reported, the nonintegral charge would presumably favor reasonable electronic conductivity. Interestingly, electrochemical conditions were claimed to influence crystal growth, as a different phase was observed in the early moments of electrolysis if a square voltage pulse was applied. It was reported later that, on changing the electrolysis conditions, $(Py)ClO_4$ ($\rho = 1$) could be formed, although details were not given [103]. To this author's knowledge, this was the first suggestion of one of the unique features of electrochemical crystal growth: the ability to adjust electrochemical parameters to direct the growth of different materials at the electrode. We shall see shortly that this feature can be exploited to rationally direct the stoichiometry of different one-dimensional materials grown at electrodes.

A report published in 1974 described the electrochemical crystal growth of metal dithiolate-perylene complexes with the formula $[Pe]_2^-$ $[MS_4C_4(CN)_4]$ (M = Ni, Cu, Pd) [104]. This growth was accomplished by controlled potential electrolysis at +1.03 V (vs. SCE) of CH_2Cl_2 solutions containing perylene and $[n-Bu_4N][MS_4C_4(CN)_4]$, resulting in the formation of needle-shaped crystals at the anode. The resulting crystals were conductive, although they displayed a positive temperature coefficient indicative of intrinsic semiconductivity.

Shortly afterward, the electrochemical preparation of highly conducting inorganic chain compounds, namely, the Krogmann's salts $K_2Pt(CN)_4X_{0.3} \cdot 3H_2O$ (X = Cl, Br), was reported [105,106]. Synthesis was accomplished by application of a small voltage across a two-electrode cell containing $K_2[Pt(CN)_4] \cdot 3H_2O$ and KBr, which resulted in oxidation of $Pt(CN)_4^{2-}$ and the growth of bronze-colored needles at the anode. These compounds are conducting by virtue of one-dimensional platinum chains (Fig. 14) composed of partially oxidized $Pt(CN)_4$ units that are more closely spaced (2.89 A) [107] than in the parent oxidized form (3.48 A) [108]. The average oxidation state of Pt in this material is $Pt^{2.33+}$ with one-third of an electron removed from each Pt site, allowing significant overlap of the Pt d_z^2 orbitals and formation of a partially filled conduction band. As a result, these materials exhibit metallic conductivity [109].

Interest in electrocrystallization as a preparative tool heightened considerably with the discovery that the one-dimensional mixed valent (ρ = 0.5) organic conductor $(TMTSF)_2X$ (X = PF_6^-, ReO_4^-, ClO_4^-) exhibited superconductivity at low temperatures [110,111]. The electrochemical preparation of an extended series of these salts with X = PF_6^-, AsF_6^-, SbF_6^-, BF_4^-, and NO_3^- was first reported in 1980 [112,113]. The synthesis was performed by oxidation of TMTSF in CH_2Cl_2 at constant current. Although no details were given, it was noted that the currents were sufficiently low to render the crystallization rate slower than diffusion of TMTSF to the electrode. This can be a rather important consideration since severe concentration polarization of TMTSF induced by higher currents at the electrode can result in electrode potentials sufficiently positive to cause formation of materials containing higher valent forms of TMTSF. The incorporation of neutral TMTSF molecules, which is necessary for formation of the mixed valent salt, would not occur if the concentration of TMTSF near the electrode was severely depleted. Also, higher electrode potentials can result in the formation of $TMTSF^{2+}$, which would also be deleterious to the formation of the conducting phase.

Electrocrystallization has also been employed to conveniently prepare one-dimensional "alloys" (probably more accurately described as solid solutions) by crystal growth in media containing structurally and chemically similar components. For example, alloys with the composition $\{(TMTSF)_{1-x}(TMTTF)_x\}_2PF_6$ (x = 0, 0.005, 0.03) were prepared by

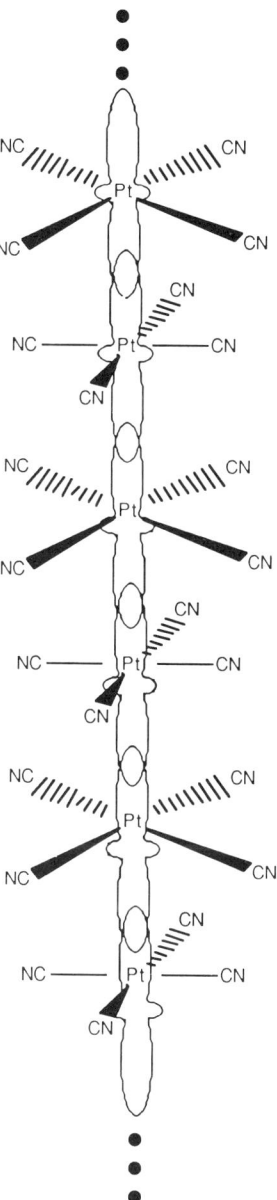

FIG. 14. Illustration of the one-dimensional chains in $K_2Pt(CN)_4Br_{0.3} \cdot 3H_2O$ showing the overlap of the Pt d_z^2 orbitals responsible for the observed metallic conductivity.

controlled potential electrolysis of TMTSF solutions containing small amounts of TMTTF and PF_6^- [114]. The metal-insulator transition of the alloy was broader and shifted to higher temperature (11.5-17 K) compared to $(TMTSF)_2(PF_6)$ prepared by identical methods. Similarly, preparation of alloys with two different anions by electrolysis in the presence of both anions was accomplished; the ambient pressure superconductivity of the chlorate salt was completely suppressed by the incorporation of IO_4^- in $(TMTSF)_2(ClO_4)_{0.95}(IO_4)_{0.05}$.

Alloys of a related series of different $(TMTTF)_2(MF_6)$ complexes have also been prepared by electrocrystallization in an effort to study the role of anion size on the electronic properties of these materials [115]. The results were somewhat similar to those found for $(TMTSF)_2(X)$ salts in that increasing the value of x in $(TMTTF)_2(SbF_6)_{1-x}(MF_6)_x$ (M = P, As) resulted in an increase in the metal-insulator transition temperature.

There have been numerous papers describing a family of segregated stack organic superconductors based on bis(ethylenedithio)tetrathiafulvalene (BEDT-TTF or ET) and inorganic counteranions with the general formula $(ET)_2(X)$ and $(ET)_2(X)(Y)$, the latter possessing two different counterions. The primary goal of these investigations has been to study the effect of different size, symmetry, and polarizability of different anions on the electronic properties. A diversity of counterions has led to a number of structural motifs. These complexes have proved to be most readily synthesized by electrocrystallization techniques, always under galvanostatic control.

The most intriguing aspect of the crystal growth of ET complexes has been the formation of multiple phases, which appears to depend on the identity of the counterion, solvent in the growth medium and current density. The structural phase is an important consideration in ET solids because it has a pronounced effect on the electronic properties, particularly superconductivity, which is thought to require significant interstack sulfur-sulfur interactions. The effect of solvent during crystal growth is evident from the markedly different crystalline habit observed for several ET-trihalide compounds when different solvents (e.g., THF, chlorobenzene, benzonitrile, nitrobenzene) are used [116]. Simultaneous growth of different phases has also been noted [117]. For example, ET complexes with ReO_4^- or BrO_4^- counterions have been reported to form crystalline phases with different stoichiometries during a single electrocrystallization experiment (Table 4) [118-121]. Multiple phase growth is also very pervasive for complexes of ET with trihalides, particularly triiodide phases in which different crystallographic phases with the same stoichiometry grow simultaneously on an electrode (Table 5). One possible explanation for this unusual behavior is the redox chemistry of the I_3^- ion, which is present as a tetraalkylammonium salt during electrocrystallization of these materials. It has been noted that the oxidation of I_3^- takes place at potentials that are cathodic of those required for oxidation of ET ($E_{p,a}$ = +0.63 V) and that therefore ET oxidation actually may be mediated by iodine radicals [122]. Indirect evidence for this behavior was the observation that more anodic electrode potentials (by 0.45 V) were required for the electrocrystallization of ET salts containing ions that were not redox-active (e.g., ClO_4^-, BF_4^-, PF_6^-). The formation of different iodine-containing phases during electrocrystallization may be due to the complex equilibriums for polyiodides in I_2/I^- solutions. The electrode potentials, concentrations of iodine species, and rates of crystallization may exert a significant influence on the relative concentrations of the different polyiodides at the electrode, subsequently affecting the type of species incorporated into the lattices of the ET complexes. The behavior appears to be very complicated, and more systematic investigations are needed for further understanding.

Multiple phase growth during electrocrystallization is not unique to the ET complexes; simultaneous growth of two different phases on oxidation of perylene in CH_2Cl_2/PF_6^- solutions was reported [130]. The composition of the two phases was reported as approximately $(PE)(PF_6)_{1.1}$·0.8 CH_2Cl_2, but the composition of each phase differed slightly with

TABLE 4

ET Complexes with Iodine

Compound	Growth conditions[a]	Conductivity	Reference
β-(ET)$_3$I$_3$	Galv (1 uA cm^{-2}), TCE	Superconducting (T_c = 1.5 K)	123, 124
α-(ET)$_2$I$_3$	Galv (uA cm^{-2})	M-I trans @ 140 K[b]	125
δ-(ET)I$_3$	Galv (65 uA cm^{-2}), TCE	M-I trans @ 130 K[b]	126
γ-(ET)$_3$(I$_3$)$_{2.5}$	Galv (65 uA cm^{-2}), TCE	Supercond (T_c = 2.5 K)	127, 126
ε-(ET)$_2$(I$_3$)(I$_8$)$_{0.5}$	Chemical	Metallic	128
(ET)$_2$(I$_3$)(I$_5$)	Chemical	Metallic	128
(ET)$_2$(I$_3$)(TII$_4$)	Galv (2.7 uA cm^{-2}), THF	Semiconducting	128
(ET)$_3$Ag$_{6.4}$I$_8$	Galv (0.5 uA cm^{-2}), MC	Semiconducting	129

[a]Galv = galvanostatic, TCE = trichloroethane, THF = tetrahydrofuran, MC = methylene chloride.
[b]M-I trans refers to a metal-insulator transition.

TABLE 5

Complexes Prepared by Electrocrystallization

Compound	σ_{300K} ($\Omega^{-1}\mathrm{cm}^{-1}$)[a]	Reference
$(2,5\text{-DM-DCNQI})_2\mathrm{Cu}$	800 5.0×10^5 (3.5 K)	153
$(\mathrm{TMTSF})_2\mathrm{PF}_6$		
$(\mathrm{TMTSF})_2\mathrm{ClO}_4$	430	114, 154
$(\mathrm{TMTSF})_2(\mathrm{ClO}_4)_{0.95}(\mathrm{IO}_4)_{0.05}$	250	
$(\mathrm{TMTSF})_{0.97}(\mathrm{TMTTF})_{0.03}\mathrm{PF}_6$		
$(\mathrm{TMTSF})_{0.995}(\mathrm{TMTTF})_{0.005}\mathrm{PF}_6$		
$(\mathrm{TSeT})_2\mathrm{Cl}$		155
$2\text{-FTSeT-Br}_{0.5}$	1.7×10^{-3}	156
$(\mathrm{TSeT})_2\mathrm{Br}$	2.0×10^{-3}	157
$(\mathrm{TSeT})_2\mathrm{I}$		
$(\mathrm{HMTP})(\mathrm{ClO}_4)$	0.3	158
$(\mathrm{HMTP})(\mathrm{ClO}_4)_{1.6}$	0.1	
$\mathrm{TTT}(\mathrm{ClO}_4)_{0.56}$	6×10^{-3}	47, 149, 159
$\mathrm{TTT}(\mathrm{PF}_6)_{0.61}$	762	
$\mathrm{TTT}(\mathrm{SbF}_6)_{0.64}$	40	
$\mathrm{TTA}(\mathrm{PF}_6)_{0.33}(\mathrm{TCE})_{0.96}$	2.3×10^{-4}	
$\mathrm{TTAI}_{1.2}$	4.2	
$\mathrm{TSA}(\mathrm{PF}_6)_{0.41}(\mathrm{TCE})_{1.47}$	2.5×10^{-5}	
$\mathrm{TSA}(\mathrm{ClO}_4)_{0.61}(\mathrm{TCE})_{1.12}$	1.7×10^{-5}	
$\mathrm{TSA}(\mathrm{TCNQ})$	15	
$\mathrm{TTN}(\mathrm{PF}_6)_{0.39}(\mathrm{TCE})_{1.07}$	1.25×10^{-7}	
$\mathrm{TTN}(\mathrm{ClO}_4)_{0.86}(\mathrm{TCE})_{0.27}$	2.56×10^{-7}	
$(\mathrm{TMTTF})_2\mathrm{X}$	60	160
$\mathrm{X} = \mathrm{BF}_4^-$	80	
ClO_4^-	10	

TABLE 5 (continued)

Compound	σ_{300K} ($\Omega^{-1}cm^{-1}$)[a]	Reference
PF_6^-	10	
Br^-	150	
$(TMTSF)_2X$		161
$X = ClO_4^-$		
BF_4^-		
PF_6^-		
$(DMDBTTF)(BF_4)$	5×10^{-8}	162
$(ET)_2(IBr_2)$		163
$(ET)_2(I_2Br)$		
$(ET)_2(X)$		164
$X = AuCl_2^-, AuBr_2^-, AuI_2^-$		
$Au(CN)_2^-, Ag(CN)_2^-$		
$(ET)_2(AuI_2)$	Superconducting (3.2 K)	165
α-$(ET)_3(NO_3)_2$	800	166
β-$(ET)_2I_3$	Superconducting (T_c = 1.5 K)	123, 124
α-$(ET)_2I_3$	M-I trans @ 140 K	125
δ-$(ET)I_3$	M-I trans @ 130 K	126
γ-$(ET)_3(I_3)_{2.5}$	Superconducting (T_c = 2.5 K)	126, 127
$(ET)_3Ag_{6.4}I_8$	Semiconducting	129
$(ET)_2(I_3)(I_5)$	Metallic	128
$(ET)_2(I_3)(TlI_4)$	Semiconducting	128
$(ET)_2AuBr_2$	Semiconducting	167
$(ET)AuBr_2Cl_2$	Semiconducting	168
$(ET)Ag_4(CN)_5$	M-I trans @ 100 K	169

TABLE 5 (continued)

Compound	σ_{300K} ($\Omega^{-1}cm^{-1}$)[a]	Reference
$(ET)_2ReO_4$	Metallic	118–121
$\alpha\text{-}(ET)_3(ReO_4)_2$	M-I trans @ 100 K	
$\beta\text{-}(ET)_3(ReO_4)_2$	Insulator	
$\gamma\text{-}(ET)_3(ReO_4)_2$	Metallic	
$(ET)_2BrO_4$	Superconducting (T_c = 2K, 4 kbar)	
$(ET)_2(BrO_4)(1,1,2\text{-trichloroethane})_{0.5}$	Metallic	
$(ET)_3(BrO_4)_2$	Metallic T > 210 K	
$TTF(X)_y$		170
$TMTTF(X)_y$		
$HMTTF(X)_y$		
$TMTSF(X)_y$		
$(TMTTF)_2MF_6$		115
M = P	40	
As	25	
Sb	10	
$(SbF_6)_{1-x}(AsF_6)_x$		
$(SbF_6)_{1-x}(PF_6)_x$		
$(TMTSF)_2X$		112
X = PF_6^-	540	
AsF_6^-	430	
SbF_6^-	500	
BF_4^-	540	
NO_3^-	780	
$TTF(X)_n$		132
X_n = $(Cl)_{1.1}$	0.06	
$(BF_4)_{0.55}$	0.6	

TABLE 5 (continued)

Compound	σ_{300K} $(\Omega^{-1}cm^{-1})^a$	Reference
$(NO_3)_{0.55}$	20	
$(Br)_{0.7}$	1.4	
$(HCO_3)_{1.5}$	0.9	
$(HSO_4)_{1.2}$	0.03	
$(C_2O_4K \cdot 3H_2O)$	0.005	
$(MeCO_2)_{0.7}$	3	
$(I)_{0.7}$	10	
$(Fumarate)_{0.25}$	0.002	
$TTF(X)_n$		133a
X = TCNQ	100	
$(N_3)_{0.74}$	0.096	
(OSO_2OMe)	5.1×10^{-8} (8.2)	
$(SCN)_{1.14}$	0.2 (or 530)	133b
$(HSO_4)_{1.2}$	4.7×10^{-2} (0.9)	
$(OCN)_2$	1.1×10^{-7}	
$(Ac)_{0.70}$	38	
$(I)_{0.70}$	10	
$(Br)_{0.70}$	1.4	
$(ClO_4)_{0.71}$	9	
$(BPh_4)_{0.86}$	2.5	
$(Fe(CN)_6)_{0.25}$	10.5	
$(AgNO_3)_{0.67}$	0.4	
$TTT(I)_{0.72}$	1.1×10^3	
$TTT(NO_3)_{0.74}$	500	
TTFTCNQ	4.2×10^2	37
$[Perylene][MS_4C_4(CN)_4]$		104
M = Ni	50	
= Cu	6	
= Pd	0.07	

TABLE 5 (continued)

Compound	σ_{300K} $(\Omega^{-1}cm^{-1})^a$	Reference
K(TNCQ)	2×10^{-4}	134
Mn(TCNQ)·0.66MeCN	4.4×10^{-3}	
$Fe_2(TCNQ)_{3.5}H_2O$	4.8×10^{-4}	
$Co(TCNQ)_{2.3}H_2O$	2.6×10^{-4}	
$Ni_2(TCNQ)_{3.6}H_2O$	7.6×10^{-4}	
Cu(TCNQ)	2.9×10^{-2}	
Cu(TCNQ)	31 (784)	
$Zn_2(TCNQ)_3(ClO_4)_{2.3} \cdot H_2O$	2.1×10^{-2}	
$Ag_2(TCNQ)_3$	8	
$Ag_4(TCNQ)_3$	44	
$Cr(TCNQ)_3$	5.1×10^{-3}	
$(Pyrene)_2(ClO_4)$		101
$(Perylene)_2(ClO_4)$		
$(Azulene)_2(ClO_4)$		
$(Naphthalene)_2(PF_6)$	0.12	171
$(Perylene)_2(PF_6)_{1.1} \cdot 8CH_2Cl_2$	2000	172
$(Perylene)_2(AsF_6)_{1.1} \cdot 7CH_2Cl_2$		
$(Perylene)_2(PF_6)_{1.1} \cdot 0.8CH_2Cl_2$	900 (880)	130
$(Perylene)_2(AsF_6)_{1.1} \cdot 0.7CH_2Cl_2$	1200 (760)	
$(Perylene)_2(PF_6)_{1.4} \cdot 0.06$ THF	70	
$(Perylene)_2(AsF_6)_{1.5} \cdot 0.5$ THF	580	
$(Perylene)_2(SbF_6)_{2.0} \cdot 7.5$ CH_2Cl_2		
$(Fluoranthene)_2(PF_6)$	0.05	173
$(B[4,5-b]PTTF)_2(BF_4)$	0.1	174
$[PcCo(CN)]_x$	57	145, 146
$[PcFe(CN)]_x$	0.4	147
$Ni(Pc)(ClO_4)_y$	700	148

TABLE 5 (continued)

Compound	σ_{300K} (Ω^{-1}cm^{-1})[a]	Reference
$K_2[Pt(CN)_4]Br_{0.3} \cdot 3H_2O$	4-1050	105
$K_2[Pt(CN)_4]Cl_{0.3} \cdot 3H_2O$	ca. 200	106
$[C_5Me_5)_2Fe]_2[TCNQ]_2$		150
$[(C_6Me_6)_2M][TCNQ]_2$ (M = Fe,Ru)	$<10^{-6}$	15
$[(C_6Me_6)_2M][TCNQCl_2]_2$ (M = Fe,Ru)	$<10^{-6}$	
$[(C_6Me_6)_2M][TCNQF_4]_2$ (M = Fe,Ru)	$<10^{-6}$	
$[(C_6Me_6)_2M][TCNQ]_4$ (M = Fe,Ru)	1.0	15
$[(C_6Me_3H_3)_2M][C_6(CN)_6]$ (M = Fe,Ru)	$<10^{-6}$	30
$[(C_6Me_6)_2M][C_6(CN)_6]$ (M = Fe,Ru)	$<10^{-6}$	
$[(C_5Me_5)Ru(2_2\text{-}1,4\text{-cyclophane})\text{-}Ru(C_5Me_5)][TCNQ]_2$	$<10^{-6}$	151
$[(C_5Me_5)Ru(2_2\text{-}1,4\text{-cyclophane})\text{-}Ru(C_5Me_5)][TCNQ]_4$	0.1	151

[a] For definition of acronyms, see Appendix.
[b] Room temperature conductivities are reported when possible; otherwise, only the gross conductivity properties are cited.

respect to crystalline habit and the actual content of PF_6^- and CH_2Cl_2. Accordingly, the formal oxidation state of perylene in the two materials was thought to be different. Only the phase with the higher formal oxidation state was observed when crystal growth was performed in THF, which probably illustrates the importance of solvent inclusion in solid-state matrices rather than any electrochemical principle.

Several papers have appeared describing the electrocrystallization of conducting salts incorporating either TTF or TCNQ, which are probably the most extensively studied compounds in the area of one-dimensional materials. The parent complex, (TTF)(TCNQ), has been synthesized by electrochemical oxidation of TTF in a solution saturated with (TTF)-(TCNQ) containing $TCNQ^-$ anion [37]. A rather extensive survey of TTF complexes with halides and a variety of other inorganic and organic counterions prepared under either potentiostatic or galvanostatic control has been reported [131–133]. Electrocrystallization of metal-TCNQ complexes has also been extensively investigated (Table 7) [134].

Electrocrystallization of nonstoichiometric $TTFBr_x$ (x = 0.7–0.8) crystals in Nafion ion-exchange films has been accomplished by ion exchange of TTF^+ into the Nafion films, followed by electrochemical cycling of the films in 1 M KBr [135,136]. Cyclic voltammograms of these films resembled those of TTF films prepared by evaporation of benzene solutions of TTF on Pt electrodes, with a very sharp cathodic peak and a somewhat broader anodic wave (Fig. 15). The large separation of the peaks is probably due to lattice energy associated with significant structural changes, as has been observed in (TTF)(TCNQ) electrodes [137, 138]. The peak shape and scan rate dependence indicated that the

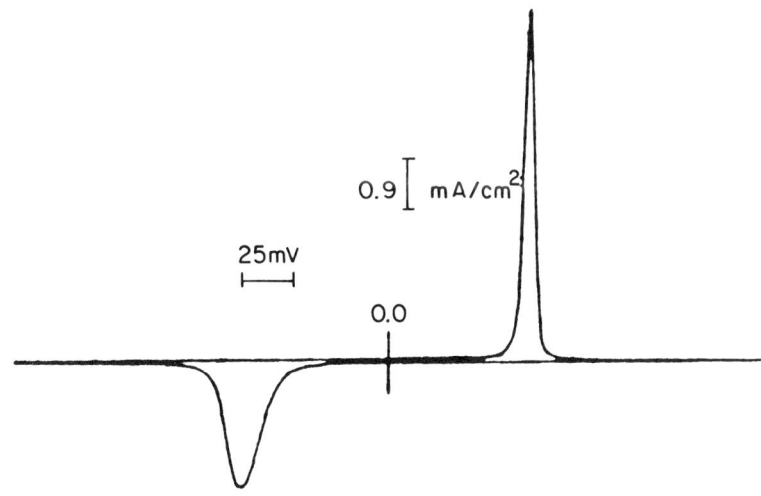

FIG. 15. Cyclic voltammogram (10 mV/sec) of a Pt/Nafion-TTF^+ electrode in 1 M KBr. (From Ref. 135.)

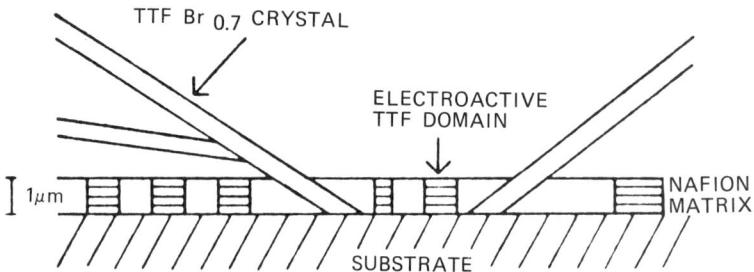

FIG. 16. Diagram of a cross section of a Pt/Nafion-TTF$^+$ electrode after formation of TTFBr$_{0.7}$ crystals. (From Ref. 136.)

electroactive TTF species were surface-bound with fairly high effective diffusion coefficients (ca 10^{-8} cm^2 sec^{-1}). The physical properties of these films is discussed in more detail in Sec. V, since only those aspects closely related to electrocrystallization are discussed here. The needle-shaped crystals of TTFBr$_{0.7}$, believed to be nonelectroactive, could be visually observed on the surface and embedded into the Nafion polymer, while electroactive domains of TTF/TTF$^+$ remaining in the polymer were responsible for the remaining cyclic voltammetry (CV) waves and the color change of the polymer upon cycling (Fig. 16). After the first few cycles, the CV behavior due to the electroactive domains was stable for several hours, and the nonelectroactive material remained in the Nafion film. It was proposed that the initial negative potential excursion resulted in reduction of the exchanged TTF$^+$ with retention of TTF in the film and insertion of K+ ion into the film to maintain electroneutrality. Subsequent oxidation of TTF on cycling then resulted in insertion of Br$^-$ into the Nafion film and growth of the TTFBr$_{0.7}$ crystals (Scheme 7). Recently, essentially identical behavior has been observed in montmorillonite clays in which TTF$^+$ is intercalated between the aluminosilicate layers [139].

The behavior of the TTFBr$_{0.7}$/Nafion films is consistent with quartz crystal microbalance (QCM) [140] recently performed in our laboratory

Scheme 7

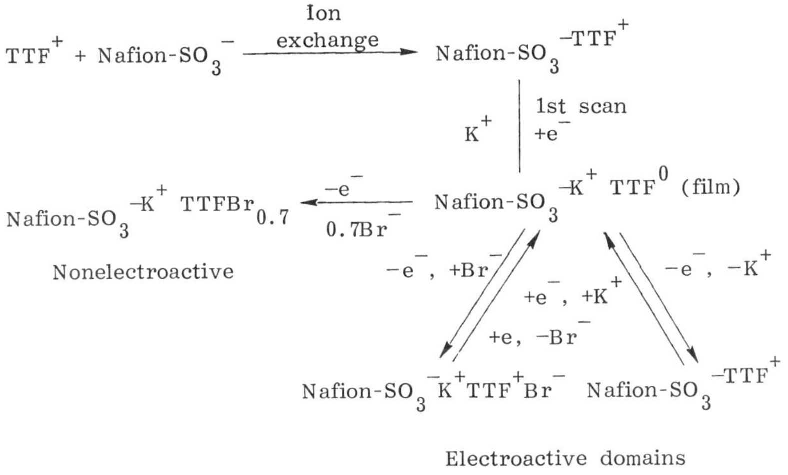

$$TTF^+ + Nafion\text{-}SO_3^- \xrightarrow{\text{Ion exchange}} Nafion\text{-}SO_3^- TTF^+$$

$$\Big\downarrow \begin{array}{c} K^+ \\ \end{array} \begin{array}{c} \text{1st scan} \\ +e^- \end{array}$$

$$Nafion\text{-}SO_3^- K^+ \; TTFBr_{0.7} \xleftarrow{-e^-}_{0.7Br^-} Nafion\text{-}SO_3^- K^+ \; TTF^0 \text{ (film)}$$

Nonelectroactive

$$-e^-, +Br^- \diagup \quad \diagdown -e^-, -K^+$$
$$\diagup +e^-, +K^+ \diagdown$$
$$\diagup +e, -Br^- \diagdown$$

$$Nafion\text{-}SO_3^- K^+ TTF^+ Br^- \qquad Nafion\text{-}SO_3^- TTF^+$$

Electroactive domains

[141]. These experiments rely on the ability to quantify mass changes in the polymer film immobilized on a piezoelectric quartz crystal from shifts in the oscillation frequency of the crystal, assuming rigid layer behavior according to the Sauerbrey relationship [142]. Recent reports describing the behavior of polyvinylferrocene films in the QCM suggest that this may be a reasonable assumption [143,144]. Addition of TTF^+Cl^- to a spin-coated Nafion film in water resulted in a frequency decrease (i.e., mass increase) corresponding to the incorporation of 2.16×10^{-9} equivalents of TTF^+ in the film. This value agreed well with coulometric determinations of active TTF that indicated 2.02×10^{-9} eq. Repeated cycling of the film in 1.0 M KBr resulted in an overall increase in mass, with progressively smaller changes in mass during each cycle (Fig. 17). Eventually, the mass of the film does not change any further, and only very small changes in mass are observed in each cycle. The overall frequency decrease of 57 Hz was close to the 63 Hz shift expected according to Scheme 7 for retention of K^+ and complete formation of $TTFBr_{0.7}$ (assuming no electroactive regions). The discrepancy is consistent with the presence of electroactive regions, which are probably responsible for the

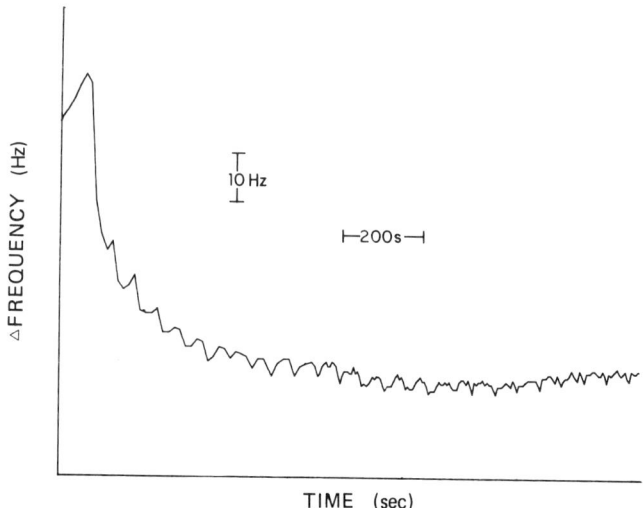

FIG. 17. Frequency response of the quartz crystal microbalance upon cycling a Nafion-TTF$^+$ film in 1.0 M KBr between 0.0 and +0.7 V (vs. SCE). The potential cycling was initiated at the positive limit. The amount of TTF$^+$ immobilized in the film coulometrically determined in the first scan was 2.02×10^{-9} eq. The decreasing frequency corresponds to an increase in mass, roughly equivalent to a total increase of 1.82×10^{-7} g.

small fluctuations in frequency that remain even after prolonged cycling. The electroactive regions may involve flux of either or both Br^- or K^+.

Several low-dimensional phthalocyanine (Pc) complexes have also been prepared by electrocrystallization techniques. Oxidation of K^+-[PcM^{III}(CN)$_2$]$^-$ (M = Co, Fe) at +1.2 V (vs. Ag/AgCl) in the absence of added electrolyte resulted in the formation of black-purple crystals of PcM^{III}(CN) at the anode [145–147]. The single crystal x-ray structure for M = Co indicated the presence of one-dimensional chains with intermolecular overlap between Pc rings that were canted 26.8° with respect to the stacking axis (Fig. 18). Although the mechanism for crystallization was proposed to involve Pc ring oxidation, the formation of the neutral solid suggests otherwise. One possible pathway for the formation

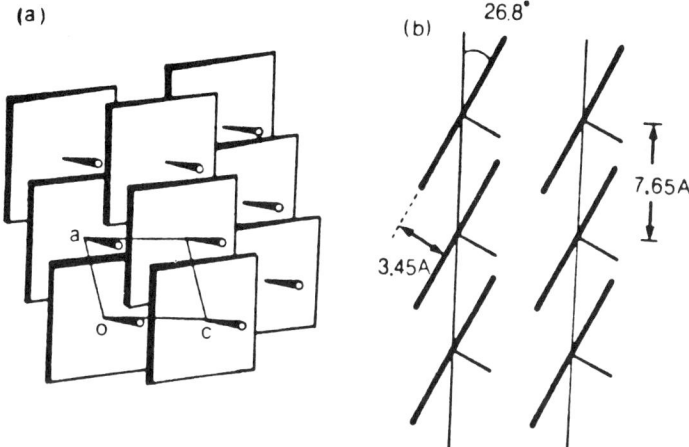

FIG. 18. Schematic drawing of the molecular packing of PcCoCN: (a) projected onto the (101) plane; (b) viewed along the (122) plane. (From Ref. 146.)

of PcM^{III}(CN) is the irreversible electrochemical oxidation of CN$^-$ to (CN)$_2$, either uncatalyzed or mediated by the complex. This would result in irreversible consumption of the [PcM^{III}(CN)$_2$]$^-$ anion, and crystallization of the insoluble neutral product at the electrode.

Galvanostatic oxidation of Ni(Pc) in the presence of ClO$_4^-$ in 1-chloronaphthalene afforded single crystals of Ni(Pc)(ClO$_4^-$)$_y$ (y = 0.39 − 0.47) [148]. Occasional batches exhibited y values as low as 0.33 and 0.50. No details of the dependence of y on growth conditions were mentioned.

From the above discussion, the reader may gain the impression that electrocrystallization is a technique limited to conducting solids. This is a reasonable assumption as conductive crystalline solids may behave as electrodes themselves, mediating electron flow from the working electrode to their surface, where the redox process and crystallization can occur. Poor conductors would be expected to passivate the electrode surface, preventing further redox processes required for crystallization. There are several examples, however, in which electrocrystallization has been utilized for growth of poorly conducting one-dimensional materials. A rather extensive list of poorly conducting ($\sigma < 10^{-4}\,\Omega^{-1}\,cm^{-1}$) compounds based on TTA, TTN, and TSA has been reported [149]. Recently, the syntheses of several poorly conducting organometallic-TCNQ charge-transfer complexes were accomplished by electrocrystallization, including [(C_5Me_5)$_2$Fe]$_2$[TCNQ]$_2$ [150]. [(C_6Me_6)$_2$M][X]$_2$ (M = Fe, Ru; X = TCNQ, TCNQCl$_2$, TCNQF$_4$) [15], [(C_5Me_5)Ru(2_2-1,4-cyclophane)-Ru(C_5Me_5)][TCNQ]$_2$ [151], and [(arene)$_2$M][$C_6(CN)_6$] (M = Fe, Ru; arene = mesitylene, hexamethylbenzene) [30]. All these materials formed on the working electrode as large high-quality crystals on reduction of the acceptor species in the presence of the organometallic cations.

It is clear from the above discussion that electrocrystallization has been used primarily as a synthetic method for rather exotic materials. However, the utility of electrocrystallization of one-dimensional solids in secondary batteries has been reported [152]. The concept was demonstrated for Li/LiClO$_4$ in THF/(Pe)ClO$_4$, Li/LiClO$_4$ in THF/(TTN)ClO$_4$, and the solid-state battery Li/polyethylene oxide + LiClO$_4$/(TTN)ClO$_4$. The best results were realized with the perylene system, whose charge and discharge reactions are described in Scheme 8. A relatively flat discharge curve (3.7 V) was reported under galvanostatic discharge (80 μA). However, coulombic efficiency was low (ca 40%), presumably because of detachment of the perylene salts from the platinum electrode. The coulombic efficiency of the liquid TTN system was very poor (ca 0.2%), but the solid-state system appeared to exhibit a very flat

Scheme 8

$$Pe + ClO_4^- \underset{discharge}{\overset{charge}{\rightleftarrows}} (Pe)_x ClO_4^- + e^-$$

$$Li^+ + e^- \underset{discharge}{\overset{charge}{\rightleftarrows}} Li$$

discharge curve. Clearly, these materials do not threaten the existence of current energy storage devices.

C. Mechanistic Aspects

Electrocrystallization has been extensively studied, and the theory explaining current-voltage behavior for different crystallization mechanisms is fairly well developed [175,176]. Recently, these principles have also been applied to the mechanism of growth of two-dimensional conducting polymers [177,178]. However, the mechanistic aspects pertaining to electrocrystallization of low-dimensional molecular solids have received scant attention, at least on a fundamental level and, currently, only some general observations have been made regarding the optimum conditions for electrochemical crystal growth.

As discussed in the previous sections, electrocrystallization of one-dimensional materials is simply analogous to conventional metathetical techniques used for crystallization in the sense that one of the reagents is introduced by the redox reaction at the working electrode. Figure 12 depicts a typical generic illustration of this process in which D is oxidized to D^+ in the presence of A^-. Any discussion would apply equally to the reduction of A to A^- in the presence of D^+. As a first-order approximation, the solubility product of the complex dictates the conditions required for crystal growth, and crystallization at the electrode occurs when $[D^+][A^-] \gg K_{sp}$ at the electrode surface. That is, when the concentration of D^+ and A^- at the electrode exceeds the solubility limit of the complex, nucleation can occur at the electrode surface, followed by crystal growth on those nuclei. One needs to furnish D^+ at a rate that allows the rate of crystallization to exceed the rate of diffusion of D^+ away from the electrode. Therefore, rapid crystallization kinetics facilitate crystal growth.

It is generally found, although not always explicitly stated, that there is an optimum current for crystal growth at the electrode.

Unfortunately, current densities, which are critical for more precise description of crystal growth conditions, are rarely given. Low yields are observed at very small currents since the formation of D^+ cannot compete with its diffusion away from the electrode, resulting in small effective concentrations of D^+ and $[D^+][A^-] \ll K_{sp}$. However, the current should also be low enough to permit D to diffuse to the electrode on a time scale competitive with crystallization. Otherwise, crystal quality can suffer at overly large currents owing to (1) high voltages induced at the electrode when the flux of D to the electrode is not sufficient to overcome concentration polarization, resulting in overoxidation or decomposition reactions, and (2) overly large nucleation rates, resulting in microcrystalline products. Because K_{sp} and crystallization kinetics vary for every complex, the optimum current levels will also differ. However, the ability to control current, and thereby the rate of introduction of one of the species, is one of the inherent advantages of electrocrystallization. This effect was briefly described for the electrocrystallization of $TMTSF_2PF_6$, where it was claimed that no crystals were formed at <1 μA but, above 10 μA, only microcrystals or powders were formed [116]. The concentrations of D and A^- in solution are also important features, the former affecting the flux of D to the working electrode and the latter ultimately affecting the product $[D^+][A^-]$ at the electrode. The concentration of A^- at the electrode is equivalent to its bulk concentration, lessened by the rate of its removal from solution by the crystallization reaction. The importance of concentration is evident from the report that electrocrystallization of TTT salts was successful only if performed in solutions saturated with TTT [159]. Also, TTF_- $(SCN)_{0.59}$ was reported to crystallize on the CV time scale only when $[TTF] > 2 \times 10^{-3}$ M.

One aspect of electrocrystallization that has not been adequately addressed is whether crystal growth occurs at the crystal-electrode interface or on the outer surfaces of existing crystalline deposits a short distance from the electrode surface. Nucleation generally appears to occur at the electrode, and the crystals remain attached to the electrode surface throughout the growth process. The observation that $(Py)_2(ClO_4)$ crystals did not form on fire-polished Pt probably indicates the need for nucleation sites (e.g., edges, kinks) [103]. If the material is conductive, once a nucleus or a macroscopic crystal forms on the electrode further propagation can result if the crystal itself behaves as an electrode itself, effecting electrochemical oxidation of D and its subsequent deposition on that crystal. This pathway is plausible, especially in light of reports that one-dimensional conductors can be used as solid electrodes (see Sec. V) and the observation that, in the case of $(Pe)_2ClO_4$, the cell conductance increased during crystallization [101]. If this is an important mechanism, the ubiquitous appearance of electrochemical crystal growth of conducting solids compared to poorly conducting materials may be due to competitive effects. When more than one phase is possible, if conductive nuclei and microcrystals are more efficient for electrodeposition compared to nonconducting nuclei by virtue of their active and increasing electrodic surface area, then the growth of conducting phases may dominate by "natural selection."

If the conductive deposits behaved as extensions of the electrode surface, it seems most likely that the redox reaction would occur on the crystal face normal to the stacking axis because of the high degree of conductivity anisotropy in one-dimensional conductors ($\sigma_\| / \sigma_\perp \approx 10^3$). Crystal growth would then be most likely along the one-dimensional axis, resulting in the needlelike habit generally observed for these materials. However, crystal habit also depends on other factors, such as temperature, impurities, intermolecular interactions between the crystal faces, and the solute, which ultimately affect the relative rates of growth of the different crystal faces [179]. Indeed, one-dimensional conductors generally assume a needle-shaped habit, even when grown by conventional methods due to the favorable π-π interactions between the crystal face normal to the one-dimensional axis and the molecular constituents in solution. It is therefore difficult to distinguish between this process and one in which the redox reaction takes place at the metal electrode, followed by diffusion of the product to the emerging crystal face. If electron transfer at the metal electrode is more favorable than at the conducting material, it is also conceivable that crystal growth may occur at the electrode-crystal interface, effectively "pushing" the crystal away from the electrode. Alternatively, D^+ may diffuse away from the electrode and crystallize on the existing crystal. One of these latter two mechanisms is certainly operative during electrocrystallization of insulating one-dimensional solids, such as TTA, TTN, and TSA compounds [151] and the series of organometallic-TCNQ complexes mentioned previously [15, 30,152,153]. In these cases, mass transport of the redox-active solute to the underlying metal electrode is clearly required for efficient crystal growth. A well-formed, large-crystalline habit, such as that observed for the organometallic-TCNQ complexes, favors this behavior, whereas denser microcrystalline deposits would be expected to impede severely flux of the redox species to the electrode surface.

Videomicroscopy experiments can be useful for examination of the macroscopic crystal growth process, particularly with regard to locating the actual crystal growth. This method has been used for monitoring growth of $(Pe)(ClO_4)$ crystals [105], although this report did not address the origin of crystal growth. Preliminary video experiments performed in our laboratory following the progress of distinctive features or defects in crystals as they grow on the electrode have suggested a variety of growth schemes. For example, the crystal growth of $(TMTSF)_2(PF_6)$ was found to occur primarily at the tip of a growing crystal and at defect sites that result from twinning or edge dislocations, as schematically represented in Fig. 19a and b. This is consistent with either conduction through the crystal or diffusion of $TMTSF^+$ generated at the platinum electrode to the tip and defect sites of the crystal. We have observed indirect evidence for the latter in experiments in which a platinum wire anode in a $TMTSF/[n-Bu_4N][PF_6]$ solution is withdrawn into a glass capillary so that the flux of TMTSF to the electrode is

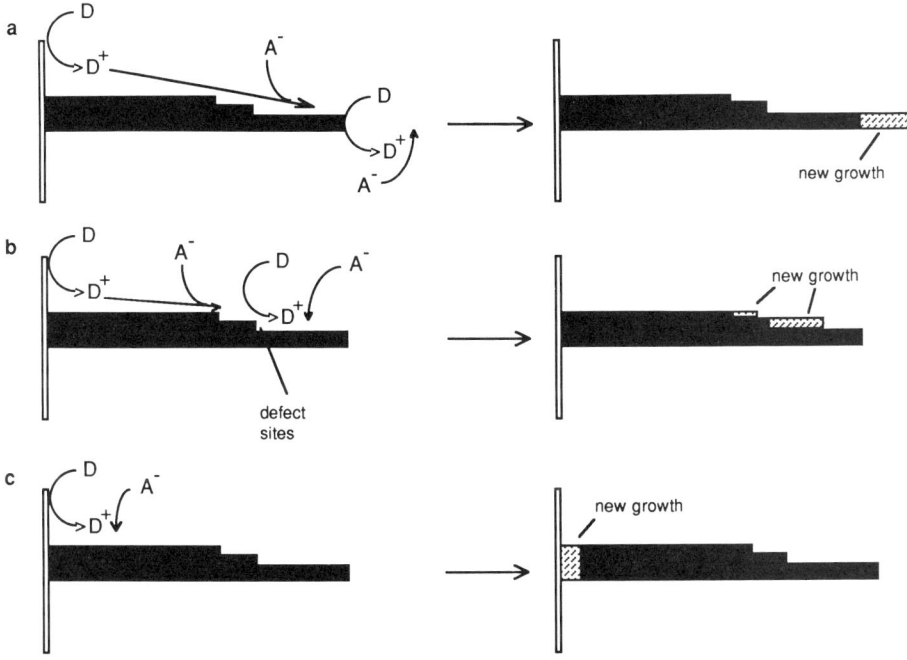

FIG. 19. Different crystal growth modes observed by videomicroscopy for electrocrystallization of $TMTSF_2PF_6$ at a platinum electrode in 0.1 M $[n\text{-}Bu_4N][PF_6]/CH_2Cl_2$ containing 5.0×10^{-3} M TMTSF (D): (a) New growth occurs at the tip of the emerging crystal, resulting from either direct oxidation of TMTSF at the crystal itself (crystal face normal to the one-dimensional axis) or by oxidation of TMTSF at the Pt electrode followed by migration to the crystal tip (b), similar to a, except that new growth occurs at high-energy defect sites, and (c) oxidation of TMTSF at the Pt electrode near the base of the emerging crystal with growth at the electrode-crystal interface. Paths a and b appear to be the major modes of crystal growth.

somewhat inhibited. Under these conditions, crystallization of TMTSF$_2$-PF$_6$ actually occurs at the solution edge of the inner capillary walls, consistent with nucleation and crystal growth on the glass surface resulting from migration of TMTSF$^+$ away from the electrode. Growth at the electrode-crystal interface was not suggested from these experiments since movement of the crystal tip and a defect were not coincident (Fig. 19c). This seems reasonable because diffusion to the base of the crystal at the electrode-crystal interface is probably not significant.

Cyclic voltammetry has been used to investigate the electrocrystallization of fluoranthenyl (FA) cation-radical salts, specifically (FA)$_2$PF$_6$ [173,55]. Cyclic voltammetry revealed electrochemical oxidation of FA to FA$^+$ and FA^{2+} (Fig. 20, peaks a and b, respectively). On the return cathodic scan, peak c was attributed to the reduction of the [FA$_2$]$^+$

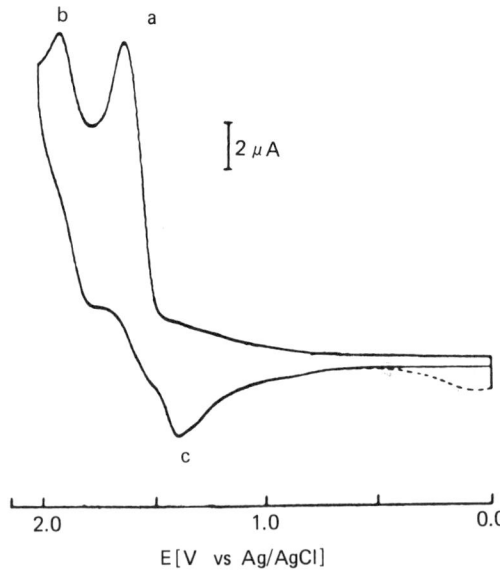

FIG. 20. Cyclic voltammogram of fluoranthene (FA) in CH$_2$Cl$_2$ (0.1 M TBAPF$_6$) at a freshly polished Pt-disk (—). Cyclic voltammogram in the presence of spurious water (---). ν = 200 mV sec^{-1}, T = -30°C. (From Ref. 55.)

radical cation dimer. The authors suggested that the shoulder and the broadness of this peak were due to reduction of $[FA_2]^+$ adsorbed on the electrode surface and in solution. The shoulder on the return cathodic peak c disappeared at high scan rates, and peak c narrowed at higher scan rates; and crystallization of the product was only observed visually at very slow scan rates. The results were interpreted as an initial oxidation of FA to the radical cation, followed by formation of $[FA_2]^+$; at slow scan rates, the dimer had sufficient time to complex with the PF_6^- counterion and form crystalline deposits (Scheme 9).

Scheme 9

$$FA - e^- \longrightarrow FA^+ \qquad \text{Peak a}$$
$$FA^+ + FA \longrightarrow [FA_2]^+$$
$$[FA_2]^+ + PF_6^- \longrightarrow [FA_2]PF_6$$
$$[FA_2]^+ + e^- \longrightarrow 2 FA \qquad \text{Peak c}$$

Cyclic voltammetry and linear sweep voltammetry at a rotating disk electrode (RDE) has also been used to investigate the electrocrystallization of $TTF(SCN)_{0.59}$ [133a]. Enhanced limiting current at the RDE upon oxidation of TTF in the presence of SCN^- was attributed to the crystallization of a conducting species. The crystallization process was visualized as a following reaction, that is, an EC process. Accordingly, cyclic voltammetry revealed that, in the presence of SCN^-, the ratio of the cathodic peak current to the anodic peak current, i_{pC}/i_{pA}, was less than unity, and a second cathodic wave, which was probably due to electrochemical reduction of the crystallized phase, was observed. In addition, the ratio of the limiting anodic current to the diffusion-controlled current in the absence of SCN^-, $i_{pA}/i_{pA°}$, decreased with increasing [TTF] until a limiting value of 0.6 was achieved at high TTF concentrations, consistent with the stoichiometry of the complex.

The formation mechanism of several other low-dimensional conducting solids dispersed in carbon paste electrodes has also been described, including TMTTF, TMTSF, and TTF compounds. Analysis of current-voltage curves of these electrodes, prepared either by oxidation of the donor in the carbon paste or by dispersion of the previously electrocrystallized phases in the carbon paste, allowed determination of the stoichiometry of these complexes. The stoichiometry of TMTTF$_2$X [180] and TMTSF$_2$X [181] complexes was thus verified by cyclic voltammetry, which indicated two successive oxidations of D or reductions of DX with equivalent charge according to Eq. (22), where D is either TMTTF or TMTSF. The observation of the partially oxidized ($\rho = 0.8$) phases TMTSFCl$_{0.8}$ and TMTSFBr$_{0.8}$ was also reported [182].

$$2D \xrightarrow[X^-]{-e^-} D_2X \xrightarrow[X^-]{-e^-} 2DX \qquad (22)$$

Investigation of the current-voltage behavior of TTF and its complexes with different anions in carbon paste electrodes also revealed that, in the presence of Cl$^-$, oxidation of TTF proceeded by stepwise removal of 0.7, 0.3, and 1 e$^-$ per TTF molecule [183], consistent with the known stoichiometry of TTFCl$_x$ phases [184]. However, the same experiments performed in Br$^-$ medium indicate three oxidation steps corresponding to removal of 0.045, 0.65, and 0.3 e$^-$ per TTF molecule (Scheme 10). The first anodic wave representing removal of only 0.045 e$^-$ was seen as a small reversible feature preceding the waves associated with formation of the known phases (Fig. 21). Although TTFBr$_x$ solids with $x = 0.59 - 0.76$ have been reported [184], the $\rho = 0.05$ phase TTFBr$_{0.05}$ has never been isolated. The low electrostatic lattice energy of this phase may prevent its isolation because of greater solubility compared to its more ionic counterparts or its facile conversion to more highly oxidized phases under chemical or electrochemical

Scheme 10

$$\text{TTF} + 0.05 \text{ Br}^- \longrightarrow \text{TTFBr}_{0.05} + 0.05 \text{ e}^-$$
$$\text{TTFBr}_{0.05} + 0.65 \text{ Br}^- \longrightarrow \text{TTFBr}_{0.70} + 0.65 \text{ e}^-$$
$$\text{TTFBr}_{0.70} + 0.30 \text{ Br}^- \longrightarrow \text{TTFBr} + 0.30 \text{ e}^-$$

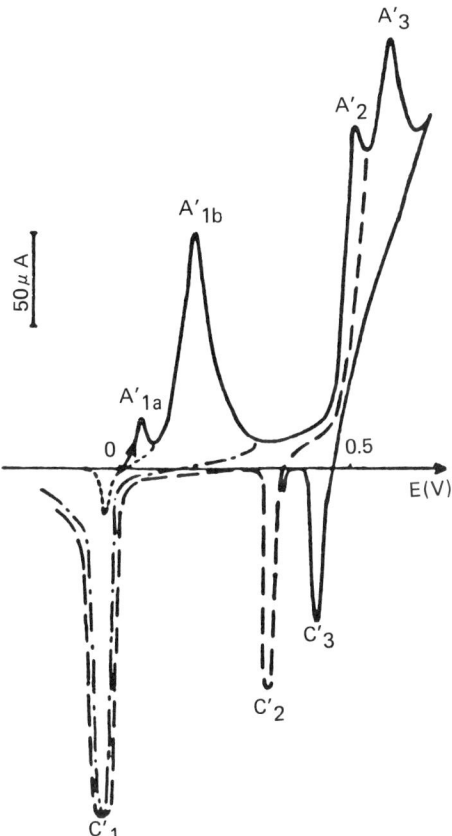

FIG. 21. Cyclic voltammogram of 1.28 mg TTF dispersed in 50 mg graphite paste in 1.0 M KBr; scan rate 2.5×10^{-5} V s^{-1}. (From Ref. 183.)

synthetic conditions. Unfortunately, scan rate dependence and lifetime studies that would elucidate the reasons for the elusive nature of this intermediate phase were not reported. However, these results demonstrate that investigations in carbon paste electrodes can be advantageous for determining stoichiometry of phases that may otherwise go undetected.

Crystallization of $TTFX_n$ may also be responsible for behavior observed in the CV of RuO_2 electrodes modified with TTF molecules covalently attached to the electrode surface by 3-(2-aminoethylamine)propyltrimethoxysilane (en silane) [185].

[structure: Ru—O and Ru—O bonded to Si(OH)/(CH$_2$)$_3$NH(CH$_2$)$_2$NHC(=O)—TTF]

electrode   en silane   TTF

The potential of the reversible wave observed for the TTF°/TTF^+ redox couple in $Et_4N^+ClO_4^-$ shifted to negative potentials by approximately 60 mV for every tenfold increase in added $Et_4N^+Cl^-$. This behavior can be attributed to stabilization of the cation relative to the neutral form, possibly owing to incipient crystallization of $TTFCl_x$ phases.

The inorganic one-dimensional conductor $K_2Pt(CN)_4X_{0.3} \cdot 3H_2O$ (X = Cl, Br) (Fig. 14) represents one of the few cases in which more detailed fundamental information regarding the electrochemical behavior of the constituents has been reported. Oxidation of $Pt(CN)_4^{2-}$ in aqueous media was reported to take place at approximately $E_{1/2}$ = +0.92 V versus SSCE, just before the onset of chlorine evolution (Fig. 22). It was proposed that the irreversible oxidation of $Pt(CN)_4^{2-}$ was accelerated in halide media compared to other electrolytes as a result of the formation of a pseudo-octahedral Pt(II) complex bound to the electrode surface via a halide bridge. Electrochemical oxidation to a short-lived Pt(III)

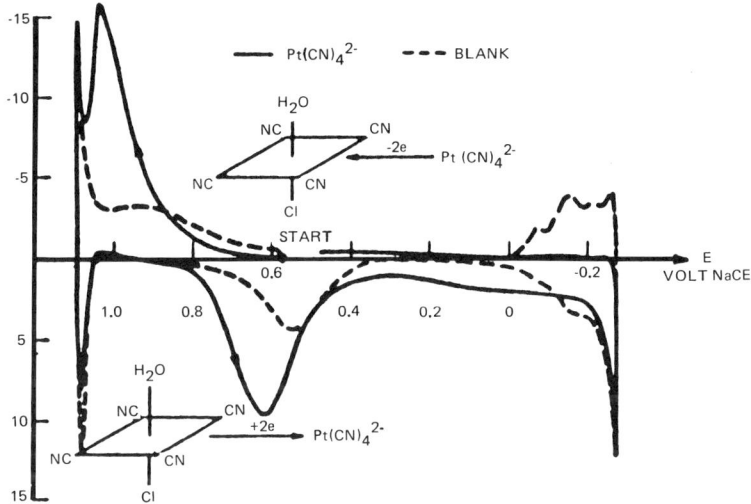

FIG. 22. Thin-layer current-potential curves for $Pt(CN)_4^{2-}$: 1 mF $K_2Pt(CN)_4$ + 10 mF NaCl + 1 F HClO4; experimental conditions: sweep rate = 2mV sec^{-1}; thin layer volume, 3.92 µl; platinum electrode area, 11.15 cm^2. Oxidation current is depicted as negative. (From Ref. 78.)

intermediate is then presumed to be followed by further rapid oxidation to the octahedral Pt(IV) species. This species is subsequently reduced at approximately +0.6 V on the return cathodic scan. The rate-limiting step was thought to involve the interaction between the adsorbed halide and the reactant complex, which suggests that choice of the electrode material, the halide, and the halide concentration may be an important consideration. The rate acceleration was proposed to coincide with expectations from the Franck-Condon principle, with the activation energy for electron transfer minimized by the presence of the distorted octahedral Pt(II) species, whose geometry is intermediate between the reactant and product.

On the basis of these results, several plausible mechanisms for the growth of $K_2Pt(CN)_4X_{0.3} \cdot 3H_2O$ can be proposed. Crystal growth may occur simply by formation of Pt(IV) species near the electrode, which then react with Pt(II) starting material; the conducting salt will grow on the electrode since this is the region of highest concentration of the Pt(IV) species. A more intriguing mechanism based on the electrochemical behavior described above would invoke trapping the Pt(III) intermediate by two sequentially added $Pt(CN)_4^{2-}$ species at the site trans to the halide bridge, resulting in an aggregate with the same oxidation state as the final conducting phase. This aggregate could then serve as a nucleus for macroscopic crystal growth (Fig. 23). Alternatively, $Pt(CN)_4^{2-}$ may assume the role of the ligand trans to the halide bridge before oxidation of the Pt(II) center. After formation of the initial nucleus, further oxidation of $Pt(CN)_4^{2-}$ may proceed with a conducting $\{Pt(CN)_4\}_x^{(2.33x)+}$ spine replacing the role of the halide bridge.

D. Role of Electrochemical Parameters in Crystallization

Recent systematic investigations of electrocrystallization has led to some insight into the role of electrochemical parameters in the crystallization process. The importance of current densities during electrocrystallization was recently demonstrated for the crystal growth of $[(C_5Me_5)_2Fe]_2$-$[TCNQ]_2$ [150], which was prepared either by oxidation of $(C_5Me_5)_2Fe$ in the presence of $TCNQ^-$ [Eq. (23)], or by reduction of TCNQ in the presence of $(C_5Me_5)_2Fe^+$ [Eq. (24)], in saturated solutions of the charge-transfer salt (to ensure precipitation at the electrode). When a constant current of +300 μA was applied

$$(C_5Me_5)_2Fe - e^- \xrightarrow{TCNQ^-} [(C_5Me_5)_2Fe]_2[TCNQ]_2 \quad (23)$$

$$TCNQ + e^- \xrightarrow{(C_5Me_5)_2Fe^+} [(C_5Me_5)_2Fe]_2[TCNQ]_2 \quad (24)$$

across two platinum electrodes in a divided H cell containing $(C_5Me_5)_2Fe$ and $[n-Bu_4N]^+[TCNQ]^-$ in the working compartment and only $[n-Bu_4N]^+$-$[TCNQ]^-$ in the counter compartment, purple crystals of $[(C_5Me_5)_2Fe]_2$-$[TCNQ]_2$ were formed in 67 percent faradaic yield. When the electrolysis was performed at +30 μA flat, octagonal single crystals were formed in

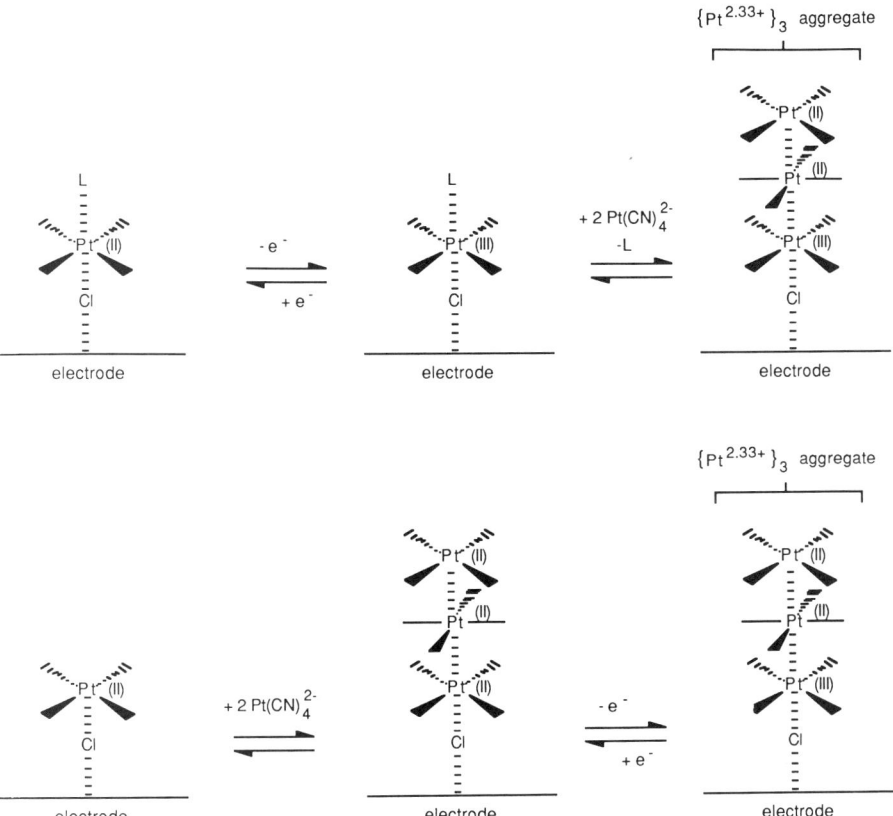

FIG. 23. Schematic representation of the possible mechanisms for electrocrystallization of $K_2Pt(CN)_4X_{0.3} \cdot 3H_2O$ (X = Cl, Br). The upper scheme illustrates oxidation of the halide-bridged $Pt(CN)_4L$ species to a Pt(III) intermediate, which subsequently loses L and is trapped by two equivalents of $Pt(CN)_4^{2-}$ to form the $\{Pt^{2.33+}\}_3$ aggregate. In the lower scheme, $L = Pt(CN)_4^{2-}$ and formation of $(Pt^{2+})_3$ aggregate precedes oxidation to the mixed valent $\{Pt^{2.33+}\}_3$ aggregate.

similar faradaic yields. However, these were more than 10x larger (0.5 × 0.2 cm) than those grown at the higher current density (Fig. 24). This behavior was attributed to the relationship between the particle size of precipitates and the relative supersaturation of the system [Eq. (25)], where Q is the momentary concentration of the complex and

$$\text{Relative supersaturation} = \frac{Q - S}{S} \qquad (25)$$

S is its equilibrium solubility under the crystallization conditions [186]. Crystal size is inversely related to the degree of relative supersaturation that exists during the crystallization process. The size dependence on current was therefore explained by the formation of more nucleation

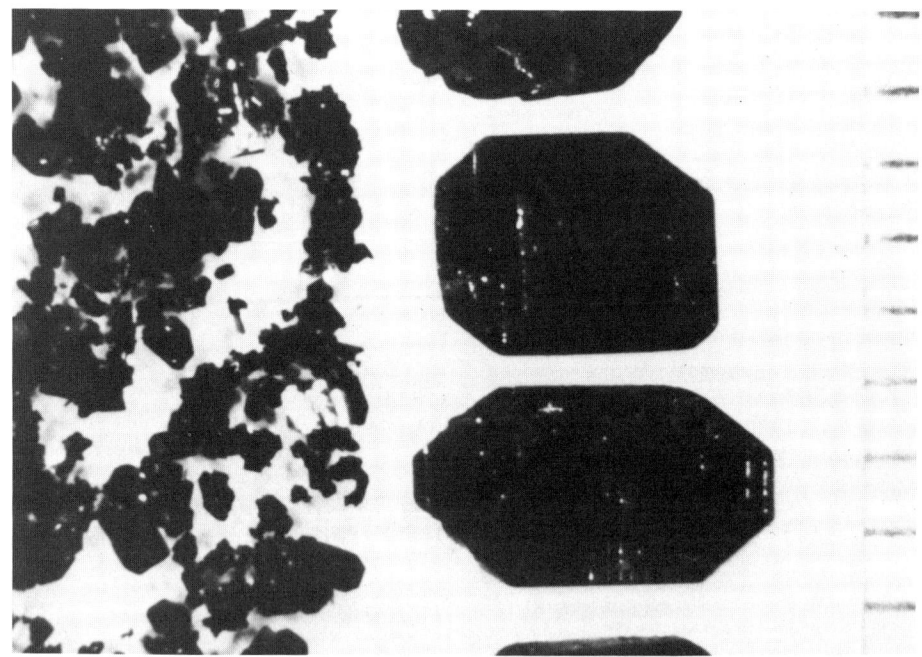

FIG. 24. Photograph of $[(\eta^5\text{-}C_5Me_5)_2Fe]_2[TCNQ]_2$ grown electrochemically by oxidation of $(\eta^5\text{-}C_5Me_5)_2Fe$ in the presence of $[n\text{-}Bu_4N]\text{-}[TCNQ]$. The crystals on the left and right were grown at +300 and 30 mA, respectively. The scale divisions represent 1.0 mm.

sites at higher current, due to a high degree of supersaturation (analogous to rapid mixing in metathesis routes). At lower currents, the degree of relative supersaturation is smaller, and crystal growth on a smaller number of nuclei is favored.

Oxidation of $(C_5Me_5)_2Fe$ in the presence of $TCNQ^-$ at significantly larger currents (>1 mA) resulted in the formation of the one-dimensional conductor $[(C_5Me_5)_2Fe][TCNQ]_2$, in which the TCNQ acceptor stack *formally* contains equivalent amounts of $TCNQ^0$ and $TCNQ^-$ and therefore possesses a nonintegral oxidation state ($\rho = 0.5$). This was attributed to concurrent electrochemical generation of $TCNQ^0$, which is subsequently incorporated into the growing crystalline lattice, owing to the more positive potential at the high current density. This was corroborated by experiments performed under constant applied potential. Electrolysis at $E_{app} \geq E°$ ($TCNQ°/TCNQ^-$), where the concentration of $TCNQ^-$ at the electrode is not large, afforded the conducting 1:2 phase, whereas more negative potentials resulted in formation of the $\rho = 1.0$ dimer phase $[(C_5Me_5)_2Fe]_2[TCNQ]_2$. Potential-dependent selectivity has also been reported in the synthesis of $[(C_6Me_6)_2M][TCNQ]_2$ and $[(C_6Me_6)_2M]$- $[TCNQ]_4$ (M = Fe, Ru) charge-transfer complexes [9]. Whereas the 1:2 phases were insulators owing to their mixed stack arrangement and integral degree of charge transfer ($\rho = 1$), the 1:4 phases were conductive as a result of the presence of segregated TCNQ acceptor stacks in which $\rho = 0.5$. This behavior has proved to be rather general, as identical results were obtained for electrocrystallization of $[C_5Me_5)Ru-(2_2-1,4-cyclophane)Ru(C_5Me_5)][TCNQ]_x$ (x = 2, 4) [151]. Similarly, the 1:2 phase was insulating, whereas the 1:4 phase ($\rho = 0.5$) was conducting. Figure 25 shows the obvious difference in the crystal morphology of these phases.

$$TCNQ + e^- \xrightarrow[E_{app} = -0.1 \text{ V}]{(C_6Me_6)_2M^{2+}} [(C_6Me_6)_2M][TCNQ]_2 \qquad (26)$$

FIG. 25. Photograph of the crystals of $\rho = 1$ $[(C_5Me_5)Ru(2_2\text{-}1,4\text{-}cyclophane)Ru(C_5Me_5)][TCNQ]_2$ (left) and $\rho = 0.5$ $[(C_5Me_5)Ru(2_2\text{-}1,4\text{-}cyclophane)Ru(C_5Me_5)][TCNQ]_4$ (right), prepared electrochemically by reduction of TCNQ in the presence of $[(C_5Me_5)Ru(2_2\text{-}1,4\text{-}cyclophane)\text{-}Ru(C_5Me_5)]^{2+}$ at -0.1 V and $+0.4$ V, respectively.

$$TCNQ + e^- \xrightarrow[E_{app} = +0.3 \text{ V}]{(C_6Me_6)_2M^{2+}} [(C_6Me_6)_2M][TCNQ]_4 \qquad (27)$$

$$TCNQ + e^- \xrightarrow[E_{app} = -0.1 \text{ V}]{(C_5Me_5)_2Ru_2(2_2\text{-}1,4\text{-cyclophane})^{2+}}$$
$$[(C_5Me_5)_2Ru_2(2_2\text{-}1,4\text{-cyclophane})][TCNQ]_2 \qquad (28)$$

$$TCNQ + e^- \xrightarrow[E_{app} = +0.3 \text{ V}]{(C_5Me_5)_2Ru_2(2_2\text{-}1,4\text{-cyclophane})^{2+}}$$
$$[(C_5Me_5)_2Ru_2(2_2\text{-}1,4\text{-cyclophane})][TCNQ]_4 \qquad (29)$$

The potential dependence of stoichiometry observed in these materials was attributed to the different relative concentrations of TCNQ and TCNQ$^-$ at the electrode during crystallization at different potentials, as described by the Nernst equation. The $\rho = 1$ phases possess TCNQ acceptor molecules that are present exclusively as the singly reduced TCNQ$^-$ anion, whereas the $\rho = 0.5$ phases *formally* contain equivalent amounts of TCNQ$^\circ$ and TCNQ$^-$. It was suggested that the concentrations of different prenucleation aggregates [179] that serve as precursors to macroscopic crystals were affected by the potential at the electrode. In this case, $[TCNQ_2]^{2-}$ preceded crystallization of the $\rho = 1$ phases that contained $[TCNQ_2]^{2-}$ in the solid state, whereas a partially reduced TCNQ aggregate such as $[TCNQ_2]^-$ or $[TCNQ_4]^{2-}$ initiated formation of the $\rho = 0.5$ phases (Scheme 11; M = organometallic cation). When

Scheme 11

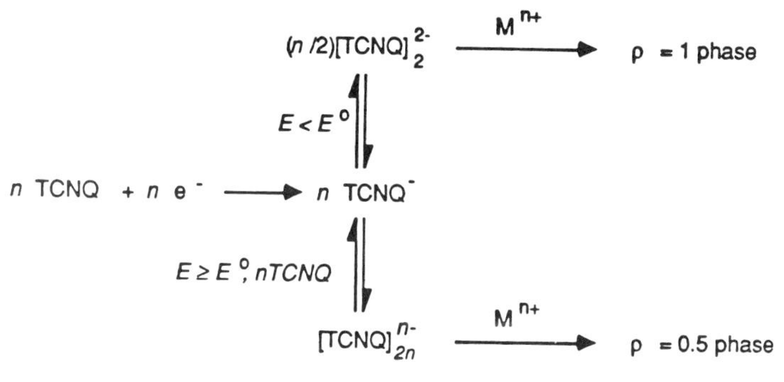

$$n\ TCNQ + n\ e^- \longrightarrow n\ TCNQ^-$$

with $(n/2)[TCNQ]_2^{2-} \xrightarrow{M^{n+}} \rho = 1$ phase via $E < E^\circ$, and $[TCNQ]_{2n}^{n-} \xrightarrow{M^{n+}} \rho = 0.5$ phase via $E \geq E^\circ, nTCNQ$

$E_{app} \geq E^\circ$, conditions are more favorable for the formation of the partially reduced aggregates at the electrode surface since [TCNQ$^\circ$] is more likely to be equal to or greater than [TCNQ$^-$] under this condition. Conversely, at very negative potentials, the predominant species at the electrode is TCNQ$^-$, favoring the presence of the fully reduced dimer dianion.

The presence of mixed valent aggregates was verified by experiments performed in an optically transparent thin-layer electrochemical cell which showed infrared bands at 2000 cm^{-1} at $E_{app} \geq E°$, attributed to [TCNQ]$_{2n}^{n-}$ aggregates. This is similar to the potential-dependent formation of mixed valent TCNQ species that has been reported for electrodes modified with a TCNQ polymer derived from 2,5-bis(2-hydroxy)-7,7,8,8-TCNQ on the basis of EPR studies [187,188].

2,5-bis(2-hydroxy)-7,7,8,8-TCNQ

The potential control of stoichiometry with the TCNQ complexes suggests that the crystallization rate constants k_1 (ρ = 1.0 phase) and k_2 (ρ = 0.5 phase) are not appreciably different (Eq. 30). If the rate constants in Eq. (30) differed significantly, it is likely that only one of the phases would be favored under all conditions. However, the solution chemistry of the molecular species is also an important factor as UV-visible spectrophotometry indicated that the concentration of [TCNQ$_2$]$^{2-}$ was not significant in nonaqueous solvents, even in concentrated solutions of TCNQ$^-$ [189]. In contrast, halogenated TCNQ anions (e.g., TCNQF$_4^-$, TCNQCl$_2^-$) were found to be extensively dimerized in solution. These anions did not form mixed valent species as only the ρ = 1 phases [(C$_6$Me$_6$)$_2$M][TCNQF$_4$]$_2$ and [(C$_6$Me$_6$)$_2$M][TCNQCl$_2$]$_2$ were formed at all potentials. Since the concentration of [TCNQ$_2$]$^{2-}$ is probably low even at negative potentials, the observation of the ρ = 1 phase at negative potentials is consistent with either very low concentrations of (TCNQ)$_{2n}^{n-}$ or $k_1 > k_2$ under these conditions. The ρ = 0.5 phase can form at the more positive potentials, even if $k_1 > k_2$, since [TCNQ]$_{2n}^{n-}$ will predominate over the disfavored [TCNQ$_2$]$^{2-}$ dimer. The correlation of stoichiometry with the relative concentrations of TCNQ and

TCNQ⁻ at the electrode implies crystallization kinetics that are fast compared to the time scale of diffusion. This probably results from large Madelung energies associated with favorable electrostatic interactions between the TCNQ⁻ and the highly charged cations. Therefore, it seems reasonable that rapid crystallization kinetics are desirable if manipulation of the selectivity toward different stoichiometric phases is to be accomplished.

$$E_{app} \ll E^o: \quad k_1[(TCNQ)_2^{2-}] > k_2[(TCNQ)_{2n}^{n-}] \quad (30a)$$

$$E_{app} \geqslant E^o: \quad k_1[(TCNQ)_2^{2-}] < k_2[(TCNQ)_{2n}^{n-}] \quad (30b)$$

Similar effects of electrochemical parameters were reported for the formation of different phases derived from HMTP [158]. Synthesis of $(HMTP)ClO_4$ ($\rho = 1$, $\sigma = 0.3 \, \Omega^{-1} \, cm^{-1}$) was accomplished upon oxidation of the neutral compound to the monocation by application of 2.5 V across cell (Scheme 12). Larger potentials resulted in formation of a less conductive phase, which roughly analyzed as $(HMTP)(ClO_4)_2$ ($\rho = 2$, $\sigma = 10^{-3} \, \Omega^{-1} \, cm^{-1}$), owing to the formation of the dication at the more positive

Scheme 12

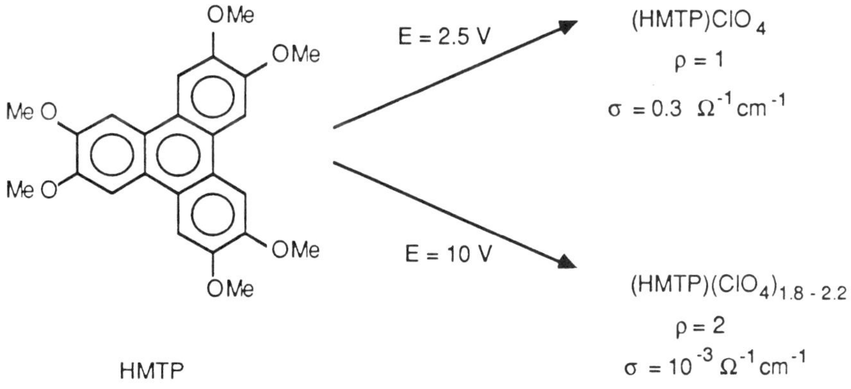

potentials. Although no structural information is available, the low conductivity of the $\rho = 2$ phase is probably due to a larger Mott transition energy associated with the higher charge on the cations. Voltage-dependent selectivity was also briefly mentioned for TTF(SCN)$_{0.59}$ and TTF(SCN)$_{0.64}$, the latter observed only at more positive electrode potentials [133b]. The higher ρ value observed for this phase is consistent with greater extent of TTF oxidation during the crystallization process.

In summary, there are several clear advantages offered by electrochemical crystal growth. The most obvious advantage is the simplicity of the preparative method; crystals are readily harvested by removal from the electrode. An element of selectivity is also introduced because only species that are electrochemically active will be produced for crystallization. Because the potential can be controlled, impurities formed at higher potentials will not interfere with the process. Other advantages originate in the ability to control the electrochemical parameters during crystal growth. Whereas the interdiffusion of D^+ and A^- in conventional metathesis routes is not easily controlled, the current during electrocrystallization can be carefully regulated, allowing precise and reproducible control of the crystallization rate. This results in rational control over the size and quality of crystalline materials, as shown for $[(C_5Me_5)_2Fe]_2[TCNQ]_2$. Therefore, electrocrystallization can be regarded as "controlled metathesis" in the preparation of DA solids.

The observation of simultaneous formation of different phases during electrocrystallization in many cases is undoubtedly due to subtle differences in the lattice energies and crystallization kinetics of these different phases. The potential-dependent selectivity described above for TCNQ complexes demonstrates that careful attention to the electrochemical crystal growth parameters may result in greater control over the crystallization process. What is clearly needed is further systematic study of the factors that influence crystal growth, including current densities and electrode potentials, electrolyte concentration, solvent, temperature, electrode material and relative concentrations of D and A in solution. The effect on kinetics of crystallization, especially the fundamental nucleation events early in the crystallization process, needs to be studied in more detail. The role that electrode surface modification exerts in nucleation and crystal growth is also a potentially interesting area.

E. Electrochemical Doping of Low-Dimensional Phthalocyanines

A rather novel electrochemical method for doping phthalocyanine polymers has recently been reported [190,191], which also demonstrates the effect of electrode potential on ρ, the degree of partial oxidation of these polymers. Although these materials, because their covalent linkages, are not molecular solids by the criteria outlined in the Introduction, their structural similarity to the general class of phthalocyanines, which crystallize in low-dimensional lattices, merits discussion. This electrochemical doping procedure is similar to electrocrystallization in that it involves the evolution of a conducting one-dimensional solid and the formation of new crystalline phases.

Oxidation of a slurry of the cofacially linked orthorhombic phase of $[Si(Pc)O]_n$ (Fig. 26) with a Pt gauze electrode in acetonitrile/n-Bu_4N^+-BF_4^- to the tetragonal phase $\{[Si(Pc)O][BF_4]_{0.50}\}_n$ was accomplished at $E_{app} > +1.80$ V (vs. SSCE). This partially oxidized ($\rho = 0.5$) product could be reduced back to the neutral tetragonal undoped polymer at -0.20 V and then reoxidized at $E_{app} > +0.35$ V. That is, oxidation of the newly formed tetragonal phase occurred much more readily than the original orthorhombic phase. Within the tetragonal structure, ρ could be varied between 0.0 and 0.5 for the BF_4^- counterion and 0.0 and 0.7 for the tosylate anion [192] by control of the applied potential (Fig. 27). The doping curves for these two anions were nearly identical, indicating that the free energy changes with doping were essentially independent of the anion. The different doping levels were manifested in dramatic differences in the room temperature conductivity and the thermopower characteristics due to the varying degree of band filling for different values of ρ, changing from a poorly conducting semiconductor at low ρ values to a metallic material at values of $\rho > 0.20$. Recent studies of

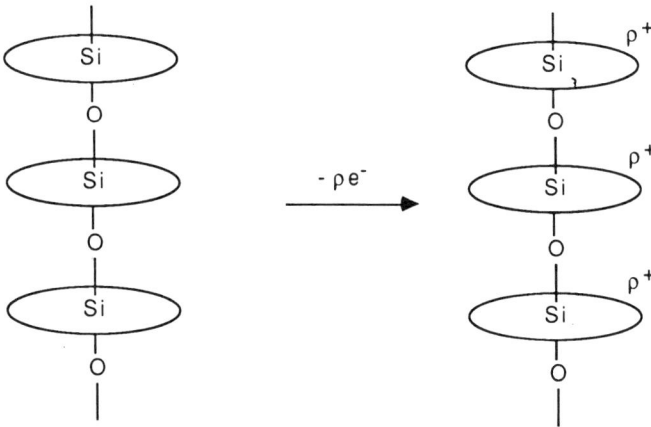

FIG. 26. (a) Figure of a generic metallophthalocyanine and (b) representation of the doping process of $[Si(Pc)O]_n$. Removal of ρ electrons per Si(Pc)O site results in an overall oxidation state of ρ^+ per site.

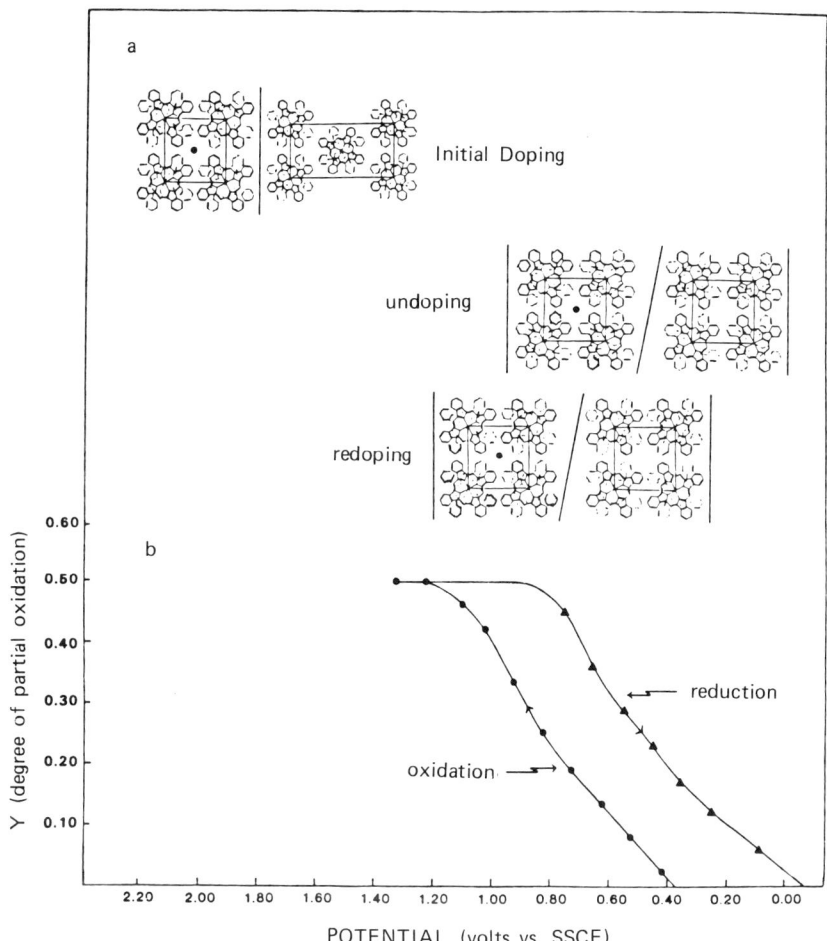

FIG. 27. (a) Structural/potential relationships in the electrochemistry of [Si(Pc)O]$_n$ in acetonitrile containing (n-Bu$_4$N)(BF$_4$). Vertical lines indicate the threshold for doping and undoping processes. Slanted lines indicate equilibrium between phases. (b) Controlled potential coulometry of tetragonal [Si(Pc)O]$_n$ in acetonitrile containing (n-Bu$_4$N)(BF$_4$). (From Ref. 190.)

thin films of H_2Pc [193] and MgPc [194], which showed reversible oxidation with attendant insertion of anions [Eq. (31)], are consistent with p-doping observed for the $[Si(Pc)O]_n$ polymers. Less detailed studies of galvanostatic doping of $[Si(Pc)O]_n$ pellets have also been reported [195].

$$MgPc + A^- \longrightarrow MgPc^+A^- + e^- \qquad (31)$$

Electrochemical "n-doping" of $[Si(Pc)O]_n$ to tetraalkylammonium and lithium salts of the partially reduced anion was also reported, where y = 0.0 − 0.25 [Eqs. (32)].

$$[Si(Pc)O]_n \xrightarrow[+ \, ny \, e^-]{+ \, ny \, Et_4N^+ClO_4^-} \{(Et_4N)_y[Si(Pc)O]\}_n \qquad (32a)$$

$$[Si(Pc)O]_n \xrightarrow[+ \, ny \, e^-]{+ \, ny \, Li^+Y^-} \{(Et_4N)_y[Si(Pc)O]\}_n \qquad (32b)$$

The ability to dope these materials either oxidatively or reductively is especially interesting when compared to the electrochemical behavior observed for the related oligomers $(n\text{-}C_6H_{13})_3SiO(SiPcO)_nSi(n\text{-}C_6H_{13})_3$ (n = 1−4) [196] and $(t\text{-}C_4H_9)Me_2SiO(SiPcO)_nSiMe_2(t\text{-}C_4H_9)$ (n = 1−3) [197]. The delocalized nature of the extended framework in these materials was demonstrated by the observation of n reduction and n oxidation waves at different potentials (Fig. 28). In the case of the hexyl-substituted oligomer, the first oxidation and reduction potentials became less energetically demanding with increasing n, consistent with decreasing electrostatic repulsion when the charge was delocalized over a larger aggregate and stabilization of the molecular orbitals by through-space interactions of the Pc rings. The $SiMe_2(t\text{-}C_4H_9)$ compound exhibits this trend only for the oxidation behavior, as very slight negative shifts of the reduction potential were observed with increasing n. Figure 28 illustrates that the potential difference between successive redox events for the hexyl compound becomes smaller and the outermost potentials appear to converge with increasing n. This behavior corresponds

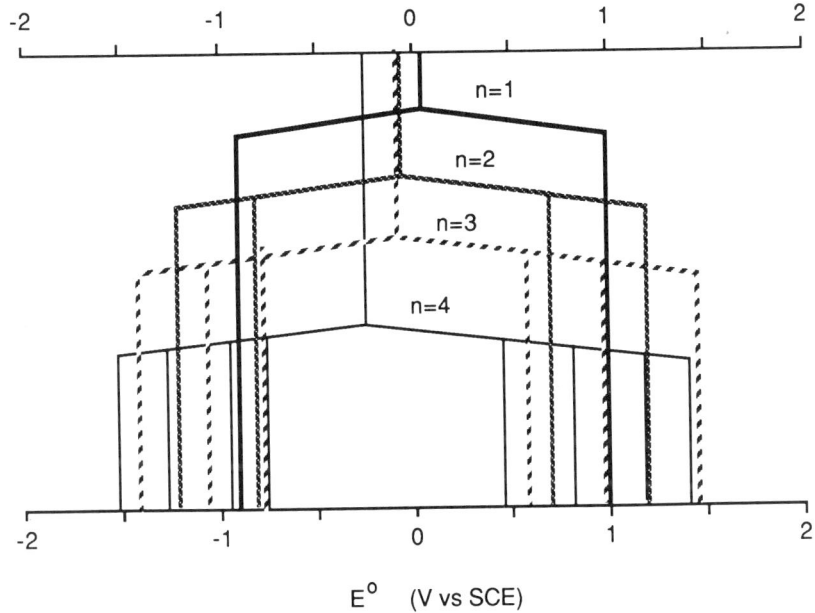

FIG. 28. Redox potentials observed for $(n-C_6H_{13})_3SiO(SiPcO)_nSi(n-C_6H_{13})_3$ (n = 1-4). The first oxidation and reduction potentials become less energetically demanding with increasing n. Also note that the averages of the first potentials, shown on the upper scale, do not shift appreciably with n. These values correspond to the Fermi energy for an oligomer with very large n.

to the progression of the $[Si(Pc)O]_n$ HOMO and LUMO to a band structure of an extended solid with increasing n. Figure 28 also suggests that the mean of the first reduction and oxidation potential, which would correspond to the Fermi energy in the extended solid, does not change significantly with n.

V. LOW-DIMENSIONAL SOLIDS AS ELECTRODE MATERIALS

A. Characteristics of Solid Electrodes

The metallic nature of some molecular materials has prompted several investigations into their properties as electrodes. Conducting polymers (e.g., polypyrrole, polythiophene, polyacetylene) have been exhaustively studied in this regard, particularly with regard to their electrochemical doping to form conductive films, and their potential utility as battery components [10]. These materials hold great promise for the development of "synthetic metal" electrodes since the polymer structure and processibility can be rationally modified. Molecular solids, that is, those lacking covalent networks according to the definition outlined in the Introduction, also possess attractive characteristics for use as electrode materials. In addition to their metallic nature, which allows their use as electrodes, the ability to modify specifically the nature of the molecular species that make up these materials suggests design of electrode surfaces that is not possible with conventional metal electrodes.

Intrinsic differences exist between electrodes derived from synthetic metals and typical metal electrodes such as platinum. A synthetic metal possesses a smaller bandwidth 4t than a metallic electrode due to a small transfer integral t, which results from less effective overlap of the molecular wavefunctions. The bandwidth for these materials is commonly less than 1 eV; for example, 4t $\sim$ 0.5 eV for (TTF)(TCNQ) [198,199]. Differences in the density of states distribution and the Fermi energy would also be expected to affect electron transfer at the surface. It has been reported that the states closest to the Fermi energy in synthetic metals derived from TCNQ salts are composed of TCNQ anion states [200]. It is also plausible that the molecular nature of these materials would result in electrodes with specific adsorption properties. This may result from π-π interactions of unsaturated substrates and the electrode or interaction of the electrode heteroatoms with functional groups of the substrate. Favorable ionic interactions between ionically charged lattice constituents and charged species in solution are also feasible (Fig. 29).

Investigations of organic solid electrodes have primarily involved materials composed of donor molecules and TCNQ that are known to exhibit reasonable conductivity at room temperature (Table 6). Some of the earliest reports described the fabrication of pellets of conducting or

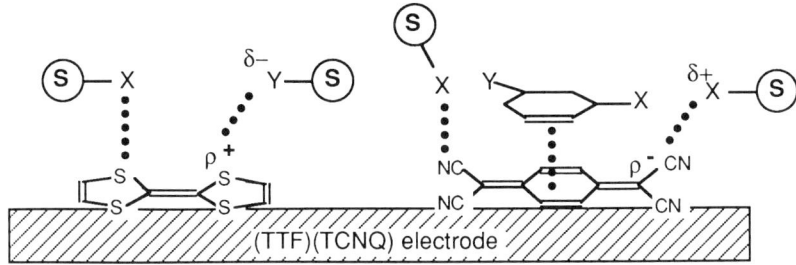

FIG. 29. Schematic representation of possible substrate-electrode interactions for a TTF-TCNQ electrode. Similar interactions would not be unexpected on other "synthetic metal" electrodes.

TABLE 6

Demonstrated Simple Redox Processes at Organic Solid Electrodes

Electrode material	Stable region (V vs. SCE)	Substrate	Reference
(TTF)(TCNQ)		$Fe(CN)_6^{3-/4-}$	204
		Cu^{2+}/Cu^0	
(TTF)(TCNQ)		TMPD	208
(TTF)(TCNQ)	+0.65--0.25 (LiAc)	$Fe(CN)_6^{3-/4-}$	207
	+0.50--0.25 (LiCl)		
	+0.49--0.25 (KBr)		
(NMP)(TCNQ)	+0.22--0.25 (NaF)		
(NMP)(TCNQ)	+0.40--0.25 (KCl)		
$NMP_{0.85}Phen_{0.15}TCNQ$	+0.41--0.25 (KCl)		
$NMP_{0.63}Phen_{0.37}TCNQ$	+0.48--0.25 (KCl)		
$(Ad)(TCNQ)_2$	+0.58--0.25 (KAc)		
$(Qn)(TCNQ)_2$	+0.40--0.25 (KAc)		
(TTT)(TCNQ)	+0.75--0.25 (KAc)		
$(TTT)(TCNQ)_2$	+0.41--0.25 (KAc)		
(2,2'-BIP)(TCNQ)	+0.42--0.25 (KAc)		
(2,2'-BIP)(TCNQ)	+0.44--0.25 (KBr)		

TABLE 6 (continued)

Electrode material	Stable region (V vs. SCE)	Substrate	Reference
(4,4'-BIP)(TCNQ)	+0.50--0.25 (KAc)		
$[(C_5H_5)_2Fe][TCNQ]$			
$TTFBr_{0.7}$/Nafion		$Fe(CN)_6^{3-/4-}$	135,136
		$Fe(EDTA)^{2-}$	
$TTFBr_{0.7}$/montmorillonite clays		$Fe(CN)_6^{3-/4-}$	144

semiconducting polycyanoanion salts, particularly TCNQ salts [201-203]. These studies focused chiefly on the role of these materials as selective potentiometric ion sensors and are discussed in more detail later in this section.

The most detailed studies of the synthetic metal electrodes themselves and interfacial electron transfer processes have involved polycrystalline (TTF)(TCNQ) electrodes prepared as pellets by compaction of the powder [204]. These electrodes were reported to behave as ideally polarizable electrodes between approximately +0.6 and −0.2 V (vs. SCE), with oxidation or reduction of the electrode occurring beyond this potential. For example, in lithium acetate (Fig. 30), the anodic currents beyond +0.65 V were attributed to oxidation of both TTF^+ and $TCNQ^-$ in the lattice [Eq. (34)]. In situ resonance Raman spectroscopy of the surface of these electrodes verified the presence of $TCNQ°$ upon oxidation past the anodic limit [205].

$$TTF\text{-}TCNQ\ (s) \xrightarrow{\text{lattice oxidation}} TTF^+\ (aq) + TTF^{2+}\ (aq) + TCNQ°\ (s) \tag{34}$$

$$TTF\text{-}TCNQ\ (s) \xrightarrow{\text{lattice reduction}} TTF°\ (s) + TCNQ^-\ (aq) + TCNQ^{2-}\ (aq) \tag{35}$$

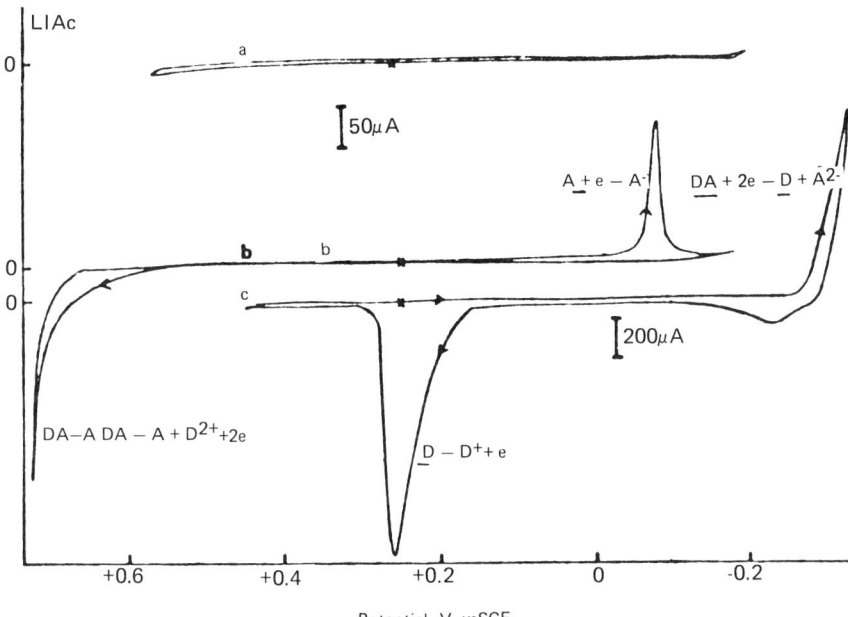

FIG. 30. Cyclic voltammogram of the TTF-TCNQ electrode in 1 M lithium acetate at a scan rate of 5 mV sec^{-1} (a) stable potential region; (b) oxidation of the electrode surface and resultant peak on reversal; (c) reduction of the electrode surface and the resultant peak on reversal. In all figures, A and D represent TCNQ and TTF, respectively. Insoluble or slightly soluble species are underlined. Scans start at potential marked by X. (From Ref. 204.)

Upon the return cathodic sweep, only a wave associated with reduction of TCNQ° was observed as the soluble acetate salts of TTF$^+$ and TTF^{2+} were lost from the electrode by convection at the extremely slow scan rates employed (5 mV sec^{-1}). Similarly, excursions from the ideal region to potentials more negative than −0.28 V resulted in reduction of both TTF$^+$ and TCNQ$^-$ [Eq. (35)], and only TTF° were reoxidized upon the return anodic sweep as the lithium salts of TCNQ$^-$ and TCNQ^{2-} were soluble. Although reduction and oxidation of the electrode was observed in all electrolytes, observation of the return peaks was strongly dependent on the solubility of the ion pairs formed during the redox process. Accordingly, in KBr, where both $TTFBr_x$ (x = 1, 2) and K^+TCNQ^- are insoluble, return waves for all the products of the electrode reaction were observed (Fig. 31).

The ideally polarizable range of this electrode in lithium acetate exceeded that expected based on the standard redox potentials for the $TTF°/TTF^+$ (+0.30 V) and $TCNQ°/TCNQ^-$ (+0.19), an effect that is probably indicative of the electrostatic lattice energy associated with the ionic solid. Indeed, the lattice oxidation and reduction potentials were essentially equivalent to these standard potentials adjusted for the enthalpy associated with formation of the crystalline lattice, reported elsewhere as 0.39 eV for (TTF)(TCNQ) [206]. The potential window is strongly dependent on electrolyte; in the presence of ions such as halides, which form insoluble species, the lattice oxidation occurs at less positive potentials compared to LiAc owing to the favorable enthalpy associated with crystallization of that new phase. Formation of nonstoichiometric phases such as $TTFBr_{0.7}$ is also likely. It was also noted that the depth of reduction and oxidation of the electrode was approximately 100 monolayers.

Subsequent studies showed similar behavior for (TTF)(TCNQ) single crystals, as well as for a series of donor-TCNQ solids (Table 6) [207]. These studies demonstrated that electrode stability with regard to lattice oxidation was dependent not only on the lattice energy of the solid but also on the effective charge on the TCNQ acceptor species. Contrary to the expected trend based on electrostatic forces, it was observed that the more highly ionic solids were less stable toward lattice oxidation. That is, for 1:1 salts, increasing values of ρ resulted in less positive lattice oxidation potentials (Fig. 32). This was attributed to the higher effective electron density on the $TCNQ^{\rho-}$ anions for larger values of ρ. This was especially evident from the more positive lattice oxidation potentials observed for $NMP_xPhen_{1-x}TCNQ$ solid solutions in which substitution of neutral phenazine into the lattice reduces ρ without substantially affecting the solid-state structure. When x = 0.85

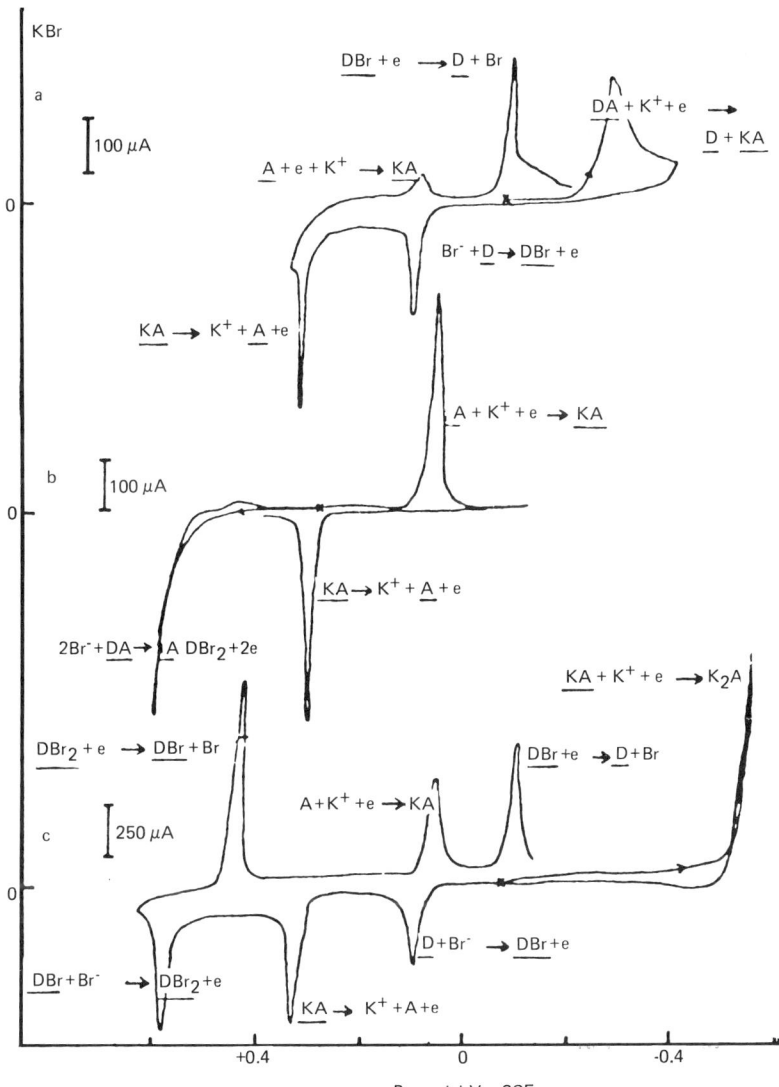

FIG. 31. Cyclic voltammogram of the TTF-TCNQ electrode in 1 M KBr: (a) initial negative scan with reduction of the electrode surface and resultant peaks on reversal; (b) initial positive scan with oxidation of the electrode surface and resultant peaks on reversal; (c) after the electrode surface has been oxidized and reduced. Scan rate, 5 mV sec^{-1}. (From Ref. 204.)

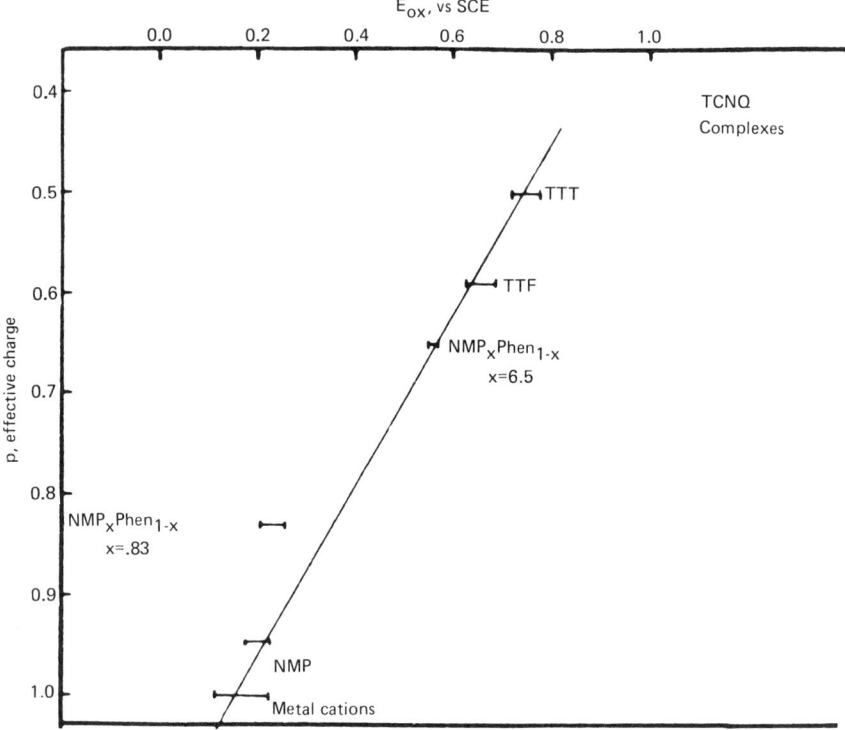

FIG. 32. Plot of the potential for lattice oxidation, E_{ox} (vs. SCE), versus the degree of charge transfer ρ. (From Ref. 207.)

($\rho = 0.85$), the lattice oxidation is +0.41 V, whereas when x = 0.63 ($\rho = 0.63$), this value shifts to +0.48 V. This suggested that ρ can be readily estimated using electrochemical methods.

The electrochemical characteristics of TTF complexes and other organic conductors dispersed in carbon paste electrodes have also been investigated [160,161,180-183]. These studies, which demonstrated that current-voltage curves could be used to determine stoichiometry of electroactive phases, indicate that these electrodes possess characteristics similar to the donor-TCNQ solid electrodes with regard to lattice oxidation limits and redox behavior. Since these were described in detail in Sec. IV.C, further elaboration is not given here.

Nearly reversible cyclic voltammograms for the $Fe(CN)_6^{3-}/Fe(CN)_6^{4-}$ and the Cu^{2+}/Cu^+ (1M KCl) redox couples have been observed on (TTF)-(TCNQ) electrodes (Fig. 33) [204]. Reversible redox behavior has also been observed on (NMP)(TCNQ) electrodes for the $TMPD/TMPD^+$ redox couple [208]. Although adsorption of TMPD onto the electrode would not have been surprising, this behavior was not observed.

Electrodes composed of synthetic metals in inert matrices have also been reported. The electrocrystallization of $TTFBr_{0.7}$ in Nafion films

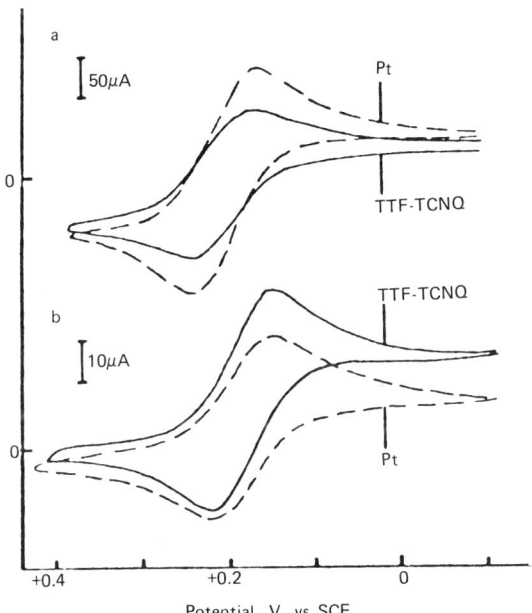

FIG. 33. Comparison of Pt and (TTF)(TCNQ) electrodes for cyclic voltammetry of soluble solution species: (a) 15 mM $K_3Fe(CN)_6/K_4Fe(CN)_6$ in acetic acid/potassium acetate (pH 6.9); (b) 10 mM $CuCl_2$ in 1 M KCl. Scan rate, 5 mV sec^{-1}. ———, (TTF)(TCNQ) electrode; ----, Pt disk electrode. (From Ref. 204.)

[135,136] and montmorillonite clays [144], mentioned in the previous section, results in films that are apparently electroactive by virtue of the conductive organic solid. The Nafion/TTFBr$_{0.7}$ films exhibited dry conductivities of approximately 3 kΩ compared to a value of 4 MΩ for TTF exchanged films without TTFBr$_{0.7}$. Cyclic voltammograms indicated a very sharp cathodic peak and a broader anodic wave, which were attributed to electroactive zones of TTF within the Nafion film (Fig. 15). It was proposed that this redox behavior was associated with the formation of TTFBr$_{0.7}$, as evidenced by the 59 mV shift of both peaks per tenfold change in Br$^-$ concentration. The large peak separation was attributed to structural transitions between the oxidized and reduced forms, based on the known crystal structure of TTF$^\circ$, which exhibits "staggered" (st) overlap of TTF molecules, whereas TTFBr$_{0.7}$ possesses stack of TTF molecules with an "eclipsed" (ec) format. On the basis of scan rate dependences, the authors proposed a "square scheme" to account for these results in which k and k' are the rate constants of following reactions in an E_r-C_i mechanism, the irreversible reactions being the very slow conversion of one structural phase to the other (Scheme 12). The

Scheme 13

$$\begin{array}{ccc} \text{TTFBr}_{0.7} \text{ (ec)} + 0.7\ e^- \rightleftharpoons \text{TTF (ec)} + 0.7\ \text{Br}^- \\ \uparrow k' & & \downarrow k \\ \text{TTFBr}_{0.7} \text{ (st)} + 0.7\ e^- \rightleftharpoons \text{TTF (st)} + 0.7\ \text{Br}^- \end{array}$$

narrow structure of the cathodic peak was interpreted as resulting from attractive interactions between the oxidized TTF molecules in the polymer that inhibit reduction; as these species are reduced, the attractive interactions decrease, resulting in more facile reduction and a steeply increasing current.

The Nafion/TTFBr$_{0.7}$ films were demonstrated to behave as electrodes for solution redox systems such as Fe(CN)$_6^{3-/4-}$ and Fe(EDTA)$^{2-}$, which were chosen because they cannot readily exchange into the Nafion film

ELECTROCHEMICAL ASPECTS OF MOLECULAR SOLIDS

due to their negative charge. Although the remaining electroactive domains of TTF^+/TTF° (Fig. 16) may mediate interfacial electron transfer, the relatively large current response, whose magnitude was nearly identical to that at a bare Pt electrode, makes this unlikely, given the small concentration of these remaining species. Attribution of this metallic behavior to the nonelectroactive conductive $TTFBr_{0.7}$ needles was corroborated by observation of Cu° electrodeposition on the $TTFBr_{0.7}$ needles, and RDE studies that showed identical intercepts in inverse Levich plots, regardless of whether the film was reduced or oxidized, support this contention. However, the irreversible cyclic voltammetric waves suggest sluggish electron transfer kinetics compared to the (TTF)-(TCNQ) solid electrodes. Montmorillonite clay electrodes demonstrate redox behavior that suggests initial intercalation of TTF^+ was qualitatively similar to the Nafion/$TTFBr_{0.7}$ electrodes. The clays exhibited mediated electron transfer to $Fe(CN)_6^{3-/4-}$ only after growth of the $TTFBr_{0.7}$ conductive phase.

Several plausible mechanisms for the interfacial redox processes observed at synthetic metal solid electrodes are summarized in Schemes 14-16, using (TTF)(TCNQ) as an example. Ideally these materials behave as true metals, with interfacial electron transfer occurring through electron exchange with the substrate via states near the Fermi

Scheme 14

Direct electron transfer

TCNQ subband

$\{(TTF^{0.59+})(TCNQ^{0.59-})\}_n + D \longrightarrow D^+ + \{(TTF^{0.59+})(TCNQ^{(0.59+1/n)-})\}_n$

$\{(TTF^{0.59+})(TCNQ^{0.59-})\}_n + A \longrightarrow A^- + \{(TTF^{0.59+})(TCNQ^{(0.59-1/n)-})\}_n$

TTF subband

$\{(TTF^{0.59+})(TCNQ^{0.59-})\}_n + D \longrightarrow D^+ + \{(TTF^{(0.59-1/n)+})(TCNQ^{0.59-})\}_n$

$\{(TTF^{0.59+})(TCNQ^{0.59-})\}_n + A \longrightarrow A^- + \{(TTF^{(0.59+1/n)+})(TCNQ^{0.59-})\}_n$

Scheme 15

Mediated electron transfer

TTF mediation

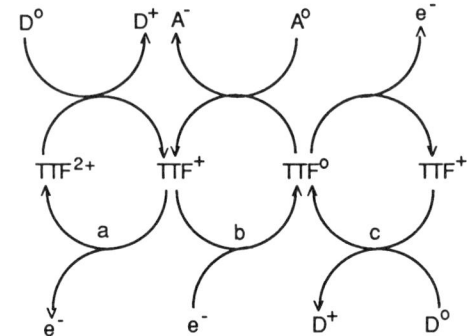

TCNQ mediation

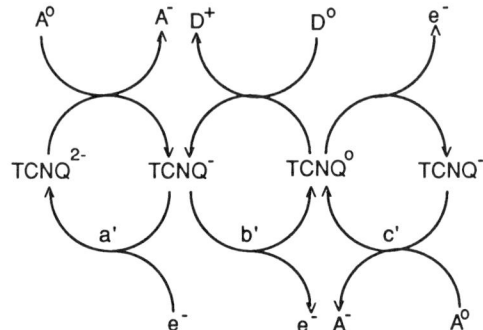

level of the electrode. The observation that the states near the Fermi level are associated with the TCNQ sites [200] is consistent with the reversible behavior of redox processes on donor-(TCNQ) electrodes. The partial charge generally present on TCNQ in these materials ($\rho < 1$) is tantamount to partial filling of the conduction band associated with the TCNQ acceptor stacks; this property may facilitate electron injection either into or out of the solid. Although the role of TCNQ species is strongly suggested, the possible role of the TTF subband cannot be ignored.

Scheme 16

retro charge-transfer prior to electron transfer

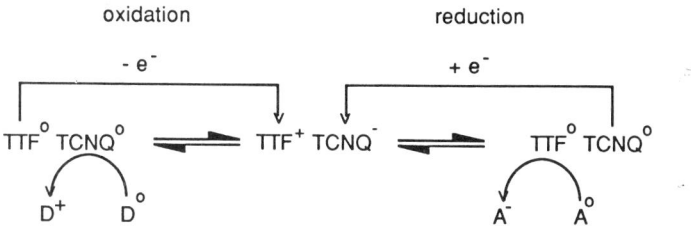

Mediated processes involving different valent states of the molecular constituents are equally plausible. The observation of TCNQ° on the surface of oxidized (TTF)(TCNQ) electrodes by Raman spectroscopy supports the possibility of mediated electron transfer. However, the lack of lattice redox reactions in the "ideally polarizable" potential region of these electrodes argues that the concentration of mediators in this range is quite small. Outside this range, or in very small mediator concentrations, electron transfer may be mediated by either TTF or TCNQ molecules in their neutral, monovalent, or divalent forms, as summarized in Scheme 15. Electrochemical reduction or oxidation of the electrode species may occur at the surface prior to electron transfer (paths a,a',b,b'), or electron transfer between the electrode species and substrate may occur prior to electrochemical regeneration of the electrode (paths c,c'). Note that, in Scheme 15, the initial steps have been drawn on the lower portion of the manifold, and TTF^+ and $TCNQ^-$ represent the ground state form of the ionic electrode. Electron transfer may also occur by reaction with either neutral TTF° or TCNQ° at the electrode surface, which may be present due to the equilibriums between these species and their partially ionized forms, depicted in Scheme 16 as resulting from "retro charge transfer." This may be due to dissolution and reprecipitation of the neutral species or simply to direct formation of the neutral species in the solid. This would be favorable only if the solvation energy of the neutral species compensated for the loss of electrostatic lattice energy accompanying formation of the neutral species. On very short time scales, one can consider the neutral forms as a "virtual" charge-transfer state of the solid. However, since this is a virtual state, this is indistinguishable from direct electron transfer into the molecular subbands. Electron exchange occurs between the substrate and one of the electrode species, followed by electrochemical oxidation or reduction of the other, regenerating the ionic forms. The nature of electron exchange depends on various factors, including the applied potential, intermolecular interactions, relative rates of electron transfer between the electrode and substrate species, electron exchange between the conductive substrate and the organic conductor, and electron exchange between the bulk organic phase and its molecular constituents.

B. Potential Applications of "Synthetic Metal" Electrodes

1. *Ion-Sensing Electrodes*

The earliest examples of electrodes prepared from low-dimensional organic solids were a series of R_4M^+ (R = C_6H_5, M = As; R = Et, M = N) and alkali metal salts of polycyano anion acceptors (Table 7) [201–203]. These were fabricated into pellets for use as ion-selective electrodes for a variety of ions, including the cations of the parent salt. The premise of this work was based on the small solubility products of these materials, which dictate the equilibrium between the ions in the solid and the solution phases. The activities of the ions in solution could then be measured by the potentiometric response of the electrode, which is based on one of the redox-active components. Electrodes prepared from Ag_2TCNQ_2 exhibited Nernstian response for Ag^+ ions according to Eq. (36), due to the primary redox process $Ag^+ + e^- \leftrightarrow Ag^°$. The $TCNQ^-$ concentration could be measured indirectly since the solution activities of Ag^+ and $TCNQ^-$ are related by the solubility product K_{sp} in Eq. (37), where a represents the activities of the designated ions and the solid. If $a(Ag_2TCNQ_2) = 1$, the response to $TCNQ^-$ could be estimated from Eq. (38),

TABLE 7

Ion-Sensing Electrodes Prepared from Organic Radical Ion Salts[a]

Electrode material	Ion detected
$Ag_2(TCNQ)_2$	Ag^+, H^+, Na^+
$Cu(TCNQ)_2$	Cu^{2+}, Ni^{2+}, H^+
$Cu_2(TCNQ)_2$	Cu^{2+}
$Pb(TCNQ)_2$	Pb^{2+}, $TCNQ^-$
$Cd(TCNQ)_2$	Cd^{2+}
$K_2(TCNQ)_2$	K^+
$Na_2(TCNQ)_2$	Na^+
$(Et_4N)(TCNQ)$	Et_4N^+, K^+
$(Et_4N)(TCNQ)_2$ ($\rho = 0.5$)	Et_4N^+, Me_2N^+, NH_4^+, Na^+, K^+, Ag^+, Mg^{2+}
$Pb(TNAP)_2$	Pb^{2+}
$Cu(TFM)_2$	Cu^{2+}
$Cu(DTF)_2$	Cu^{2+}
$(C_6H_5)_4As(DTF)$	$(C_6H_5)_4As^+$, H^+, K^+, Et_4N^+

[a]From Refs. 201-203.

where E is the measured electrode potential. The analysis was based on the assumption that $TCNQ^-$ was extensively dimerized as $(TCNQ)_2^{2-}$, based on the reported behavior of $TCNQ^-$ in aqueous solutions [189]. The detection limit for Ag^+ was estimated as 10^{-9} M.

$$E = E°(Ag/Ag+) + RT/F \log a(Ag^+) \qquad (36)$$

$$K_{sp} = \frac{[a(Ag^+)]^2[a(TCNQ_2^{2-})]}{a(Ag_2TCNQ_2)} \qquad (37)$$

$$E = E°(Ag/Ag^+) + 29.6 \log K_{sp} - 29.6 \log a(TCNQ_2^{2-}) \qquad (38)$$

Because of the relationship of K_{sp} and the activities of the cations and anions, the potential response to cations in solution could be measured even when the metal redox couple was not directly involved. That is, when the primary redox event of the electrode involved a more energetically accessible reduction or oxidation of the anions, Eq. (39) could be used to measure $[M^{n+}]$, where X is the polycyanoanion. For example, the response to Pb^{2+} was found to follow this relationship with Nernstian response for both Pb^{2+} and $TCNQ^-$ between 10^{-5} – 10^{-1} M, assuming that the $TCNQ_2^{2-}/2TCNQ^{2-}$ couple was involved. The validity of this relationship was also demonstrated by good agreement between potentiometrically determined K_{sp} values and those determined by atomic absorption.

$$E(M^{n+}) = E°(X^{n-}/X^{(n\pm1)-}) - RT/F \ln K_{sp} - RT/F \ln a(M^{n+}) \qquad (39)$$

The response toward redox-inert Et_4N^+, using either the simple salt $(Et_4N)(TCNQ)$ ($\rho = 1.0$) or the complex salt $(Et_4N)(TCNQ)_2$ ($\rho = 0.5$), indicated that different primary redox events were involved in the potentiometric detection. For a given concentration between 10^{-2}-1.0 M Et_4N^+, where Nernstian behavior was observed, the electrode potential of the complex salt was 130 mV more positive than the simple salt. This was proposed to reflect the role of the $TCNQ_2^{2-}/2TCNQ^0$ and $TCNQ_2^{2-}/2TCNQ^{2-}$ redox couples in the complex and simple salts, respectively, since the former couple possesses the more positive standard potential. However, this explanation ignores the role of differences in K_{sp} that may result from different lattice energies for the two phases. Indeed, it was noted that the 130 mV difference was not as large as the difference in standard redox potentials for the two couples and, therefore, it is not entirely clear that the fundamental redox processes responsible for the electrode behavior are different. The Et_4N^+ electrodes were also found to exhibit a near-Nernstian pH dependence of 46.3 mV decade^{-1}

between pH = 1-5. This was attributed to the formation of active $TCNQH_2/TCNQ$ layers on the electrode formed from protonation of the lattice $TCNQ^-$ anion under these conditions [Eq. (40)].

$$2\,TCNQ^- + 2\,H^+ \rightleftharpoons TCNQH_2 + TCNQ^0 \qquad (40)$$

The selectivity of these electrodes toward different cations depends on the relative stability of salts of those cations and, as a result, the electrodes cannot be considered specific. Indeed, significant response toward cations not present in the prefabricated electrode was commonly observed, presumably due to ion exchange with the solid [Eq. (41)]. Thus the selectivity is dependent primarily on differences in K_{sp} of different cations, and interference from other ions will be observed if Eq. (41) is operative. This was consistent with different working ranges and selectivities observed in comparisons of electrodes prepared with 11,11,-12,12-tetracyano-2,6-quinodimethane (TNAP), 9-dicyanomethylene-2,4,7-trinitrofluorene (DTF), 2,4,5,7-tetranitrofluorenemalanonitrile (TFM) and TCNQ, whose salts would be expected to have different K_{sp} values. These interferences represent the major impediment to the use of these electrodes as ion-selective electrodes and, accordingly, it does not appear that the materials are viable candidates for this application in the absence of ion-selective membranes.

$$M_1^+X^-(s) + M_2^+\,(aq) \rightleftharpoons M_2^+X^-(s) + M_1^+\,(aq) \qquad (41)$$

TNAP   DTF   TFM

2. Enzymatic Electrode Processes

Amperometric biosensors are generally composed of conventional metal electrodes and mediators that can shuttle electrons from the electrode to the active center commonly buried deep within the polypeptide structure of the enzyme. For example, detection of glucose via O_2 and $(C_5H_5)_2Fe$ mediators has been reported (Fig. 34) [209–213].

Recently, synthetic metals have also attracted considerable attention as electrochemical transducers for a variety of enzyme-based sensor devices. Significantly, these electrodes have been reported to effect electron transfer to enzymes without purposely added mediator species. In addition to simplifying construction, this approach potentially obviates mediator-limited currents (owing either to low mediator concentrations or sluggish kinetics) which are deleterious to precise detection. The

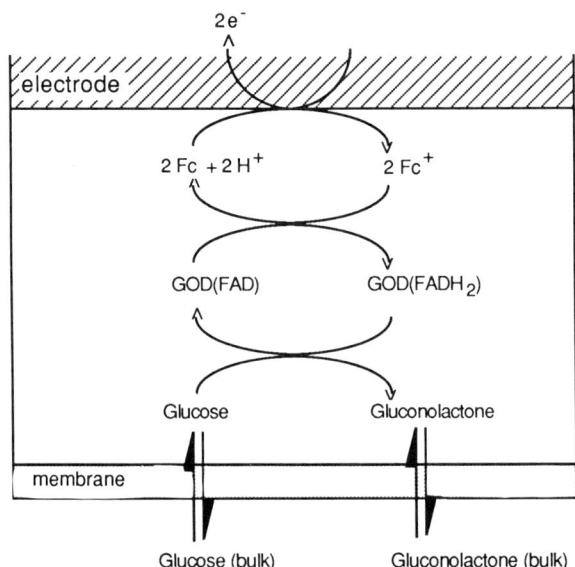

FIG. 34. Schematic representation of an enzymatic electrode in which homogeneous mediation is operative. In this particular example, the mediator is the $(C_5H_5)_2Fe/(C_5H_5)_2Fe^+$ couple (Fc/Fc^+), glucose oxidase the enzyme, and glucose the substrate.

ability to prepare different conducting electrodes with various functional groups may allow design of electrode surfaces that are selective toward substrates due to substrate-electrode interactions as well as ionic interactions involving lattice ions, as previously illustrated in Fig. 29. The redox activity of these electrodes and the solubility of the products of these reactions should also allow generation of a fresh surface if contamination of the electrode surface results. This is not feasible for electrodes modified with covalently attached species or polymer layers. It has also been claimed that the lower surface tension of organic electrodes compared to metal electrodes makes denaturation of enzymes less likely [214].

The oxidation of reduced β-nicotinamide adenine dinucleotide coenzyme (NADH) has been demonstrated on (NMP)(TCNQ) solid electrodes prepared by either cold-pressing the CT salt into a pellet, dip-coating a glassy carbon electrode, or filling a cavity with a slurry prepared from the salt and polyvinylchloride [215]. The oxidation of NADH was of interest because more than 250 enzymes use this cofactor for oxidation of substrates.

$$\text{NAD}^+ \xrightleftharpoons[\text{2e}^-,\ \text{H}^+\ (\text{substrate})]{-2\text{e}^-,-\text{H}^+\ (\text{electrode})} \text{NADH}$$

TABLE 8

Enzymatic Processes at Organic Solid Electrodes

Electrode material	Enzyme/coenzyme	Substrate	Reference
(CuDPA)(TCNQ)	Glucose oxidase[b]	Glucose	218–220
(TTF)(TCNQ)	L-aminoacid oxidase[b]	L-phenylalanine	223
$[(C_5H_5)_2Fe][(TCNQ)]$	D-aminoacid oxidase[b]	D-alanine	
$[(C_2H_5)_3NH][TCNQ]$	Choline oxidase[b]	Betaine aldehyde	
[Qn][TCNQ]	Xanthine oxidase[b]	Xanthine	
	Monoamine oxidase[b]		
	Pyruvate oxidase[b]	Phosphate	
(NMP)(TCNQ)	$FADH_2/FAD$		218
(NMP)(TCNQ)	$NADH/NAD^+$		215
(NMP)(TCNQ)	Ethanol dehydrogenase[c]	Ethanol	217
(TTF)(TCNQ)	Glucose oxidase[b]	Rat brain glucose[b]	225, 230
(TTF)(TCNQ)	Xanthine oxidase[b]	Purine	227
		Hypoxanthine	
		Xanthine	
(NMP)(TCNQ)	Glucose oxidase	Glucose	221, 222
(NMad)(TCNQ)			
(NMP)(TCNQ)	Horseradish peroxidase	Hydrogen peroxide	231
(NMad)(TCNQ)	Cytochrome b_2	L-lactate	

[a]If reported.
[b]Mediated by the $FAD/FADH_2$ cofactor.
[c]Mediated by the $NAD^+/NADH$ coenzyme.

It was found that NADH adsorbed on the surface according to a Langmuir isotherm, which could be equivalently expressed according to Michaelis-Menten kinetics [216]. Rotating disk experiments were used to determine the equilibrium constant for adsorption of NADH, K_1, and the rate constant for its oxidation at the electrode, k_2 [Eq. (42)]. The heterogenous rate constant for oxidation was reported as 10^{-3} cm sec^{-1}, consistent with fairly rapid electrode kinetics.

$$\text{NADH (sol)} \underset{}{\overset{K_1}{\rightleftharpoons}} \text{NADH (ads)} \xrightarrow{-e^-, \; k_2} \text{NAD}^+ \quad (42)$$

The oxidation of NADH was involved in a subsequent demonstration of amperometric ethanol detection, using an (NMP)(TCNQ) electrode in the presence of ethanol dehydrogenase [Eqs. (43) and (44)] (SH_2 = ethanol), the electrode assembly contained within a dialysis membrane (Fig. 35) [217]. Thus the enzyme is contained within a thin-layer region between the electrode and the dialysis membrane, and the substrate must diffuse through the membrane. The rate-limiting step was determined to be the enzymatic oxidation of ethanol, consistent with the fast electrode kinetics. This work demonstrated the advantage of fast electrode kinetics as rapid electrochemical consumption of NADH prevented back reaction and product inhibition of the enzyme was obviated.

$$SH_2 + NAD^+ \xrightarrow{\text{enzyme}} S + NADH + H^+ \quad (43)$$

$$NADH \xrightarrow{\text{(NMP)(TCNQ) electrode}} NAD^+ + H^+ + 2e^- \quad (44)$$

It has also been reported that various reduced flavoenzymes can be directly oxidized using (donor)(TCNQ) electrodes via the flavin adenine dinucleotide active site [Eqs. (45) and (46)] [218–223]. Some of the most detailed investigation in this category has involved glucose oxidase [221–223] using a variety of organic electrodes including (TTF)(TCNQ), which was most desirable because it possessed the lowest background currents. Rapid electrode kinetics were observed for these processes,

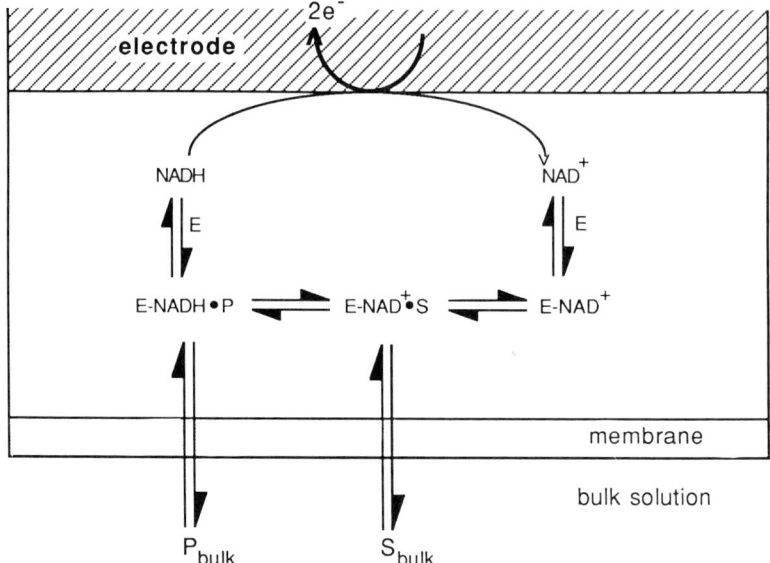

FIG. 35. Schematic representation of a generic synthetic metal enzymatic electrode that uses the $NADH/NAD^+$ coenzyme. This scheme is operative in the enzyme catalyzed electrochemical oxidation of ethanol.

and the heterogeneous electrochemical rate constants for different electrode materials were similar for the different electrode materials [223]. This was consistent with rate-limiting substrate diffusion through the dialysis membrane (Fig. 36), which is desirable since sensor response is less significantly affected by electrode and enzyme kinetics under this condition. Investigations with different enzyme/substrate systems found that only choline oxidase/betaine aldehyde exhibited kinetics that were partially rate-limited by both the enzymatic reaction and substrate transport [225].

The rapid electrode kinetics were confirmed by a rigorous kinetic analysis of the reactions employing the organic enzyme electrodes [224]. However, the actual role of the organic electrode in electron transfer is still not completely understood. It was originally proposed that the

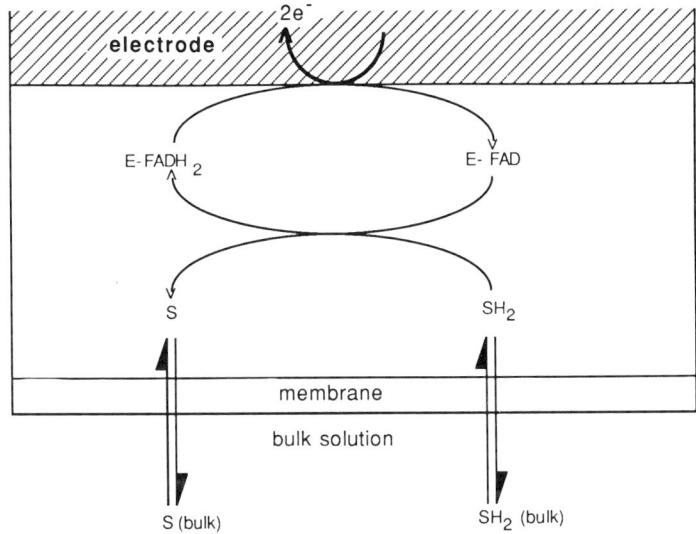

FIG. 36. Schematic representation of a generic synthetic metal enzymatic electrode that uses the $FADH_2/FAD$ coenzyme. This scheme is operative in electrochemical oxidation of substrates catalyzed by enzymes possessing this coenzyme.

mechanism for electron transfer was dependent on the nature of the enzyme: enzymes containing flavin mononucleotide and heme as the active site (cytochrome b_2) involved direct electron transfer with the electrode, whereas the flavoenzymes, including glucose oxidase, proceeded by homogeneous mediation of the molecular constituents produced by slight dissolution of the organic electrode [214,222]. However, rotating ring-disk experiments indicated that, in the region where amperometric detection was performed, dissolution of the organic electrode could not be detected as ring currents $<10^{-7}$ A for (NMP)(TCNQ) were observed between -0.1 and 0.3 V [223]. Also insertion of the membrane between the enzyme and the electrode resulted in inactive responses. These observations ruled out homogeneous mediation, but they did not eliminate the possibility of heterogeneous mediation (Scheme 17). The latter was

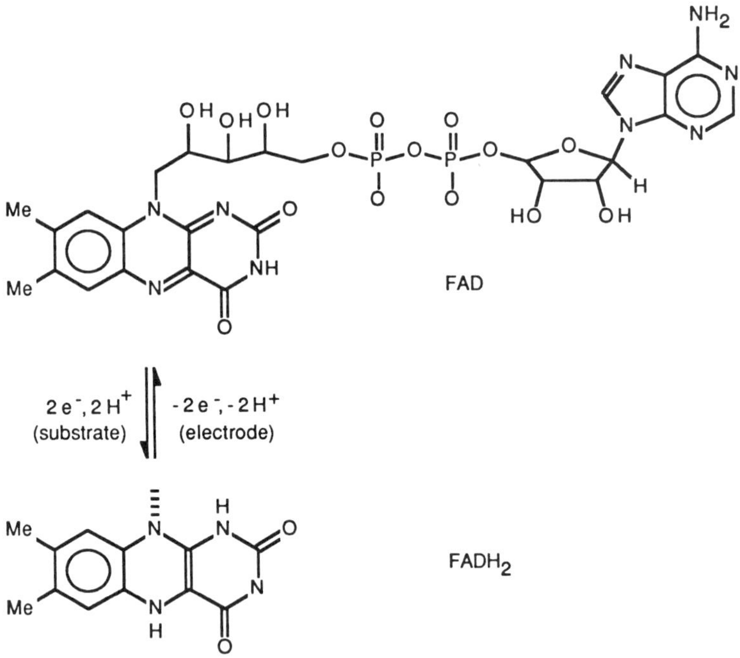

[Structural diagram of FAD showing flavin adenine dinucleotide with methyl groups, hydroxyls, phosphate groups, ribose, and adenine]

$2e^-, 2H^+$ (substrate) ⇌ $-2e^-, -2H^+$ (electrode)

[Structural diagram of FADH$_2$]

$SH_2 + E\text{-}FAD \longrightarrow S + E\text{-}FADH_2$ \hfill (45)

$E\text{-}FADH_2 \xrightarrow{\text{organic electrode}} E\text{-}FAD + 2H^+ + 2e^-$ \hfill (46)

Scheme 17

$TTF^+TCNQ^-(s) \underset{}{\overset{K}{\rightleftharpoons}} TTF^0(s) + TCNQ^0(s)$
$\downarrow \qquad\qquad\qquad \downarrow \; FADH_2 \to FAD$
$TTF^+(s) + TCNQ^-(s)$

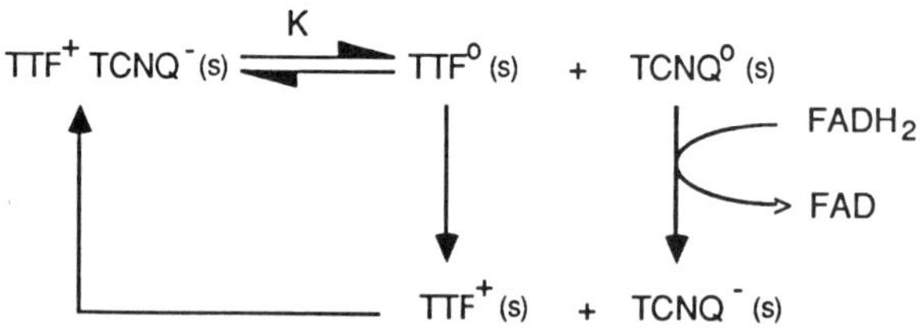

proposed to account for the glucose oxidase results on the basis of a half-order dependence on enzyme concentration and a transfer coefficient α equal to 0.3. According to the kinetic model, direct electron transfer would give first-order dependence on enzyme concentration and $\alpha = 1/2$, and homogeneous mediation would give only values of $\alpha = 0$, $1/2$, or 1 [225].

Regardless of the mechanism operative in these enzymatic electrode processes, the success of synthetic metal electrodes may be due to rate enhancements resulting from favorable π-π interactions of the electrode constituents with the active redox site on either the NADH or $FADH_2$ cofactor. Indeed, the rapid oxidation of flavin enzymes by TCNQ has been claimed [226]. Further investigations are obviously required to resolve the microscopic mechanistic issues surrounding these electrodes.

Amperometric detection of purine, hypoxanthine, and xanthine via xanthine oxidase has also been accomplished using (TTF)(TCNQ) electrodes [227]. These three reagents are involved in the stepwise reaction of xanthine oxidase with purine (Scheme 18). Accordingly, the sensitivity toward these compounds decreased in the order purine > hypoxanthine > xanthine because of lesser amplification as one proceeds along this series. As with the previously mentioned studies, kinetic analysis indicated that the substrate permeation through the dialysis membrane was rate-determining. The effective Michaelis-Menten constants were larger than those reported for the homogeneous enzymatic reactions [228,229]. This was attributed to the better electron-accepting properties of the (TTF)-(TCNQ) electrode, resulting in more efficient oxidation of the reduced enzyme compared to oxygen, which serves as the oxidant in the homogeneous reactions. It was also demonstrated that hypoxanthine and purine could be detected in blood plasma with linear responses in the 10^{-3}–10^{-4} M concentration range.

Enzymatic detection with synthetic metal electrodes has also been demonstrated in the absence of a membrane, owing to the irreversible and

Scheme 18

purine $\xrightarrow{\text{xanthine oxidase}}$ hypoxanthine $\xrightarrow{\text{xanthine oxidase}}$ xanthine $\xrightarrow{\text{xanthine oxidase}}$ uric acid

robust adsorption of the enzyme on the electrode, facilitating the use of single crystal (TTF)(TCNQ) microelectrodes for in vivo applications [225,230]. The microelectrodes were prepared by packing (TTF)(TCNQ) crystals into a 2 mm deep cavity formed by withdrawing a 0.25 mm diameter silver wire into Teflon insulation, followed by soaking the crystals in glucose oxidase. They were then directly implanted in the brains of freely moving live rats to detect glucose levels. Fairly linear response curves and only minor decreases in activity were observed for glucose detection, even after 10 days in vivo. After a period of 28 days, the activity dropped to approximately half its original value. Insulin injected into the subject gave the expected smaller currents at an implanted electrode due to reduced glucose concentration (Fig. 37).

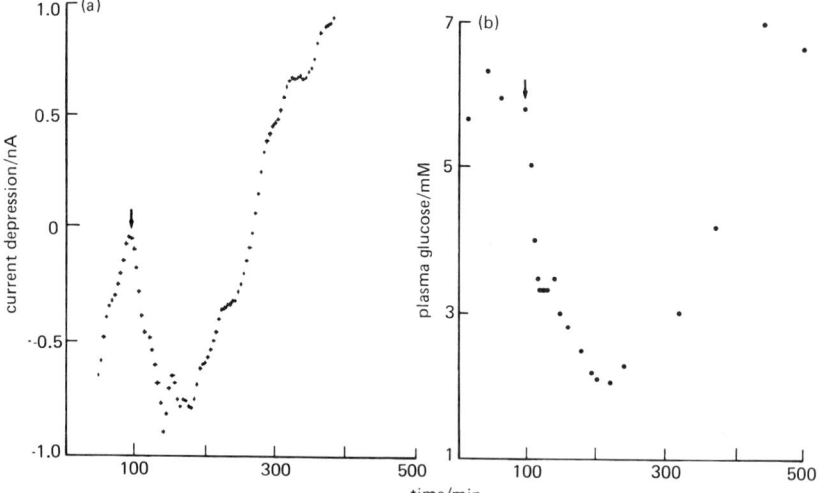

FIG. 37. Comparison of the response to an injection of insulin (25 units kg^{-1}) of (a) an implanted glucose electrode with (b) blood glucose as determined from blood samples. Subject: live rat. (From Ref. 225.)

C. Other Electrode Materials

Single crystal structures of numerous metallophthalocyanines indicate the presence of one-dimensional stacks, with significant overlap of the π orbitals of adjacent rings [232-234]. Accordingly, these materials possess electronic band structures, which results in either semiconducting or metallic behavior, the latter generally observed only after partial oxidation of the phthalocyanine. The p-type semiconducting behavior has led to numerous investigations of these materials as photoelectrodes, as extensively reviewed elsewhere [235]. Thin films of phthalocyanines, which presumably retain the extended one-dimensional structures of single crystals, are well known to effect electrocatalytic reduction of oxygen, although it is not always clear whether the actual faradaic process takes place by direct electron transfer at the phthalocyanine or at the conductive substrate underneath the films. Kinetic data for the electroreduction of oxygen have been reported that support the contention that faradaic processes occur at the phthalocyanine [236]. Reversible redox behavior for $Fe(CN)_6^{4-}/Fe(CN)_6^{3-}$ was reported on ZnPc and FePc films [237,238], which had working ranges of -0.8-+1.1 V and -0.75-+0.5 V, respectively. The observation that redox reactions proceeded readily in the positive potential regime, whereas reductive processes were difficult, has been attributed to the p-type nature of phthalocyanines.

A rather unusual electrode, although not rigorously a solid electrode, which consists of anisotropic electron donors and acceptors incorporated into thin bilayer lipid membranes (ca 50 A), has been reported [239, 240]. Reversible redox behavior was observed for $Fe(CN)_6^{4-}/Fe(CN)_6^{3-}$ when the bilayer membranes contained TTF, TCNQ, $(C_5Me_5)_2Fe$, I_2, or mesotetraphenylporphyrin (Fig. 38a). Application of an electric field across the membrane containing TCNQ in the absence of redox couples resulted in linear I/V curves, consistent with ohmic behavior. However, in the presence of $Fe(CN)_6^{4-}/Fe(CN)_6^{3-}$, the bilayer behaved as a metallic

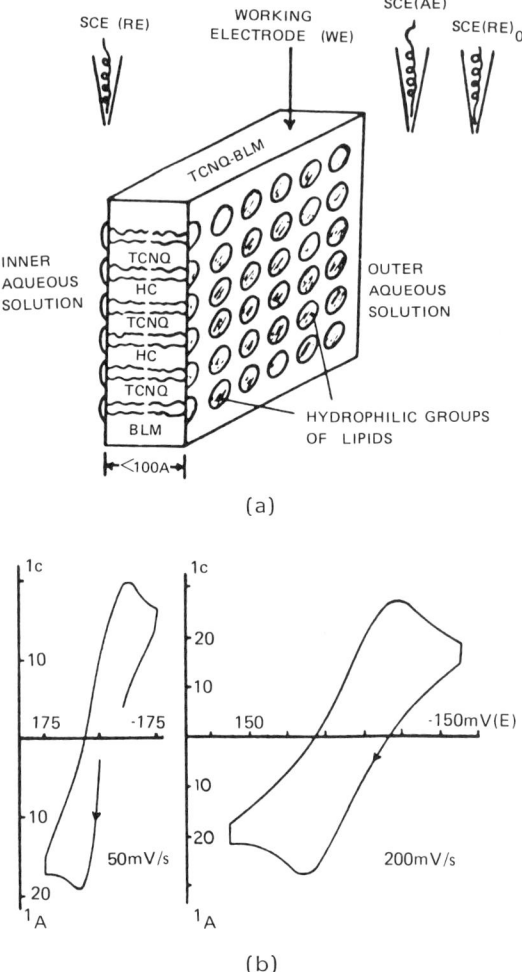

FIG. 38. (a) Schematic representation of a portion of a bilayer lipid membrane (BLM) containing TCNQ used in cyclic voltammetry. One side of the BLM (WE) indicates the working electrode, and AE and RE_0 are SCEs that serve as auxiliary and reference electrodes, respectively. The RE_1 reference electrode was used to apply a field across the BLM, as well for potentiometric experiments. (b) Cyclic voltammograms at a TCNQ-containing BLM working electrode in 5×10^{-4} M quinhydrone in 1 M KCl (V vs. SCE). The BLM was formed from a lipid mixture of 5 ml of natural lecithin and 45 ml of oxidized cholesterol in n-octane. (From Ref. 240.)

electrode with diffusion-shaped waves observed for the expected redox process, except for smaller currents compared to platinum (Fig. 38b). Presumably, a redox couple was present to mediate charge transfer on the side of the membrane opposite the working side, although this was not expressly stated in these reports. The actual mechanism of electron transfer in the membrane is not clearly established, as it is unclear whether these electrodes simply mediate electron transfer by electron hopping or actually behave as metallic species. It is perhaps intriguing to attribute this behavior to the formation of ordered TCNQ aggregates that can support current flow across the membrane.

Conducting Langmuir-Blodgett (LB) films containing CT complexes have been prepared by I_2 doping of N-docosylpyridinium-TCNQ films. Later, it was discovered that *intrinsic* conductivity could be observed in LB films of (TMTTF)-(octadecylTCNQ) (10^{-1} Ω^{-1} cm^{-1}) [241,242] and (N-docosylpyridinium)-(TCNQ)$_2$ (10^{-2} Ω^{-1} cm^{-1}) [243,244].

(TMTTF)-(octadecylTCNQ)   (N-docosylpyridinium)-(TCNQ)$_2$

The structure of these films is not rigorously determined, but the observation of high-conductivity anisotropy ($\sigma_{\parallel}/\sigma_{\perp} = 10^{13}$ Ω^{-1} cm^{-1}) and EPR spectral features similar to (TTF)(TCNQ) suggests low-dimensional structure parallel to the substrate surface with intermolecular interactions similar to those observed in the better-defined crystalline solids. Although the electrodic behavior of these films has not yet been reported, the demonstration of electrode modification with LB films [245–150] and the observation of unique electrode characteristics of crystalline TCNQ solids suggest interesting possibilities for conducting LB films.

VI. SUMMARY

The unique properties and behavior of low-dimensional solids are generally dependent on the physical characteristics of their constituents, which facilitates understanding of structure-function relationships in these materials. The electrochemical properties of the molecular species play a significant and somewhat predictive role in the solid-state properties of low-dimensional solids, including conductivity, magnetism, and optical absorbance. The redox properties of these compounds also allow ready synthesis of CT complexes by electrochemical methods, which are inherently advantageous since the crystallization process can be precisely regulated. In many cases, this allows reproducible control of crystal size, quality, and stoichiometry, the last having a significant influence on the conductivity of low-dimensional solids. The fundamental aspects of electrocrystallization of low-dimensional materials on metal electrodes have received scant attention and require further investigation. Initial investigations in which TTF and TCNQ species were detected on Ag using surface-enhanced Raman spectroscopy [251] suggest that this technique may be useful for investigating the events leading to crystallization. That many low-dimensional materials form multiple phases also indicates that more careful attention to conditions is required.

The characteristics of low-dimensional solids have resulted in some interesting potential applications as electrode materials, including ion sensors and enzyme electrodes. Their metallic nature allows them to be utilized in the absence of binders such as carbon paste, and their organic nature may facilitate electrode processes by specific electrode-substrate interactions. This suggests the possibility of making electrodes that are composed of organic conductors with molecularly engineered functional groups to effect specific reactions. This concept of electrode modification by actually designing the electrode material itself differs from the conventional approach of modifying metal electrodes with electroactive polymers or covalently attached redox species.

Also, the ability to electrocrystallize conducting salts in inert matrices suggests the possibility of microelectrode preparation in this fashion; possibly arrays of synthetic metal microelectrodes designed on this principle could replace compressed pellets. Clearly, further work is required in this area to exploit these novel materials fully.

ACKNOWLEDGMENTS

The author wishes to thank R. E. Putscher for his assistance in literature searches, and P. N. Bartlett, J. Gaudiello, and T. J. Marks for the generous donation of reprints.

APPENDIX
Definitions of Acronyms for Molecular Components

Ad	Acridine
2,2'-BIP	2,2'-bipyridine
4,4'-BIP	4,4'-bipyridine
B[4,5-b]PTTF	Bis[4,5-b]pyridino-1,1',3,3'-tetrathiafulvalene
DBTTF	Dibenzotetrathiafulvalene
DDQ	Dichlorodicyanoquinodimethane
DEDMTSeF	Diethyldimethyltetraselenafulvalene
DMDBTTF	Dimethyldibenzotetrathiafulvalene
2,5-DM-DCNQI	2,5-dimethyl-N,N'-dicyanoquinonediimine
DMP	Dimethylphenazine
DPA	Dipyridylamine
DTDSeF	Diselenadithiafulvalene
DTF	9-dicyanomethylene-2,4,7-trinitrofluorene
ET	Bis(ethylenedithio)tetrathiafulvalene
FA	Fluoranthene
2-FTSeT	2-fluorotetraselenafulvalene
HAB	Hexaaminobenzene

HCTMM	Hexacyanotrimethylenemethane
HCTMCP	Hexacyanotrimethylenecyclopropane
HET	2,3,6,7,10,11-tris(N,N'-ethylenediamino)triphenylene
HMTP	Hexamethoxytriphenylene
HMTSeF	Hexamethylenetetramethyltetraselenafulvalene
HMTTeF	Hexamethylenetetramethyltetratellurafulvalene
HOC	Hexaazaoctadecahydrocoronene
MV	Methylviologen or N,N'-dimethylbipyridine
NMAd	N-methylacridine
NMP	N-methylphenazine
OCNAQ	11,11,12,12,13,13,14,14-octacyano-1,4,5,8-Anthradiquinotetramethane
PDCB	Paradicyanobenzene
Pe	Perylene
PMDA	Pyromelliticdianhydride
Py	Pyrene
Qn	Quinoline
TAE	1,1',3,3'-diethylene-2,2'-bibenzimidazolidene
TCNE	Tetracyanoethylene
TCNQ	Tetracyanoquinodimethane
TDMAB	1,2,4,5-tetrakis(dimethylamino)benzene
TFM	2,4,5,7-tetranitrofluorenemalononitrile
TMDAP	Tetramethyldiaminopyrene
TMPD	Tetramethyl-p-phenylenediamine
TMSA	Tetramethoxyselenaanthracene
TMTSF	Tetramethyltetraselenafulvalene
TMTTF	Tetramethyltetrathiafulvalene
TNAP	11,11,12,12-tetracyano-2,6-quinodimethane (or tetracyanonaphthaquinone)
TSeA	Tetraselenanthracene
TSeF	Tetraselenafulvalene
TSeN	Tetraselenanaphthacene

TTA	Tetrathioanthracene
TTF	Tetrathiafulvalene
TTN	Tetrathionaphthacene
TTT	Tetrathiotetracene
TTeF	Tetratellurafulvalene

REFERENCES

1. (a) J. M. Lehn and C. W. Rees (eds.), *Molecular Semiconductors*, Springer-Verlag, New York, 1982; (b) *Molecular Electronic Devices*, F. Carter, ed., Marcel Dekker, New York, 1982.

2. (a) J. Simon, J.-J. Andre, and A. Skoulios, J. Nouv. Chim. *10*: 295 (1986). (b) J. Simon, F. Tournilhac, and A. Skoulios, J. Nouv. Chim. *11*:383 (1986).

3. (a) J. Ferraris, D. O. Cowan, V. Walatka, and J. H. Perlstein, J. Am. Chem. Soc. *95*:948 (1973). (b) For a general review, see D. O. Cowan, Chem. Eng. News *64*:28, July 21, 1986.

4. E. Ehrenfreund, E. F. Rybaczewski, A. F. Garito, and A. J. Heeger, Phys. Rev. Lett. *28*:873 (1972).

5. D. Jerome and H. J. Schutz, Adv. Phys. *31*:299 (1982).

6. J. M. Williams, T. J. Emge, H. H. Wang, M. A. Beno, P. T. Copps, L. N. Hall, K. D. Carlson, and G. W. Crabtree, Inorg. Chem. *23*:2558 (1984).

7. S. S. Parkin, E. M. Engler, R. R. Schumaker, R. Lagier, V. Y. Lee, J. C. Scott, and R. L. Greene, Phys. Rev. Lett. *50*:270 (1983).

8. G. Satio, T. Enoki, T. Toriumi, H. Inokuchi, Solid State Commun. *42*:557 (1982).

9. M. D. Ward and D. C. Johnson, Inorg. Chem., *26*:4213 (1987).

10. For a general review of the area of conducting polymers, see (a) *Handbook of Conducting Polymers*, Vols. I and II, T. A. Skotheim, ed., Marcel Dekker, New York, 1986; and (b) J. E. Frommer and R. R. Chance, *Encyclopedia of Polymer Science and Engineering*, M. Grayson and J. Kroschwitz, eds., Wiley, New York, 1985, p. 462, and references therein.

11. D. F. Eaton, A. G. Anderson, W. Tam, and Y. Wang, J. Am. Chem. Soc. *109*:1886 (1987).

12. D. F. Eaton, A. G. Anderson, W. Tam, and Y. Wang, ACS Symposium Ser. Polymers for High Technol. *346*:381 (1987).

13. M. D. Ward, P. J. Fagan, and J. C. Calabrese, submitted for publication.
14. J. S. Miller, P. J. Krusic, D. A. Dixon, W. M. Reiff, J. H. Zhang, E. A. Anderson, and A. J. Epstein, J. Am. Chem. Soc. 108:4459 (1986).
15. J. S. Miller, J. C. Calabrese, A. J. Epstein, R. W. Bigelow, J. H. Zhang, and W. M. Reiff, J. Chem. Soc. Chem. Comm., 1026, 1986.
16. J. S. Miller, A. H. Reis, Jr., E. Gebert, J. J. Ritsko, W. R. Salaneck, L. Kovnat, T. W. Cape, and R. P. Van Duyne, J. Am. Chem. Soc. 101:7112 (1979).
17. R. W. Murray, Electroanal. Chem. 13:191 (1984).
18. J. D. Corbett, Chem. Rev. 85:383 (1985).
19. J. S. Miller (Ed.), Extended Linear Chain Compounds, Vols. 1-3, 1981-1983.
20. F. H. Herbstein, Perspectives in Structural Chemistry, Vol. IV, J. D. Dunitz and J. A. Ibers, Eds., Wiley, New York, 1971, p. 166, and references therein.
21. A. I. Kitaigorodsky in Molecular Crystals and Molecules, Academic Press, New York, 1973.
22. (a) J. S. Miller and A. J. Epstein, Prog. Inorg. Chem. 20:1 (1976); (b) A. F. Garrito and A. J. Heeger, Acc. Chem. Res. 7:232 (1973).
23. J. S. Miller (ed.), Extended Linear Chain Compounds, Vols. 1-3, 1981-1983.
24. D. B. Chestnut and P. Arthur, Jr., J. Chem. Phys. 36:2969 (1962).
25. K. Bechgaard, Mol. Cryst. Liq. Cryst. 79:1 (1982).
26. R. S. Mulliken and W. B. Pearson, Molecular Complexes: A Lecture and Reprint Volume, Wiley, New York, 1969.
27. P. Batail, S. J. LaPlaca, J. J. Mayerle, and J. B. Torrance, J. Am. Chem. Soc. 103:951 (1981).
28. W. H. Bentley and H. G. Drickamer, J. Chem. Phys. 42:1573 (1965).
29. J. J. Mayerle, J. B. Torrance, and J. I. Crowley, Acta Crystallogr. Sect. B 35:2988 (1979).
30. M. D. Ward, Organometall. 6:754 (1987).
31. M. D. Ward and J. C. Calabrese, Organometallics, in press.

32. K. Nakamura, Y. Kai, N. Yasuoka, and N. Kasai, Bull. Chem. Soc. Jpn. 54:3300 (1981).
33. M. D. Ward, unpublished results.
34. K. Bechgaard and V. D. Parker, J. Am. Chem. Soc. 94:4749 (1972).
35. R. Breslow, Pure and Applied Chem. 54:927 (1982).
36. J. A. McCleverty, Prog. Inorg. Chem. 10:49 (1968).
37. R. C. Wheland and J. L. Gillson, J. Am. Chem. Soc. 98:3916 (1976).
38. E. M. Engler, F. B. Kaufman, D. C. Green, C. E. Klots, and R. N. Compton, J. Am. Chem. Soc. 97:2921 (1975).
39. R. D. McCullough, G. B. Kok, K. A. Lerstrup, and D. O. Cowan, J. Am. Chem. Soc. 109:4115 (1987).
40. S. Hunig, G. Kiesslich, H. Quast, and D. Scheutzow, Justus Liebigs Ann. Chem. 310 (1973).
41. These redox properties were measured by the author in acetonitrile containing 0.1 M [n-Bu$_4$N][BF$_4$].
42. T. Fukanaga, J. Am. Chem. Soc. 98:610 (1976).
43. Redox potentials were determined by the author in methylene chloride containing 0.1 M TBAClO$_4$.
44. S. Tanaka, J. A. Bruce, and M. S. Wrighton, J. Phys. Chem. 85:3778 (1981).
45. R. Breslow, P. Maslak, and J. S. Thomaides, J. Am. Chem. Soc. 106:6453 (1984). The oxidation of HOC to the trication and tetracation was reported to occur at E° = 0.87 and 1.25 V versus SCE, respectively.
46. These values, determined by the author in 0.1 M [n-Bu$_4$N][BF$_4$] in acetonitrile, differ significantly from those reported in Ref. 45. HOC also exhibits reversible electrochemical oxidation to the trication (E° = 0.52) and the tetracation (E° = 0.91).
47. T. Nogami, H. Tanaka, S. Ohnishi, Y. Tasaka, and H. Mikawa, Bull. Chem. Soc. Jpn. 57:22 (1984).
48. F. Wudl, D. E. Schafer, and B. Miller, J. Am. Chem. Soc. 98:252 (1976).
49. A. Yamahira, T. Nogami, and H. Mikawa, J. Chem. Soc. Chem. Comm. 904 (1983).
50. J. C. Stark, R. Reed, L. A. Acampora, D. J. Sandman, S. Jansen, M. T. Jones, and B. M. Foxman, Organometall. 3:732 (1984).

51. H. Endres, M. Hiller, H. J. Keller, K. Bender, E. Gogu, I. Heinen, and D. Schweitzer, Z. Naturforsch. 40b:1664 (1985).

52. F. Wudl and E. A. Shalom, J. Am. Chem. Soc. 104:1154 (1982).

53. R. D. McCullough, G. B. Kok, K. A. Lerstrup, and D. A. Cowan, J. Am. Chem. Soc. 109:4115 (1987).

54. E. S. Pysh and N. C. Yang, J. Am. Chem. Soc. 85:2124 (1963).

55. V. Enkelmann, B. S. Morra, C. Krohnke, G. Wegner, and J. Heinze, Chem. Phys. 66:303 (1982).

56. Two-electron oxidation; K. Elbl, C. Kreiger, and H. A. Staab, Angew. Chem. Int. Ed. Engl. 25:1023 (1986).

57. J. D. L. Holloway and W. E. Geiger, J. Am. Chem. Soc. 101:2038 (1979).

58. J. L. Robbins, N. Edelstein, B. Spencer, and J. C. Smart, J. Am. Chem. Soc. 104:1882 (1982).

59. U. Koelle and F. Khouzami, Angew. Chem. Int. Ed. Eng. 19:640 (1980).

60. (a) R. J. Wilson, L. F. Warren, and M. F. Hawthorne, J. Am. Chem. Soc. 91:758 (1969); (b) R. P. Van Duyne and C. N. Reilley, Anal. Chem. 44:158 (1972).

61. L. I. Denisovich, N. V. Zakurin, A. A. Bezrukova, and S. P. Gubin, J. Organometall. Chem. 81:207 (1974).

62. S. P. Gubin, S. A. Smirnova, L. I. Denisovich, and A. A. Lubovich, J. Organometall. Chem. 30:243 (1971).

63. J. C. Smart and J. L. Robbins, J. Am. Chem. Soc. 100:3936 (1978).

64. U. Kolle and J. Grub, J. Organometall. Chem. 289:133 (1985).

65. W. F. Little, C. N. Reilley, J. D. Johnson, K. N. Lynn, and A. P. Sanders, J. Am. Chem. Soc. 86:1376 (1964).

66. J.-R. Hamon, D. Astruc, and P. Michaud, J. Am. Chem. Soc. 103:758 (1981).

67. W. Pukacki, M. Pawlak, A. Graja, M. Lequan, and R. M. Lequan, Inorg. Chem. 26:1328 (1987).

68. L. P. Yur'eva, S. M. Peregudova, L. N. Nekrasov, A. P. Korotkov, N. N. Zaitseva, N. V. Zakurin, and A. Yu. Vasil'kov, J. Organometall. Chem. 219:43 (1981).

69. R. M. Markle and J. J. Lagowski, Organometall. 5:595 (1986).

70. O. W. Webster, J. Am. Chem. Soc. 86:2898 (1964).

71. M. E. Peover, J. Chem. Soc. (London), 4540 (1962).

72. A. F. Garito and A. J. Heeger, Acc. Chem. Res. 7:232 (1974).

73. T. Mitsuhashi, M. Goto, K. Honda, Y. Maruyama, T. Sugawara, T. Inabe, and T. Watanabe, J. Chem. Soc. Chem. Comm. 810 (1987).

74. OCNAQ also exhibits reversible electrocehmical reduction to the trianion ($E° = -0.44$ V) and the tetraanion ($E° = -0.53$ V).

75. C. L. Bird and A. T. Kuhn, Chem. Soc. Rev. 10:49 (1981).

76. R. G. Finke, R. H. Voegeli, E. D. Laganis, and V. Boekelheide, Organometall. 2:347 (1983).

77. Redox potential determined by author in acetonitrile containing 0.1 M $TBABF_4$; compound supplied by V. Boekelheide.

78. C.-N. Lai and A. T. Hubbard, Inorg. Chem. 11:2081 (1972).

79. J. S. Chappell, A. N. Bloch, W. A. Bryden, M. Maxfield, T. O. Poehler, and D. O. Cowan, J. Am. Chem. Soc. 103:2442 (1981).

80. N. F. Mott and E. A. Davis, Electronic Processes in Non-Crystalline Materials, Clarendon Press, Oxford, 1971.

81. P. P. Edwards and M. J. Sienko, J. Phys. Rev. 17:2575 (1978).

82. R. M. Hedges and F. A. Matsen, J. Chem. Phys. 28:950 (1958).

83. E. S. Pysh and N. C. Yang, J. Am. Chem. Soc. 85:2124 (1963).

84. J. O. Howell, J. M. Goncalves, C. Amatore, L. Klasinc, R. M. Wightman, and J. K. Kochi, J. Am. Chem. Soc. 106:3968 (1984).

85. B. W. Sullivan and B. M. Foxman, Organometall. 2:187 (1983).

86. R. C. Wheland, J. Am. Chem. Soc. 98:3926 (1976).

87. J. B. Torrance, Acc. Chem. Res. 12:79 (1979).

88. J. Diekmann, W. R. Hertler, and R. E. Benson, J. Org. Chem. 28:2719 (1963).

89. D. J. Sandman and A. F. Garito, J. Org. Chem. 39:1165 (1974).

90. C. J. Bender, Chem. Soc. Rev. 15:475 (1986).

91. J. B. Torrance, J. E. Vazquez, J. J. Mayerle, and V. Y. Lee, Phys. Rev. Lett. 46:253 (1981).

92. T. J. LePage and R. Breslow, J. Am. Chem. Soc. 109:6412 (1987).

93. T. P. Radhakrishnan, Z. Soos, H. Endres, and L. J. Azevedo, J. Chem. Phys. 85:1126 (1986).

94. J. S. Miller, J. H. Zhang, and W. M. Reiff, J. Am. Chem. Soc. 109:4584 (1987).

95. J. S. Miller, J. C. Calabrese, R. W. Bigelow, A. J. Epstein, R. W. Zhang, and W. M. Reiff, J. Chem. Soc. Chem. Comm. 1986:1026.

96. J. S. Miller, J. C. Calabrese, H. Rommelmann, S. R. Chittapeddi, R. W. Zhang, W. M. Reiff, and A. J. Epstein, J. Am. Chem. Soc. 109:769 (1987).

97. J. S. Miller and A. J. Epstein, J. Am. Chem. Soc. 109:3850 (1987).

98. H. M. McConnell, Proc. R. A. Welch Found. Chem. Res. 11:144 (1967).

99. F. B. Kaufman, E. M. Engler, D. C. Green, and J. Q. Chambers, J. Am. Chem. Soc. 98:1596 (1976).

100. B. A. Scott, S. J. La Placa, J. B. Torrance, B. D. Silverman, and B. Welber, J. Am. Chem. Soc. 99:6631 (1977).

101. T. C. Chiang, A. H. Reddoch, and D. F. Williams, J. Chem. Phys. 54:2051 (1971).

102. P. L. Nordio, Z. G. Soos, and H. M. McConnell, Annu. Rev. Phys. Chem. 17:273 (1966).

103. D. F. Williams, Science 197:1194 (1977).

104. L. Alcacer and A. H. Maki, J. Phys. Chem. 78:215 (1974).

105. J. S. Miller, Science 194:189 (1976).

106. J. S. Miller and A. J. Epstein, Prog. Inorg. Chem. 20:1 (1976).

107. J. M. Williams, M. Iwata, S. W. Peterson, K. A. Leslie, and H. J. Guggenheim, Phys. Rev. Lett. 34:1653 (1975).

108. D. M. Washecheck, S. W. Peterson, A. H. Reis, Jr., and J. M. Williams, Inorg. Chem. 15:74 (1976).

109. For a general review of these materials, see J. M. Williams, A. J. Schultz, A. E. Underhill, and K. Carneiro, *Extended Linear Chain Compounds*, Vol. 1, J. S. Miller, Ed., Plenum, New York, 1982, p. 73.

110. S. S. P. Parkin, M. Ribault, D. Jerome, and K. Bechgaard, J. Phys. C 14:L445:5305 (1981).

111. K. Bechgaard, K. Carneiro, F. B. Rasmussen, M. Olsen, and C. S. Jacobsen, Phys. Rev. Lett. 46:852 (1981).

112. D. Jerome, A. Mazaud, M. Ribault, and K. Bechgaard, J. Phys. (Paris) Lett. 41:L95 (1980).

113. K. Bechgaard, C. S. Jacobsen, K. Mortensen, H. J. Pedersen, and N. Thorup, Solid State Comm. 33:1119 (1980).

114. E. M. Engler, R. Greene, P. Haen, Y. Tomkiewicz, K. Mortensen, and J. Berendzen, Mol. Crystl. Liq. Cryst. 79:15 (1982).

115. R. Laversanne, C. Coulon, B. Gallois, J. P. Pouget, and R. Moret, J. Phys. Lett. 45:L393 (1984).

116. H. Anzai, M. Tokumoto, T. Ishiguro, G. Saito, H. Kobayashi, R. Kato, and A. Kobayashi, Synth. Met. 19:611 (1987).

117. J. M. Williams and K. Carneiro, Adv. Inorg. Chem. Radiochem. 29:249 (1985).

118. J. M. Williams, M. A. Beno, H. H. Wang, P. E. Reed, L. J. Azevedo, and J. E. Schirber, Inorg. Chem. 23:1790 (1984).

119. S. S. P. Parkin, E. M. Engler, R. R. Schumaker, R. Lagier, V. Y. Lee, J. C. Scott, and R. L. Greene, Phys. Rev. Lett. 50:270 (1983).

120. M. A. Beno, G. S. Blackman, P. C. W. Leung, K. D. Carlson, P. T. Copps, and J. M. Williams, Mol. Cryst. Liq. Cryst. 119:409 (1985).

121. S. S. P. Pakin, E. M. Engler, V. Y. Lee, and R. R. Schumaker, Mol. Cryst. Liq. Cryst. 119:375 (1985).

122. H. Endres, M. Hiller, H. J. Keller, K. Bender, E. Gogu, I. Heinen, and D. Schweitzer, Z. Naturforsch. 40b:1664 (1985).

123. E. B. Yagubskii, I. F. Shchegolev, V. N. Laukhin, P. A. Kononovich, M. V. Kartsovnik, A. V. Zvarykina, and L. I. Buravov, JETP Lett. (Engl. Trans.) 39:12 (1984).

124. J. M. Williams, T. J. Emge, H. H. Wang, M. A. Beno, P. T. Copps, L. N. Hall, K. D. Carlson, and G. W. Crabtree, Inorg. Chem. 23:2558 (1984).

125. K. Bender, I. Henning, D. Schweitzer, K. Deitz, H. Endres, and H. J. Keller, Mol. Cryst. Liq. Cryst. 119:361 (1984).

126. R. P. Shibaeva, V. F. Kaminskii, and E. B. Yaqubskii, Mol. Cryst. Liq. Cryst. 119:361 (1985).

127. E. B. Yagubskii, I. F. Shchegolev, S. I. Pesotskii, V. N. Laukhin, P. A. Kononivich, M. V. Kartsovnik, and A. V. Zvarykina, JETP Lett. (Engl. Trans.) 39:328 (1984).

128. M. A. Beno, U. Geiser, K. L. Kostka, H. H. Wang, K. S. Webb, M. A. Firestone, K. D. Carlson, L. Nunez, M.-Y. Whangbo, and J. M. Williams, Inorg. Chem. 26:1912 (1987).

129. U. Geiser, H. H. Wang, K. M. Donega, B. A. Anderson, J. M. Williams, and J. F. Kwak, Inorg. Chem. 25:402 (1986).

130. H. J. Keller, D. Nothe, H. Pritzkow, D. Wehe, M. Werner, P. Koch, and D. Schweitzer, Mol. Crystl. Liq. Cryst. 62:181 (1980).

131. P. Kathirgamanathan and D. Rosseinsky, J. Chem. Soc. Chem. Comm. 356 (1980).

132. P. A. C. Gane, P. Kathirgamanathan, and D. Rosseinsky, J. Chem. Soc. Chem. Comm. 378 (1981).

133. (a) P. Kathirgamanathan, S. A. Mucklejohn, and D. R. Rosseinsky, J. Chem. Soc. Chem. Comm. 86 (1979); (b) M. Lamache and K. E. Kacemi, Mol. Cryst. Liq. Cryst. 120:255 (1985).

134. P. Kathirgamanathan and D. R. Rossiensky, J. Chem. Soc. Chem. Comm. 839 (1980).

135. T. P. Henning, H. S. White, and A. J. Bard, J. Am. Chem. Soc. 103:3937 (1981).

136. (a) T. P. Henning, H. S. White, and A. J. Bard, J. Am. Chem. Soc. 104:5862 (1982). (b) T. P. Henning and A. J. Bard, J. Electrochem. Soc. 130:613 (1983).

137. C. D. Jaeger and A. J. Bard, J. Am. Chem. Soc. 101:1690 (1979).

138. C. D. Jaeger and A. J. Bard, J. Am. Chem. Soc. 102:5435 (1980).

139. M. T. Carter and A. J. Bard, J. Electroanal. Chem. 229:191 (1987).

140. S. Bruckenstein and M. Shay, J. Electroanal. Chem. 188:131 (1985).

141. M. D. Ward, unpublished results.

142. G. Z. Sauerbrey, Phyzik 155:206 (1959).

143. P. T. Varineau and D. A. Buttry, Phys. Chem. 91:1292 (1987).

144. M. D. Ward, J. Phys. Chem. 92:2049 (1988).

145. Y. Orihashi, N. Kobayashi, E. Tsuchida, H. Matsuda, H. Nakanishi, and M. Kato, Chem. Lett. 1617 (1985).

146. E. Tsuchida, Y. Orihashi, N. Kobayashi, and H. Ohno, Synth. Met. 15:201 (1986).

147. Y. Orihashi, H. Ohno, E. Tsuchida, H. Matsuda, H. Nakanishi, and M. Kato, Chem. Lett. 601 (1987).

148. M. Almeida, M. G. Kanatzidas, L. M. Tonge, T. J. Marks, H. O. Marcy, W. J. McCarthy, and C. R. Kannewurf, Solid State Commun. 63:457 (1987).

149. T. Nogami, H. Tanaka, S. Ohnishi, Y. Tasaka, and H. Mikawa, Bull. Chem. Soc. Jpn. 57:22 (1984).

150. M. D. Ward, Inorg. Chem. 25:4444 (1986).
151. M. D. Ward and P. J. Fagan, submitted for publication.
152. T. Nogami, S. Ohnishi, Y. Tasaka, and H. Mikawa, Mol. Cryst. Liq. Cryst. 101:367 (1983).
153. A. Aumiller, P. Erk, G. Klebe, S. Hunig, J. V. von Schutz, and H.-P. Wemer, Angew. Chem. Int. Ed. Eng. 25:740 (1986).
154. K. Bechgaard, C. S. Jacobsen, K. Mortensen, H. J. Pedersen, and N. Thorup, Solid State Commun. 33:1119 (1980).
155. A. V. Zvarykina, P. A. Kononovich, V. N. Laukhin, A. G. Khomenko, and E. B. Yagubskii, Izv. Akad. Nauk SSSR, Ser. Khim. 11:2624 (1983).
156. B. Hilti, C. W. Mayer, G. Rihs, H. Loeliger, and P. Baltzer, Mol. Cryst. Liq. Cryst. 120:267 (1985).
157. B. Hilti, C. W. Mayer, and G. Rihs, Helv. Chim. Acta. 61:1462 (1978).
158. L. Y. Chiang, D. C. Johnston, J. B. Stokes, and A. N. Bloch, Synth. Met. 19:697 (1987).
159. T. Nogami, J. Ohnishi, H. Tanaka, and H. Mikawa, J. Phys. 44:1253 (1983).
160. M. Lamache and R. Najean, Electrochim. Acta. 29:273 (1984).
161. M. Lamache, S. Wuryanto, and F. Benhamou, Electrochim. Acta. 30:817 (1985).
162. C. Tanaka, J. Tanaka, K. Dietz, C. Katayama, and M. Tanaka, Bull. Chem. Soc. Jpn. 56:405 (1983).
163. H. Endres, M. Hiller, H. J. Keller, K. Bender, E. Gogu, I. Heinen, and D. Schweitzer, Z. Naturforsch. 40b:1664 (1985).
164. H. Anzai, M. Tokumoto, T. Ishiguro, H. Kobayashi, R. Kato, A. Kobayashi, Physica 143B:293 (1986).
165. E. Amberger, H. Fuchs, and K. Polborn, Angew. Chem. Int. Ed. Eng. 24:968 (1985).
166. A. Weber, H. Endres, H. J. Keller, E. Gogu, I. Heinen, K. Bender, and D. Schweitzer, Z. Naturforsch. 40b:1658 (1985).
167. T. Mori, F. Sakai, G. Saito, and H. Inokuchi, Chem. Lett. 1589 (1986).
168. L. C. Porter, H. H. Wang, M. A. Beno, K. D. Carlson, C. M. Pipan, R. B. Proksh, and J. M. Williams, Solid State Comm. 64:387 (1987).

169. U. Geiser, H. H. Wang, J. M. Williams, E. L. Venturini, J. F. Kwak, and M. H. Whangbo, Synth. Met. 19:599 (1987).

170. H. Anzai, M. Tokumoto, and G. Saito, Mol. Cryst. Liq. Crystl. 125:385 (1985).

171. H. P. Fritz, H. Gebauer, P. Friedrich, P. Ecker, R. Artes, and U. Schubert, Z. Naturforsch. 33b:498 (1978).

172. R. Wilckens, H. P. Geserich, W. Ruppel, P. Koch, D. Schweitzer, and H. J. Keller, Solid State Commun. 41:615 (1982).

173. C. Krohnke, V. Enkelmann, and G. Wegner, Angew. Chem. 92:941 (1980).

174. G. C. Papavasiliou, Chim. Chron., New Ser. 15:161 (1986).

175. J. O'M. Bockris and G. A. Razumney, *Fundamental Aspects of Electrocrystallization*, Plenum, New York, 1967, Chap. 8.

176. M. Fleischmann and H. R. Thirsk, Adv. Electrochem. Electrochem. Engin. 3:123 (1963).

177. A. R. Hillman and E. F. Mallen, J. Electroanal. Chem. 220:351 (1987).

178. R. E. Noftle and D. Pletcher, J. Electroanal. Chem. 227:229 (1987).

179. For a general discussion of factors that can influence crystal growth, see L. Addadi, Z. Berkovitch-Yellin, I. Weissbuch, J. van Mil, L. J. W. Shimon, M. Lahav, and L. Leiserowitz, Angew. Chem. Int. Ed. Engl. 24:466 (1985).

180. M. Lamache and R. Najean, Electrochim. Acta 29:273 (1984).

181. M. Lamache, S. Wuryanto, and F. Benhamou, Electrochim. Acta 30:817 (1985).

182. M. Lamache, S. Wuryanto, and F. Benhamou, Electrochim. Acta 29:1055 (1984).

183. M. Lamache, H. Menet, and A. Moradpour, J. Am. Chem. Soc. 104:4520 (1982).

184. B. A. Scott, S. J. LaPlaca, J. B. Torrance, B. D. Silverman, and B. Welber, J. Am. Chem. Soc. 99:6631 (1977).

185. K. Kuo, P. R. Moses, J. R. Lenhard, D. C. Green, and R. W. Murray, Anal. Chem. 51:745 (1979).

186. D. A. Skoog and D. M. West, *Analytical Chemistry*, Holt, Rinehart and Winston, New York, 1974, p. 147.

187. H. Karimi and J. Q. Chambers, J. Electroanal. Chem. 217:313 (1987).

188. G. Inzelt, R. W. Day, J. F. Kinstle, and J. Q. Chambers, J. Phys. Chem. 87:4592 (1983).

189. R. H. Boyd and W. D. Phillips, J. Chem. Phys. 43:2927 (1965).

190. J. G. Gaudiello, M. Almeida, T. J. Marks, W. J. McCarthy, J. C. Butler, and C. R. Kannewurf, J. Phys. Chem. 90:4917 (1986).

191. J. G. Gaudiello, H. O. Marcy, W. J. McCarthy, M. K. Moguel, C. R. Kannewurf, and T. J. Marks, Synth. Met. 15:115 (1986).

192. M. Almeida, J. Gaudiello, T. J. Marks, J. C. Butler, H. O. Marcy, and C. R. Kannewurf, Synth. Met. 21:261 (1987).

193. H. Djellab and F. Dalard, J. Electroanal. Chem. 221:105 (1987).

194. J. L. Kahl, L. R. Faulkner, K. Dwarakanath, and H. Tachikawa, J. Am. Chem. Soc. 108:5434 (1986).

195. E. A. Orthmann, V. Enkelmann, and G. Wegner, Makromol. Chem. Rapid Comm. 4:687 (1983).

196. B. L. Wheeler, G. Nagasubramanian, A. J. Bard, L. A. Schechtman, D. R. Dininny, and M. E. Kenney, J. Am. Chem. Soc. 106:7404 (1984).

197. T. M. Mezza, N. R. Armstrong, G. W. Ritter, II, J. P. Iafalice, and M. E. Kenney, J. Electroanal. Chem. 137:227 (1982).

198. A. A. Bright, A. F. Garito, and A. J. Heeger, Solid State Comm. 13:943 (1973).

199. P. M. Grant, R. L. Greene, G. C. Wrighton, and G. Castro, Phys. Rev. Lett. 31:1311 (1973).

200. P. Nielsen, A. J. Epstein, and D. J. Sandman, Solid State Commun. 15:53 (1974).

201. M. Sharp, Anal. Chim. Acta 85:17 (1976).

202. M. Sharp, Anal. Chim. Acta 59:137 (1972).

203. M. Sharp, Anal. Chim. Acta 54:13 (1971).

204. C. D. Jaeger and A. J. Bard, J. Am. Chem. Soc. 101:1690 (1979).

205. W. L. Wallace, C. D. Jaeger, and A. J. Bard, J. Am. Chem. Soc. 101:4840 (1979).

206. R. Metzger, J. Chem. Phys. 66:2525 (1977).

207. C. D. Jaeger and A. J. Bard, J. Am. Chem. Soc. 102:5435 (1980).

208. D. J. Sandman, G. D. Zoski, L. Samuelson, and W. A. Burke, Synth. Metals 4:249 (1982).

209. L. B. Wingard, Fed. Proc. Fed. Am. Soc. Exp. Biol. 42:288 (1983).

210. P. W. Carr and L. D. Bowers, *Immobilized Enzymes in Analytical and Clinical Chemistry*, Wiley, New York, 1980, p. 460.

211. A. E. G. Cass, G. Davis, G. D. Francis, H. A. O. Hill, W. J. Aston, I. J. Higgins, E. V. Plotkin, L. D. L. Scott, and A. P. F. Turner, Anal. Chem. 56:667 (1984).

212. L. C. Clark and C. Lyons, Ann. N.Y. Acad. Sci. 102:29 (1962).

213. G. G. Guilbault and G. O. Lubrano, Anal. Chim. Acta. 64:436 (1973).

214. J. J. Kulys, Biosensors 2:3 (1986).

215. W. J. Albery and P. N. Bartlett, J. Chem. Soc. Chem. Comm. 1984:234.

216. L. Michaelis and M. L. Menten, Biochem. Z. 49:333 (1913).

217. W. J. Albery, N. P. Bartlett, A. E. G. Cass, and K. W. Sim, J. Electroanal. Chem. 218:127 (1987).

218. W. J. Albery, P. N. Bartlett, A. E. G. Cass, D. H. Craston, B. G. D. Haggett, J. Chem. Soc. Farad. Trans. 1, 82:1033 (1986).

219. W. J. Albery, N. P. Bartlett, D. H. Craston, M. Bycroft, and C. P. Jones, European Patent Appl. No. 85308277.4 or GB84/28599.

220. W. J. Albery, N. P. Bartlett, M. Bycroft, D. H. Craston, and B. J. Driscoll, J. Electroanal. Chem. 218:119 (1987).

221. J. J. Kulys, M. V. Pesliakiene, and A. S. Samalius, Bioelectrochem. Bioenerg. 8:81 (1981).

222. N. K. Cenas and J. J. Kulys, Bioelectrochem. Bioenerg. 8:103 (1981).

223. W. J. Albery, P. N. Bartlett, and D. H. Craston, J. Electroanal. Chem. 194:223 (1985).

224. W. J. Albery and P. N. Bartlett, J. Electroanal. Chem. 194:211 (1985).

225. W. J. Albery, P. N. Bartlett, and A. E. G. Cass, Phil. Trans. R. Soc. Lond. B. 316:107 (1987).

226. J. J. Kulys and N. K. Cenas, Biochim. Biophys. Acta 744:57 (1983).

227. K. McKenna and A. Brajter-Toth, Anal. Chem. 59:954 (1987).

228. F. Bergmann and L. Levene, Biochim. Biophys. Acta 429:672 (1976).

229. L. Greenlee and P. Handler, J. Biol. Chem. 239:1090 (1964).

230. M. G. Boutelle, C. Stanford, M. Fillenz, W. J. Albery, and P. N. Bartlett, Neurosci. Lett. 72:283 (1986).

231. J. J. Kulys and A. S. Samalius, Bioelectrochem. Bioenerg. 10:385 (1983).

232. B. Honigmann, H.-U. Lenne, and R. Schrodel, Z. Kristallogr. Kristallgeom. Kristallphys. Kristallchem. 122:185 (1965).

233. C. J. Brown, J. Chem. Soc. A 2488 (1968).

234. C. J. Brown, J. Chem. Soc. A 2494 (1968).

235. J. Simon and J.-J. Andre, *Molecular Semiconductors*, J. M. Lehn and C. W. Rees, Eds., Springer-Verlag, Berlin, 1985, p. 73.

236. A. J. Appleby and M. Savy, Electrochim. Acta 21:567 (1976).

237. H. Tachikawa and L. R. Faulkner, J. Am. Chem. Soc. 100:4379 (1978).

238. F.-R. Fan and L. R. Faulkner, J. Am. Chem. Soc. 101:4779 (1979).

239. H. T. Tien, Bioelectrochem. Bioenerg. 13:299 (1984).

240. H. T. Tien, J. Phys. Chem. 88:3172 (1984).

241. T. Nakamura, F. Takei, M. Tanaka, M. Matsumoto, T. Sekiguchi, E. Manda, Y. Kawabata, and G. Saito, Chem. Lett. 323 (1986).

242. Y. Kawabata, T. Nakamura, M. Matsumoto, M. Tanaka, T. Sekiguchi, H. Komizu, E. Manda, and G. Saito, Synth. Met. 19:663 (1987).

243. T. Nakamura, M. Matsumoto, F. Takei, M. Tanaka, T. Sekiguchi, E. Manda, and Y. Kawabata, Chem. Lett. 709 (1986).

244. M. Matsumoto, T. Nakamura, F. Takei, M. Tanaka, T. Sekiguchi, M. Mizuno, E. Manda, and Y. Kawabata, Synth. Met. 19:675 (1987).

245. J. S. Facci, P. A. Falcigno, and J. M. Gold, Langmuir 2:732 (1986).

246. M. Fujihara and T. Araki, Bull. Chem. Soc. Jpn. 59:2375 (1986).

247. M. Fujihara and S. Poosittisak, J. Electroanal. Chem. 199:481 (1986).

248. H. Daifuku, K. Aoki, K. Tokuda, and H. Matsuda, J. Electroanal. Chem. 183:1 (1985).

249. H. Daifuku, I. Yoshimura, I. Hirata, K. Aoki, K. Tokuda, and H. Matsuda, J. Electroanal. Chem. *199*:47 (1986).
250. K. Aoki, K. Tokuda, and H. Matsuda, J. Electroanal. Chem. *199*:69 (1986).
251. A. Girlando and G. Sandona, Surf. Sci. *160*:87 (1985).

Author Index

Numbers in parentheses are reference numbers and indicate that an author's work is referred to although his name is not cited in the text. Underlined numbers give the page on which the complete reference is listed.

Abel, A. U., 165(295), <u>180</u>
Abruna, H. D., 56(128), <u>59</u>
  (132, 133), 60(133), 73(161),
  80(172), <u>86</u>, <u>87</u>, <u>88</u>
Acampora, L. A., 197(50), <u>301</u>
Adams, R. H., 72(159), 116(107),
  118(126, 127, 129), 127(126),
  <u>88</u>, <u>172</u>, <u>173</u>
Addadi, L., 245(179), 259(179),
  <u>308</u>
Adeloju, S. B., 24(33), 32(33), <u>82</u>
Akaponkw, U., 115(103), <u>172</u>
Akiba, V., 71(156), <u>88</u>
Alacer, L., 225(104), 234(104), <u>304</u>
Alawi, M. A., 130(203), <u>176</u>
Albery, W. J., 95(11), 99(11, 19),
  100(11), 101(32), 103(45),
  105(45), 109(11), 114(32), 122
  (32), 125(32), 152(243), 285
  (215), 286(215, 217, 218, 219,
  220, 223, 225, 236), 287(217,
  218, 219, 220, 223), 288(223,
  224, 225), 289(223), 291(225),

292(225, 230), <u>168</u>, <u>169</u>, <u>178</u>,
  <u>310</u>, <u>311</u>
Alexander, P. W., 115(103), <u>172</u>
Almeida, M., 235(148), 241(148),
  263(190, 192), 265(190), <u>306</u>,
  <u>309</u>
Amatore, C., 207(84), <u>303</u>
Amberger, E., 232(165), <u>307</u>
Anderson, A. G., 182(11, 12),
  <u>299</u>
Anderson, B. A., 230(129),
  232(129), <u>305</u>
Anderson, E. A., 182(14), <u>300</u>
Anderson, J. E., 119(135), 129
  (135), <u>173</u>
Anderson, J. L., 98(15), 103
  (42, 43), 112(82), 119(135,
  136, 137), 129(135, 136, 137),
  130(197), 156(197), 154(253,
  254), 155(253, 254), <u>168</u>, <u>169</u>,
  <u>171</u>, <u>176</u>, <u>178</u>
Anderson, J. R., 24(34), <u>82</u>
Ando, T., 62(137), <u>87</u>

Andre, J.-J., 182(2a), 293(235), 299, 311
Andrews, R. W., 123(176), 175
Ang, K. P., 99(17, 18), 101(26, 27, 33), 106(48), 109(27), 112(33, 70), 114(33), 115(102), 122(17, 33), 123(33, 174, 175), 124(174), 125(33, 174), 126(18, 33, 174), 128(133), 129(26), 130(26, 27, 210), 131(102), 132(102), 133 (102), 134(27, 210), 135(27), 136(27), 137(213, 214), 144 (102, 233), 145(102), 147(233), 155(214), 156(213), 157(268), 167(33, 233), 168, 169, 172, 175, 176, 177, 179
Anson, F. C., 5(37), 9(12), 10(12), 35(70), 53(3), 68(150), 69(153), 73(164), 74(164), 81, 84, 87, 88
Anzai, H., 229(116), 232(164), 233(170), 244(116), 305, 307, 308
Aoki, K., 103(44), 105(44), 295 (248, 249, 250), 169, 311, 312
Appleby, A. J., 293(236), 311
Appelqvist, R., 157(264), 179
Araki, T., 295(246), 311
Aren, K., 123(178), 175
Ariel, M., 114(96), 117(118), 122(169), 125(169, 183), 171, 172, 175
Armentrout, D. N., 119(134), 173
Armstrong, N. R., 266(197), 309
Artes, R., 235(171), 308
Arthur, P. Jr., 187(24), 300
Ashton, W. J., 157(267), 179
Asslett, L. H., 112(73), 170
Astafeva, V. V., 23(31), 82
Aston, W. J., 284(211), 310
Astruc, D., 199(66), 302
Aumiller, A., 231(153), 246(153), 307
Austin, D. S., 116(110), 172
Azevedo, L. J., 218(93), 229(118), 233(118), 303, 305

Baldwin, R. D., 77(170, 171), 79(171), 88
Baldwin, R. P., 49(83), 52(124), 58(130), 63(130), 64(130, 140), 119(149, 150), 120(149, 150), 150(239), 151(239), 157(269, 270, 271), 158(269), 159(269), 77(168), 84, 86, 87, 88, 174, 177, 179
Baltzer, P., 231(156), 307
Bard, A. J., 9(11), 72(158), 73(162, 163), 87(155), 157 (262), 160(262), 237(135, 136, 137, 138, 139), 238(136), 266(196), 270(204, 205), 271 (204), 272(207), 273(204), 274(207), 275(204), 276(135, 136), 81, 85, 87, 88, 179, 306, 309
Barros, A. A., 39(96), 85
Bartlett, P. N., 285(215), 286 (215, 217, 218, 219, 220, 223, 225, 230), 287(217, 218, 219, 220, 223), 288(223, 224, 225), 289(223), 291(225), 292(225, 230), 310, 311
Batail, P., 190(27), 300
Bately, G. E., 2(1), 122(167), 81, 174
Batrakov, V. V., 7(4), 81
Batycka, H., 46(113, 114), 86
Bechgaard, K., 187(25), 189(25), 197(34), 201(34), 226(110, 111, 112, 113), 231(154), 233(112), 300, 301, 304, 307
Bednarkiewicz, E., 45(112), 86
Belew, W. L., 26(51), 83
Benadikova, H., 39(95), 85
Bender, C. J., 214(90), 303
Bender, K., 197(52), 229(122), 230(125), 232(125, 163, 166), 302, 305, 307
Benhamou, F., 232(161), 250 (181, 182), 274(161), 274 (181, 182), 307, 308

AUTHOR INDEX

Beno, M. A., 182(6), 229(118, 120), 230(124, 128), 232(124, 128, 168), 233(118, 120), 299, 305, 307
Benson, R. E., 213(88), 303
Bentley, W. H., 190(28), 300
Benton, C. S., 35(72), 84
Berendzen, J., 228(114), 231(114), 305
Berger, T. A., 120(156), 125(181), 174
Bergmann, F., 291(228), 311
Berkovitch-Yellin, Z., 245(179), 259(179), 308
Bermejo, E., 39(118a), 86
Bernsteiner, A., 42(106a), 85
Bezrukova, 198(61), 302
Bigelow, R. W., 182(15), 218(95), 236(15), 242(15), 246(15), 257(15), 300, 304
Bird, C. L., 201(75), 303
Bird, E., 140(225), 177
Bixler, J. W., 154(256), 155(256), 178
Blackman, G. S., 229(120), 233(120), 305
Blaedel, W. J., 11(18), 95(8), 107(497), 113(87), 116(276), 82, 168, 169, 179
Bloch, A. N., 203(79), 231(158), 261(158), 303, 307
Bockris, J. O'M., 243(175), 308
Boekelheide, V., 201(76, 77), 303
Boguslaski, R. C., 161(276), 179
Bollett, C., 116(108), 172
Bonakder, M., 35(71), 36(79), 37(71), 43(109), 44(109), 50(119), 65(144), 84, 85, 86, 87
Bond, A. M., 24(33), 32(33), 34(68), 114(98), 122(79), 107(54), 112(54, 77, 78, 79, 80, 81, 82, 83), 113(54), 127(54), 131(54), 133(54, 78), 138(54), 149(54), 153(54), 125(187), 151(241), 154(256), 155(256), 156(260), 82, 84, 171, 172, 175, 178

Bos, P., 113(88, 92), 114(93), 171
Boublikova, P., 40(99, 102), 85
Boutelle, M. G., 286(230), 292(230), 311
Bowers, L. D., 161(274), 164(274), 284(210), 179, 310
Boyd, R. H., 260(189), 281(189), 309
Brajter-Toth, A., 67(148), 286(227), 291(227), 87, 310
Bratin, K., 115(104), 131(104), 133(104), 137(215), 142(215), 148(234, 236), 172, 176, 177
Braun, H., 25(37), 82
Breitwieser, C., 116(116), 172
Breslow, R., 197(35, 45), 201(35), 218(92), 219(45), 301, 303
Brett, C. M. A., 95(11), 99(11), 100(11), 101(32), 109(11), 114(32), 122(32), 125(32), 168, 169
Brewster, J. D., 130(197), 156(197), 176
Briggs, M. H., 24(33), 32(33), 112(80), 82, 171
Bright, A. A., 268(198), 309
Brinkman, A. T., 116(113), 172
Brinkman, U. A., 152(242), 178
Brodbelt, J. S., 138(219), 139(219), 177
Brown, A. P., 9(12), 10(12), 35(70), 81, 84
Brown, C. J., 293(233), 293(234), 311
Brown, S. D., 125(186), 175
Bruce, J. A., 197(44), 301
Bruckenstein, S., 238(140), 306
Bruins, C. H. P., 103(40), 169
Brunt, K., 103(40), 127(192), 169, 176
Bruntlet, C. S., 115(104), 131(104), 133(104), 172
Bryden, W. A., 203(79), 303
Buchberger, W., 116(116), 172
Buffle, J., 122(173), 175

Buravov, L. I., 230(123), 232(123), 305
Burke, W. A., 275(208), 309
Butler, J. C., 263(190), 265(190), 309
Buttry, D. A. J., 239(143), 306
Bycroft, M., 286(219, 220), 287(219, 220), 310

Cabelka, T. D., 116(110), 172
Cabral, J. O., 39(96), 85
Calabrese, J. C., 182(13, 15), 218(95, 96), 236(15), 242(15), 246(15), 257(15), 300, 304
Caliguri, E. J., 118(128), 173
Caohuu, T., 163(289), 167(289), 169
Cape, T. W., 182(16), 300
Cardoza, B., 163(289), 167(289), 180
Carlson, K. D., 182(6), 229(120), 230(124, 128), 232(124, 128, 168), 233(120), 299, 305, 307
Carneiro, K., 226(109, 111), 229(117), 304, 305
Carr, P. W., 161(274), 164(274), 284(210), 179, 310
Carter, M. T., 238(139), 306
Cass, A. E. G., 157(267), 284(211), 286(217, 218, 225), 287(217, 218), 288(225), 291(225), 292(225), 179, 310
Castner, J. F., 157(265), 179
Castro, G., 268(199), 309
Caude, M., 116(108), 172
Caudill, W. L., 117(124), 119(139), 143(226), 144(226), 156(258, 259), 173, 177, 178
Cenas, N. K., 286(222), 287(222), 291(226), 310
Chambers, C. A., 43(107), 17
Chambers, C. A. H., 126(189), 175
Chambers, J. Q., 221(99), 260(187, 188), 304, 308, 309
Chan, H. K., 121(159), 174

Chance, R. R., 182(10b), 183(10b), 268(10b), 299
Chaney, E. N., 49(83), 52(124), 84, 86
Chappell, J. S., 203(79), 303
Cheek, G. T., 65(143), 87
Cheng, H. Y., 39(98), 117(121), 85, 173
Chesney, D. J., 119(135, 136), 129(135, 136), 173
Chestnut, D. B., 187(24), 300
Chia, V. K. F., 35(73), 84
Chiang, L. Y., 231(158), 261(158), 307
Chiang, T. C., 223(101), 224(101), 235(101), 245(101), 304
Chiavari, G., 130(200), 176
Chin, D. T., 95(13), 168
Chittapeddi, S. R., 218(96), 304
Christensen, K., 58(130), 63(130), 64(130), 86
Christie, J. H., 125(188), 175
Clark, L. C., 284(212), 310
Clemens, A. H., 165(294), 180
Colin, H., 127(191), 175
Colwell, J. A., 164(293), 165(293), 180
Compton, R. N., 197(38), 301
Concialini, V., 130(200), 176
Cope, D. K., 155(255), 178
Copeland, T. R., 121(165), 174
Copps, P. T., 182(6), 229(120), 230(124), 232(124), 233(120), 299, 305
Corbett, J. D., 183(18), 300
Coulon, C., 228(115), 233(115), 305
Cowan, D. O., 182(3a, 3b), 197(39), 198(53), 203(79), 299, 301, 302, 303
Cowlard, F. C., 117(125), 118(125), 173
Cox, J. A., 68(151), 69(151), 160(272), 87, 179
Crabtree, G. W., 182(6), 230(124), 232(124), 299, 305

Craston, D. H., 286(218, 219, 220, 223), 287(218, 219, 220, 223), 288(223), 289(223), 310
Crowley, J. I., 190(29), 300
Curran, D. J., 107(49), 119(133, 143, 144), 169, 173, 174

Daifuku, H., 295(248, 249), 311, 312
Dalard, F., 266(193), 309
Dalgard, L., 154(250), 178
Damaskin, B. B., 7(4, 6), 46(115), 81, 86
Darnall, D., 77(169), 88
Davenport, 116(114), 172
Davis, E. A., 205(80), 303
Davis, G., 157(267), 284(211), 179, 310
Davis, J. M., 107(56), 108(56), 170
Day, R. W., 260(188), 309
de Abreu, M. A., 156(261), 178
Dean, J. A., 25(40), 83
Debowski, J., 113(90), 171
De Castro, E. S., 66(147), 67(147), 69(147), 87
Deitz, K., 230(125), 232(125), 305
De Jong, H., 144(230), 177
deJong, H. G., 106(47), 169
Delahay, P., 17(27), 82
Denisovich, L. I., 198(61, 68), 302
Deshmukh, B. K., 37(91), 43(109), 44(109), 84, 85
deVries, W. T., 123(179), 175
Dewald, H. D., 101(30), 107(51), 114(30), 122(30, 170), 125(30, 185), 144(232), 169, 170, 175, 177
Dieker, J. H., 121(158), 174
Diekmann, J., 213(88), 303
Dietz, K., 232(162), 307
Ding, X.-D., 148(234), 177
Dininny, D. R., 266(196), 309
Dixon, D. A., 182(14), 300
Djellab, H., 266(193), 309

Donega, K. M., 230(129), 232(129), 305
D'Oruzio, P. A., 165(294), 180
Dreiling, R., 118(126), 127(126), 173
Drickamer, H. G. J., 190(28), 300
Driscoll, B. J., 286(220), 287(220), 310
Duke, P. D., 109(64), 116(109), 170, 172
Durst, R. A., 53(127), 108(63), 131(206), 86, 170, 176
Dwara Kanath, K., 266(194), 309

Eaton, D. F., 182(11, 12), 299
Ecker, P., 235(171), 308
Edelstein, N., 198(58), 302
Edgerton, T. R., 130(202), 176
Edwards, P. P., 206(81), 303
Ege, D., 73(163), 88
Ehrenfreund, E., 182(4), 299
Elbicki, J. M., 96(22), 100(22), 104(22), 148(237, 238), 149(237), 150(237), 168, 177
Elbl, K., 198(56), 302
Elferink, F., 114(99), 172
Elving, P. J., 117(117), 172
Emge, T. J., 182(6), 230(124), 232(124), 299, 305
Endres, H., 197(51), 218(93), 229(122), 230(125), 232(125, 163, 166), 302, 303, 305, 307
Engler, E. M., 182(7), 197(38), 221(99), 228(114), 229(119, 121), 231(114), 233(119, 121), 299, 301, 304, 305
Engstrom, R. C., 66(145), 119(145, 146), 174
Enkelmann, V., 198(55), 235(173), 248(55, 173), 266(195), 302, 309
Enoki, T., 182(8), 299
Epstein, A. J., 182(14, 15), 185(22a), 212(22a), 218(95, 96),

219(97), 225(106), 236(15, 106), 242(15), 246(15), 251(15), 268 (200), 278(200), 300, 304, 309
Erk, P., 231(153), 246(153), 307
Equey, J. F., 73(163), 88
Eskilsson, H., 21(30), 25(30), 82
Espenschied, M. W., 56(129), 86
Ewing, A. G., 53(127), 143(226), 144(226), 156(259), 86, 177, 178

Facci, J. S., 295(245), 311
Fagan, D. T., 119(152), 174
Fagan, P. J., 182(13), 236(151), 242(151), 246(151), 257(151), 300, 307
Falat, L., 39(98), 117(121), 85, 173
Falcigno, P. A., 295(245), 311
Fan, F.-R., 293(238), 311
Fan, F. R. F., 72(158), 88
Farias, P. A., 26(43), 29(43), 83
Farias, P. A. M., 10(13), 13(21), 31(60), 32(60), 35(75), 36(78, 80), 37(21, 73, 75, 78), 48(21, 75), 49(21), 81, 82, 83, 84, 85
Farlee, R. D., 190(31), 214(31), 216(31), 300
Faulkner, L. R., 9(11), 53(126), 266(194), 293(237, 238), 81, 86, 309, 311
Feher, Z., 113(89), 171
Fenn, R. J., 119(133), 173
Ferraris, J., 182(3a), 299
Fillenz, M., 286(230), 292(230), 311
Finke, R. G., 201(76), 303
Firestone, M. A., 230(128), 232 (128), 305
Fischer, U., 165(295), 180
Fleet, B., 101(23, 24, 29), 109 (23), 112(70, 77), 117(120), 118(23), 120(23), 122(29), 128 (23), 130(24), 133(24), 163(289), 167(289), 168, 169, 171, 172, 180
Fleischmann, M., 243(176), 308
Flora, C. J., 20(28), 23(28), 82

Florence, T. M., 26(51), 31(66), 114(100), 121(100), 122(167), 83, 172, 174
Flores, J. R., 42(106), 85
Fogg, A. G., 35(76), 39(96), 113(84), 121(159), 84, 85, 171, 174
Forcier, G., 148(234), 177
Forte, A. J., 130(199), 176
Fosdick, L. E., 154(254), 155 (254), 178
Foxman, B. M., 197(50), 207(85), 301, 303
Francis, G. D., 157(267), 284 (211), 179, 310
Frank, J., 130(198), 176
Frei, R. W., 113(88, 92), 114 (93), 116(113), 152(242), 171, 172, 178
French, W. G., 43(108a), 85
Friedrich, M., 32(64), 83
Friedrich, P., 235(171), 308
Frieha, B. A., 11(20), 37(91), 39(97), 43(108), 50(121), 51(121, 122, 123), 52(123), 101(31), 122(31), 125(31), 127(90), 82, 84, 85, 86, 169, 175
Frieha, B., 157(263), 179
Fritz, H. P., 235(171), 308
Frommer, J. E., 182(10b), 183 (10b), 268(10b), 299
Frumkin, A. N., 7(6), 81
Fuchs, H., 232(165), 307
Fujihara, M., 295(246, 247), 311
Fukanaga, T., 197(42), 199(42), 301

Galiasatos, C., 66(147), 67(147), 69(147), 87
Gallois, B., 228(115), 233(115), 305
Galus, Z., 118(129), 173
Gamache, P. H., 140(225), 177
Gammelgaard, B., 24(34), 82

AUTHOR INDEX

Gane, P. A. C., 233(132), 237(132), 306
Gardea, J., 77(169), 88
Garito, A. F., 182(4), 199(72), 213(89), 268(198), 299, 303, 309
Garrito, A. F., 185(22b), 212(22b), 300
Gatford, C., 25(36), 82
Gaudiello, J. G., 260(190), 263 (190, 191), 265(190), 281(190), 309
Gebauer, H., 235(171), 308
Gebert, E., 182(16), 300
Gehron, M. J., 67(148), 87
Geiger, W. E., 198(57), 302
Geiser, U., 230(128, 129), 232, (128, 129, 169), 305, 308
Geno, P. W., 77(168), 88
Gerhardt, G. A., 72(159), 88
Geserich, H. P., 235(172), 308
Ghosh, P. K., 73(162, 163), 88
Giddings, J. C., 107(56), 108(56), 170
Gillson, J. L., 197(37), 198(37), 200(37), 210(37), 234(37), 237 (37), 301
Giner, J., 116(106), 172
Girlando, A., 296(251), 312
Glauert, M. B., 99(16), 168
Go, W., 125(184), 175
Gogu, E., 197(51), 229(122), 232(163, 166), 302, 305, 307
Gold, J. M., 295(245), 311
Golimowski, J., 31(58), 32(58), 83
Goncalves, J. M., 207(84), 303
Goncalves, M., 122(173), 175
Gorenc, B., 50(120), 86
Gorton, L., 157(266), 179
Goto, M., 201(73), 303
Goton, L., 157(264), 179
Gough, D. A., 161(277, 280), 162 (285), 179, 180
Grabaric, B. S., 125(187), 175
Graja, A., 199(67), 302
Grant, P. M., 268(199), 309
Gratzl, M., 100(21), 136(21), 168

Green, D. C., 197(38), 221 (99), 252(185), 301, 304, 308
Greene, B., 58(131), 75(131), 76(131), 86
Greene, G., 123(177), 175
Greene, R., 228(114), 231(114), 305
Greene, R. L., 182(7), 229(119), 233(119), 268(199), 299, 305, 309
Greenlee, L., 291(229), 311
Grub, J., 198(64), 302
Guadalupe, A. R., 56(128), 59 (132, 133), 60(133), 73(161), 80(172), 86, 87, 88
Gubin, S. P., 198(61, 62), 302
Guggenheim, H. J., 226(107), 304
Guibault, G. G., 161(275), 179
Guilbault, G. G., 284(213), 310
Guiochon, G. A., 127(191), 175
Gunasingham, H., 95(12), 97(12), 98(12), 99(12, 17), 100(12), 101(24, 25, 26, 27, 29, 33), 105(12), 106(12, 48), 108(25), 109(12, 25, 27), 112(33, 70), 113(12), 114(33), 115(102), 117(119), 120(117), 122(17), 123(33, 174, 175), 124(174), 125(33, 174), 126(18, 33, 174), 128(25, 33), 129(26), 130 (24, 26, 27, 210), 131(102), 132(102), 133(102), 134(24, 27, 210), 135(25, 27), 136 (27), 137(213, 214), 144(102, 233), 145(102), 147(233), 155(214), 156(213), 157(268), 162(286), 167(233), 168, 169, 172, 175, 176, 177, 179, 180
Gupta, Das S., 163(289), 167 (289), 180
Guthrie, E. J., 119(138), 155 (257), 173, 178
Guy, R. D., 72(160), 88

Haen, P., 228(114), 231(114), 305

Haggett, B. G. D., 286(218), 287 (218), 310
Halbert, M. K., 157(270), 179
Hall, L. N., 182(6), 230(124), 232 (124), 299, 305
Hameka, H. F., 161(278), 179
Hamon, J.-R., 199(66), 302
Hanawa, H., 62(136), 87
Handler, P., 291(229), 311
Hanekemp, H. B., 106(46, 47), 108(46), 113(46, 88, 92), 114 (93), 169, 171
Hansen, E. H., 111(68, 69), 112 (68), 120(157), 170, 174
Haraldsson, C., 21(30), 25(30), 82
Harrow, J. F., 112(71), 170
Hawkridge, F. M., 157(265), 179
Hawthorne, M. F., 198(60a), 302
Hedges, K. M., 206(82), 303
Heeger, A. J., 182(4), 185(22b), 199(72), 212(22b), 268(198), 299, 300, 303, 309
Heineman, W. R., 37(90), 38(38), 43(110), 66(147), 67(147), 69 (147), 104(38), 114(38), 122(38), 127(193), 160(273), 84, 85, 87, 169, 176, 179
Heinen, I., 197(51), 229(122), 232 (163, 166), 301, 305, 307
Heinze, J., 198(55), 248(55), 301
Henning, I., 230(125), 232(125), 305
Henning, T. P., 237(135, 136), 238 (136), 276(135, 136), 306
Hepler, B. R., 117(123), 173
Herath, V. S., 157(268), 179
Herbstein, F. H., 184(20), 190(20), 207(20), 300
Heritage, I. D., 112(81, 82), 171
Hernandez, L., 49(118a), 86
Hertler, W. R., 213(88), 303
Hill, H. A. O., 157(267), 284(211), 179, 310
Hillman, A. R., 243(177), 308
Hiller, M., 197(51), 229(122), 232(163), 302, 305, 307
Hilti, B., 231(156, 157), 307

Higgens, I. J., 157(267), 284 (211), 179, 310
Hikume, M., 164(292), 180
Hirata, I., 295(249), 312
Hirata, Y., 103(41), 169
Hiroshima, O., 129(195), 176
Hojabri, H., 150(240), 178
Holloway, J. D. L., 198(51), 302
Honda, K., 201(73), 303
Honigmann, B., 293(232), 311
Houck, G. P., 117(124), 173
Hough, D. L., 165(294), 180
Horvoi, G., 113(89), 171
Howell, J. O., 119(139), 154 (252), 156(258), 207(84), 173, 178, 303
Hubbard, A. T., 13(23), 35(72, 73), 201(78), 253(78), 82, 84, 303
Huber, C. O., 116(112, 115), 172
Huber, E. W., 66(147), 67(147), 69(147), 87
Huber, J. F. K., 133(208), 176
Hudson, H. A., 112(83), 114(98), 171, 172
Hung, M. A., 40(101), 85
Hughes, S., 152(245, 246), 178
Hunig, S., 197(40), 231(153), 246(153), 301, 307
Hutchins, L. D., 119(148), 174
Hutchins-Kumar, L. D., 138 (222), 140(222), 141(222), 177
Hu, I. F., 14(24a), 119(152), 82, 174
Hui, B. S., 116(112), 172
Humphrey, D. W., 130(201), 176

Ikariyama, Y., 160(273), 179
Ikenoya, S., 129(195), 176
Inabe, T., 201(73), 303
Inokuchi, H., 182(8), 232(167), 299, 307
Inzelt, G., 260(188), 309

Ishiguro, T., 229(116), 232(307), 244(116), 305, 307
Iuaska, A., 121(160), 174
Iwata, M., 226(107), 304
Izutsu, K., 62(136, 137), 87

Jacobsen, C. S., 226(111, 113), 231(154), 304, 307
Jaeger, C. D., 237(137, 138), 270(204, 205), 271(204), 272(207), 273(204), 274(207), 306, 309
Jagner, D., 21(30), 25(30), 123(178), 125(185), 82, 175
Janata, J., 112(71, 162), 121(162), 143(162), 170, 174
Jansen, S., 197(50), 301
Jarbawi, T. B., 37(90), 43(110), 84, 85
Josefson, M., 123(178), 175
Jayaweera, P., 107(53), 170
Jehring, H., 44(111), 86
Jelen, F., 40(99), 85
Jerome, D., 182(5), 226(110, 112), 299, 304
Jhaveri, S. S., 80(172), 88
Jin, Z., 131(204), 133(204), 176
Johansson, G., 157(264, 266), 161(282), 162(282), 179
Johnson, D. C., 9(14), 98(14), 107(54), 112(54), 113(54), 116(54), 123(176), 127(54), 131(54), 133(54), 138(54), 149(54), 152(244, 245, 246, 247, 248, 249), 153(54, 249), 182(9), 168, 170, 172, 175, 178, 299
Johnson, J. D., 199(65), 302
Jones, C. P., 286(219), 287(219), 310
Johnston, D. C., 231(158), 261(158), 307
Jones, M. T., 197(50), 301
Jones, R. D., 112(82), 171
Jones, S., 143(226), 144(226), 156(259), 177, 178

Jordan, J., 95(6), 167
Jorgenson, J. W., 119(138), 143(228, 229), 144(229), 146(229), 155(257), 173, 177

Kacemi, K. E., 234(133b), 262(133b), 306
Kahl, J. L., 266(194), 309
Kai, Y., 190(32), 217(32), 301
Kalcher, K., 64(141), 75(166, 167), 78(166), 87, 88
Kakutani, T., 11(14), 81
Kalvoda, R., 36(74, 77), 39(95), 46(116, 117), 49(74), 84, 85, 86
Kamau, G. N., 120(155), 174
Kaminskii, V. F., 230(126), 305
Kanatzidas, M. G., 235(148), 241(148), 306
Kannewurf, C. R., 235(148), 241(148), 263(190, 191, 192), 265(190), 306, 307, 309
Kapauan, A. F., 112(74), 170
Karimi, H., 260(187), 308
Kartsovnik, M. V., 230(123, 127), 232(123, 127), 305
Karube, I., 164(292), 180
Karweik, D., 14(24a), 82
Kasai, N., 190(32), 217(32), 301
Katayma, C., 232(162), 307
Kathirgamanathan, P., 233(132), 234(133a), 235(134), 237(131, 132, 134), 249(133a), 306
Kato, M., 235(145, 147), 240(145, 147), 306
Kato, R., 229(116), 224(116), 305
Kaufman, F. B., 197(38), 221(99), 301, 304
Kawabata, Y., 295(241, 243, 244), 311
Kawabe, K., 129(195), 176
Kazee, B., 119(154), 120(154), 174

Keller, H. J., 197(51), 229(122, 130), 230(125), 232(125, 163, 166), 235(130, 172), 302, 305, 306, 307, 308
Keller, R., 118(127), 173
Kemula, W., 90(2), 113(90), 114 (95), 167, 171
Kenney, M. E., 266(196, 197), 309
Khomenko, A. G., 231(155), 307
Khoo, S. B., 137(213, 214), 155 (214), 156(213), 176
Khouzami, F., 198(59), 302
Khoyetsian, R., 163(289), 167(289), 180
Kinstle, J. F., 260(188), 309
Kiesslich, G., 197(40), 301
Kihara, S., 122(168), 174
Kimla, A., 113(85), 171
Kisse, T. R., 161(276), 179
Kissinger, P. T., 38(38), 101(36, 37), 104(38), 106(36), 109(37), 112(75), 114(38), 115(104), 118(126), 122(38), 127(36, 37, 126, 193), 130(37), 131(104), 132(104), 137(215), 138(216, 218, 219, 233), 140(224), 142 (215), 169, 170, 172, 173, 176, 177
Kitaigorodsky, A. I., 184(21), 190(21), 300
Klasinc, L., 207(84), 303
Klatt, L. N., 95(8), 168
Klebe, G., 231(153), 246(153), 307
Klots, C. E., 197(38), 301
Knecht, L. A., 119(138), 155(257), 173, 178
Ko, W. H., 165(296), 180
Kobayashi, A., 229(116), 232(164), 244(116), 305, 307
Kobayashi, H., 229(116), 232(164), 244(116), 305, 307
Kobayashi, N., 235(145, 146), 240(145, 146), 241(146), 306
Koch, P., 229(130), 235(130), 235(172), 306, 308
Kochi, J. K., 207(84), 303
Koelle, U., 198(59), 302

Koen, J. G., 133(208), 176
Kok, G. B., 197(39), 198(53), 301, 302
Kok, U. T., 152(242), 178
Kok, W. T., 116(113), 172
Kolle, U., 198(64), 302
Kolthoff, I. M., 113(86), 171
Komizu, H., 295(242), 311
Komy, Z., 24(35), 82
Kononovich, P. A., 230(123, 127), 231(155), 232(123, 127), 305, 307
Kopanica, M., 39(26, 94), 49 (26, 94), 50(94), 52(125), 82, 85, 86
Korfhage, K. M. 157(268), 158 (268), 159(268), 179
Korotov, A. P., 199(68), 302
Kosta, L., 121(161), 174
Kostka, K. L., 230(128), 232 (128), 305
Kounaves, S. P., 122(173), 175
Koval, C., 35(70), 84
Kovnat, L., 182(16), 300
Kowalski, B. R., 125(186), 175
Kowalski, Z., 114(97), 171
Kreiger, C., 198(56), 302
Kreuzig, F., 130(198), 176
Krohnke, C., 198(55), 235(173), 248(55, 173), 302, 308
Krstulovic, A. M., 127(191), 175
Krull, I. S., 148(234, 235, 236), 177
Krusic, P. J., 182(14), 300
Kryger, L., 58(130), 63(130), 64(130, 140), 125(185), 86, 87, 175
Kuawana, T., 119(152), 174
Kubiak, W., 114(97), 171
Kublik, Z., 45(112), 114(101), 86, 172
Kubo, T., 164(292), 180
Kuhn, A. T., 201(75), 303
Kulesza, P. J., 68(151), 69(151), 87
Kulkarni, K. R., 160(272), 179

Kulys, J. J., 161(279), 285(214), 286(221, 222, 231), 287(221, 222), 289(214, 221), 291(226), 179, 310, 311
Kuo, K., 252(185), 308
Kutner, W., 75(165), 113(90), 114(95), 88, 171
Kuwana, T., 14(24a), 43(108a), 117(122), 119(141, 142, 153, 154), 120(154), 82, 85, 172, 173, 174
kVasil'kov, A. P., 199(68), 302
Kwak, J. F., 230(129), 232(129, 169), 305

LaCourse, W. R., 148(235, 236), 177
Lafalice, J. P., 266(197), 309
Laganis, E. D., 201(76), 303
Lagier, R., 182(7), 229(119), 233(119), 299, 305
Lagowski, J. J., 199(68), 302
Lahav, M., 245(179), 249(179), 308
Lai, C. N., 201(78), 253(78), 303
Lam, N. K., 39(94), 49(94), 50(94), 85
Lamache, M., 231(160), 232(161), 234(133b), 250(180, 181, 182, 183), 262(133b), 274(160, 161, 180, 181, 182, 183), 306, 308
Langlais, P., 140(225), 177
Lankelma, J., 187(55), 170
LaPlaca, S. J., 190(27), 221(100), 250(184), 300, 304, 308
LaQue, F. L., 165(297), 180
Laser, D., 117(118), 172
Latimer, G. W., 25(41), 26(41), 83
Laukhin, V. N., 230(123, 127), 231(155), 232(123, 127), 305, 307
Laversanne, K., 228(115), 233(115), 305
Lavin, A. G., 150(240), 178
Laviron, E., 8(7), 9(7, 9, 10), 11(15), 81

Lee, J. K., 43(107), 126(189), 85, 175
Lee, V. Y., 182(7), 214(91), 229(119, 121), 233(119, 121), 299, 303, 305
Leiserowitz, L., 245(179), 249(179), 308
Lemar, M., 129(194), 176
Lenhard, J. R., 252(185), 308
Lenne, H. U., 293(232), 311
LePage, T. J., 218(92), 303
Lequan, M., 199(67), 302
Lequan, R. M., 199(67), 302
Lerstrup, K. A., 197(39), 198(53), 301, 302
Leslie, K. A., 226(107), 304
Leung, P. C. W., 229(120), 233(120), 305
Levene, L., 291(228), 311
Levich, V. G., 94(4), 97(4), 98(4), 167
Lewis, J. C., 117(125), 118(125), 173
Lewis, J. M. 35(76), 84
Lexa, J., 63(139), 87
Leypold, J. K., 161(277, 280), 179
Li, R. L., 39(98), 85
Lieberman, S. H., 122(172), 175
Lin, M. S., 14(24), 36(81, 82), 37(88), 38(84, 85, 86, 87), 41(24), 42(24), 47(63, 87), 48(63, 87), 82, 83, 84
Lin, P. T., 103(41), 169
Lindquist, J., 130(130), 173
Little, C. J. 101(23), 109(23), 111(67), 118(23), 120(23), 128(23), 133(23), 168, 170
Little, W. F., 199(65), 302
Liu, H. Y., 73(164), 74(164), 88
Liu, K. E., 80(172), 88
Loeliger, H., 231(156), 307
Long, M. W., 119(134), 173
Lopex, J. A. P., 49(118a), 86
Lores, E. M., 130(202), 176
Lovric, M., 11(19), 82

Lubert, K. H., 62(135), 63(138), 87
Lubovich, A. A., 198(62), 302
Lubrano, G. O., 284(213), 310
Lucisano, J. Y., 162(285), 180
Luecke, G. R., 95(14), 98(14), 152(248), 168, 178
Lukaszewski, Z., 46(113, 114), 86
Lund, W., 2(1), 81
Lundback, H., 163(289), 167(289), 180
Lundbaek, H., 161(282), 162(282), 179
Lunte, C. E., 140(224), 177
Luo, D. B., 13(21), 35(75, 92), 37(21, 75), 48(21, 75), 49(21), 82, 84, 85
Lyle, S. J., 114(94), 171
Lynn, K. N., 199(65), 302
Lyons, C., 284(212), 310

Mabbott, G. A., 11(18), 82
MacCrehan, W. A., 131(206), 176
MacDonald, A., 116(110), 172
Magno, M. C., 112(74), 170
Mahmoud, J. S., 26(42, 43, 47, 49), 27(49), 28(52), 29(42), 31(60), 32(49), 33(47), 35(75), 36(78, 80), 37(75, 78, 89, 93), 48(75), 83, 84, 85
Mairanovskii, S. G., 47(118), 86
Majda, M., 69(154), 87
Maki, A. H., 225(104), 234(104), 304
Mallen, E. F., 243(177), 308
Maloy, J. T., 161(281), 179
Manda, E., 295(241, 242, 243, 244), 311
Marcy, H. D., 235(148), 241(148), 263(191, 192), 306, 309
Mark, J. E., 66(147), 67(147), 69(147), 87
Mark, K. A., 140(225), 177
Markle, R. M., 199(69), 302
Marko-Varga, G., 157(264), 179

Marks, T. J., 235(148), 241(148), 263(190, 191, 192), 265(190), 306, 309
Mart, L., 123(180), 175
Martin, C. R., 56(129), 66(146), 71(146), 72(157), 86, 87, 88
Martinez, T., 34(69), 75(164a), 84, 88
Marciyama, Y., 201(73), 303
Maslak, P., 197(45), 219(45), 301
Matsen, F. A., 206(82), 303
Matson, W. R., 140(225), 177
Matsuda, H., 95(5, 7, 10), 99 (10), 103(44), 105(44), 235 (145, 147), 240(145, 147), 295(248, 249, 250), 167, 168, 169, 306, 311, 312
Matsue, T., 71(156), 88
Matsumoto, K., 162(284), 180
Matsumoto, M., 295(241, 242, 243, 244), 311
Maxfield, M., 203(79), 303
Mayer, C. W., 231(156, 157), 307
Mayer, G. S., 138(217), 177
Mayerle, J. J., 190(27, 29), 214(91), 300, 303
Mayfield, R. K., 164(293), 165 (293), 180
Mazaud, A., 226(112), 233(112), 304
McCarthy, W. J., 235(148), 241 (148), 263(190, 191), 265 (190), 306, 309
McCintock, S. A., 138(221), 177
McCleverty, J. A., 201(35), 301
McCrehan, W. A., 108(63), 133 (207), 138(207), 170, 176
McConnell, H. M., 219(98), 223 (102), 304
McCormick, M. J., 112(81), 171
McCullough, R. D., 197(39), 198(53), 301, 302
McDonald, A., 109(64), 170
McKenna, K., 286(227), 291(227), 310

AUTHOR INDEX

McLachlan, N. N., 112(77, 78), 133(78), 171
McLean, J. D., 119(134), 173
Mefford, I. N., 118(128), 129 (196), 130(196), 173, 176
Mell, L. D., 161(281), 179
Menet, H., 250(183), 251(183), 274(183), 308
Menten, M. L., 287(216), 310
Merkei, K. E., 152(244), 178
Mertens, M. J. M., 143(227), 177
Meschi, P. L., 95(14), 98(14), 168
Metzgar, M., 25(37), 82
Metzger, R., 272(206), 309
Meyer, T. J., 75(165), 88
Mezza, T. M., 266(197), 309
Michaelis, L., 287(216), 310
Michaud, P., 199(66), 302
Michel, L., 113(91), 171
Mikawa, H., 197(48, 49), 231(48, 149, 159), 242(149, 152), 244 (159), 246(152), 301, 306, 307
Mikkelson, S. R., 137(212), 176
Miller, B., 197(48), 301
Miller, C. J., 69(154), 87
Miller, J. S., 182(14, 15, 16), 184 (19), 185(22a), 187(23), 212 (22a), 218(94, 95, 96), 219(97), 225(105, 106), 236(15, 105, 106), 242(15), 246(15, 105), 257(15), 300, 303, 304
Miner, D. J., 107(59), 170
Mitsuhashi, T., 201(73), 303
Mizuno, M., 295(244), 311
Moghaddam, B., 72(159), 88
Moguel, M. K., 263(191), 309
Moldaveanu, J., 103(42, 43), 154(253), 155(253), 169, 178
Montgomery, D. D., 69(153), 87
Moody, G. L., 101(35), 169
Moradpour, A., 250(183), 251(183), 274(183), 308
Moret, R., 228(115), 233(115), 305
Morgan, C., 35(71), 37(71), 58(131), 75(131), 76(131), 84, 86

Morgan, D. M., 96(22), 100(22), 104(22), 107(57), 108(57), 148(237, 238), 149(237), 150(237), 168, 170, 177
Mori, T., 232(167), 307
Morra, B. S., 198(55), 248(55), 302
Morrison, T. N., 116(115), 172
Mortensen, K., 226(113), 228 (114), 231(114, 154), 304, 305, 307
Moseman, R. F., 130(202), 176
Moses, P. R., 252(185), 306
Mott, N. F., 205(80), 303
Motu, A. C. M. Almedia, 122(173), 175
Mucklejohn, S. A., 234(133a), 237(133a), 306
Muller, A., 165(295), 180
Muller, O. H., 2(1), 90(1), 167
Mulliken, R. S., 190(26), 300
Murphy, K., 31(59), 83
Murray, P. T., 66(147), 67(147), 69(147), 87
Murray, R. M., 157(262), 160 (262), 179
Murray, R. W., 8(8), 53(127), 57(8), 66(145), 75(165), 183(17), 252(185), 81, 86, 87, 88, 300, 308

Nagaoka, T., 117(119), 172
Nagasubramanian, G., 266(196), 309
Nagy, F., 72(159), 88
Nagy, G., 113(89), 171
Najean, R., 231(160), 235(180), 274(160), 307, 308
Nakamura, T., 62(136, 137), 190(32), 217(32), 295(241, 242, 243, 244), 87, 301, 311
Nakanishi, H., 235(145, 147), 240(145, 147), 306
Namaratne, S., 72(160), 88
Naotsuka, M., 162(284), 180

Naturforsch, Z., 197(51), 229(122), 232(163, 166), 235(171), 302, 305, 307, 308
Neeb, R., 25(38), 82
Nekrasov, L. N., 199(68), 302
Nelson, R. F., 65(143), 87
Nelson, S. D., 130(199), 176
Nentwig, J., 161(282), 162(282), 179
Nernst, W., 93(3), 167
Neuberger, G. C., 152(249), 153(249), 178
Ngo, C. C., 99(17, 18), 101(33), 112(33), 114(33), 122(17, 33), 123(33, 174, 175), 124(174), 125(33, 174), 126(18, 33, 174), 128(33), 167(33), 168, 169, 172, 175
Nieboer, E., 20(28), 23(28), 82
Nielsen, P., 268(200), 278(200), 309
Nimmo, M., 28(57), 30(57), 50(57), 83
Noftle, R. E., 243(178), 308
Nogumi, T., 197(47, 49), 23(47, 149, 159), 242(149, 152), 244(159), 246(152), 301, 306, 307
Nomura, T., 162(284), 180
Nonidez, W. K., 130(197), 156(197), 176
Nordio, P. L., 223(101), 304
Nothe, D., 229(130), 235(130), 306
Novotny, L., 46(116), 86
Novothy, M., 103(41), 169
Nunez, L., 230(128), 232(128), 305
Nürnberg, H. W., 13(22), 18(22), 22(32), 25(22), 31(58), 41(103), 82, 83, 85

O'Dea, J., 144(231), 177
O'Dea, J. J., 101(34), 114(34), 169
Ohmae, M., 129(195), 176

Ohnishi, J., 231(159), 244(159), 307
Ohnishi, S., 197(47), 231(47, 149), 242(149, 152), 246(152), 301, 306, 307
Ohno, H., 235(146, 147), 240(146, 147), 241(146), 306
Oke, A., 72(159), 118(127), 88, 173
Okinaka, Y., 113(86), 171
Oliva, P., 116(108), 172
Olsen, M., 226(111), 304
Olsson, B., 161(282), 162(282), 179
Orihashi, Y., 235(145, 146, 147), 240(145, 146, 147), 241(146), 306
O'Riordan, D. M. T., 61(134), 87
Orthmann, E. A., 266(195), 309
Osa, T., 71(156), 88
Osteryoung, J., 4(17), 9(17), 11(17), 37(17), 101(34), 114(34), 144(23), 82, 169, 177
Osteryoung, R. A., 125(184, 188), 133(209), 175, 176
Ou, T. Y., 130(197), 156(197), 176
Ouziel, E., 114(96), 171
Oyama, N., 68(150), 87

Pacakova, V., 119(131, 132, 140), 173
Pack, M., 50(119), 86
Packett, D. L., 77(170), 88
Pakin, S. S. P., 229(121), 233(121), 305
Palecek, E., 40(99, 100, 101, 102), 42(100), 85
Palziel, J. A., 11(16), 82
Panzer, R. E., 117(117), 172
Papavasiliou, G. C., 235(174), 308
Parkin, S. S., 182(7), 226(110), 299, 304

Parkin, S. S. P., 229(119), 233 (117), 305
Parker, V. D., 197(34), 201(34), 301
Pawlak, M., 199(67), 302
Pearson, W. B., 190(26), 300
Pedersen, H. J., 226(113), 231 (154), 304, 307
Peng, T., 37(88), 119(151), 120 (151), 84, 174
Penner, R. M., 72(157), 88
Peover, M. E., 199(71), 200(71), 303
Peregudova, S. M., 199(68), 302
Perlstein, J. H., 182(3a), 299
Persson, B., 157(266), 179
Pesliakiene, M. V., 286(221), 287 (221), 289(221), 310
Pesotskii, S. I., 230(127), 232(127), 305
Peterson, S. W., 226(107, 108), 304
Peterson, W. M., 108(61), 170
Petric, D., 50(120), 86
Petri, O. A., 7(4), 81
Philar, B., 23(32), 82
Phillips, W. D., 260(189), 281(189), 309
Pihalar, B., 121(161), 174
Pihlar, B., 13(22), 18(22), 25(22), 50(120), 82, 86
Pinedo, H. M., 114(99), 172
Pipan, C. M., 232(168), 307
Pletcher, D., 243(178), 308
Ploegmakers, H. H. J. L., 143 (227), 177
Plotkin, E. V., 157(267), 284(211), 179, 310
Podolak, 119(140), 173
Poehler, 203(79), 303
Polborn, K., 232(165), 307
Polta, J. A., 116(110, 111), 152 (244, 247, 248), 172, 178
Poppe, H., 107(55), 108(60), 119 (147), 170, 174
Poosittisak, S., 295(247), 311
Porter, L. C., 232(168), 307

Porthault, M., 129(194), 176
Postbieglova, I., 40(100), 42(100), 85
Pouget, J. P., 228(115), 233(115), 305
Poulson, B., 165(298), 180
Prabhu, S., 98(15), 168
Prabu, S. V., 64(140), 87
Price, J. F., 77(170, 171), 79 (171), 88
Pritzkow, H., 229(130), 235(130), 306
Prokhorova, G. V., 23(31), 82
Proksh, R. B., 232(168), 307
Pukacki, W., 199(67), 302
Pungor, E., 100(21), 113(89), 136(21), 168
Purdy, W. C., 117(123), 137(212), 138(220, 221), 156(261), 173, 176, 177, 178
Pysh, E. S., 198(54), 207(83), 302, 303

Quast, H., 197(40), 301

Radhakrishnan, T. P., 218(93), 303
Radzik, D. M., 138(219), 139 (219), 177
Ramaley, L., 11(16), 107(53), 82, 170
Ramsing, A. U., 111(68), 112 (68), 170
Rappaport, S. M., 131(204), 133(204), 176
Rasmussen, F. B., 226(111), 304
Ravichandran, K., 77(168, 170), 119(149, 150), 120(149, 150), 150(239), 151(239), 157(269), 158(269), 159(269), 88, 174, 177, 179
Razumney, G. A., 243(175), 308

Rechnitz, G. A., 161(278), 164 (291), 179, 180
Reddoch, A. H., 223(101), 224 (101), 235(101), 245(101), 304
Reed, P. E., 229(118), 233(118), 305
Reed, R., 197(50), 301
Refshauge, C., 118(126), 127(126), 173
Reggers, G., 24(35), 82
Reiff, W. M., 182(14, 15), 218(94, 95, 96), 236(15), 242(15), 246(15), 257(15), 300, 303, 304
Reilley, C. N., 198(60b), 199(65), 302
Reim, R. E., 131(205), 176
Reis, A. H. Jr., 182(16), 226(108), 300, 304
Reussel, H. A., 130(203), 176
Ribault, M., 226(110, 112), 233 (112), 304
Rice, M. E., 118(129), 173
Rihs, G., 231(156, 157), 307
Riley, C., 112(72, 73), 170
Riley, J. P., 31(59), 83
Ritsko, J. J., 182(16), 300
Ritter, G. W. II, 266(197), 309
Riviello, J. M., 150(240), 178
Robbins, J. L., 198(58, 63), 302
Roberts, J. L. Jr., 109(65), 170
Rocks, B., 112(72), 170
Rocks, B. F., 112(73), 170
Roekens, E., 24(35), 82
Rommelmann, H., 218(96), 304
Rooijen, H. W., 119(147), 174
Rosseinsky, D., 233(132), 237 (131, 132), 306
Rosseinsky, D. R., 234(133a), 235(134), 237(133a, 134), 262 (133b), 306
Roston, D. A., 138(216, 218, 223), 176, 177
Rubinstein, I., (155), 87
Ruf, H., 33(64), 83
Ruppel, W., 235(172), 308
Rusling, J. F., 120(155), 174
Rutzel, H., 123(180), 175

Ruzicka, J., 111(68, 69, 162), 112(68), 120(157), 121(162), 143(162), 170, 174
Rybaczewski, E. F., 182(4), 299

Saito, G., 229(116), 232(167), 233(170), 244(116), 295(241, 242), 305, 307, 308, 311
Sakai, F., 232(167), 307
Salaneck, W. R., 182(16), 300
Saleh, M. I., 114(94), 171
Salikhdzhanova, R. M. F., 23(31), 82
Samalius, A. S., 286(221, 231), 287(221), 289(221), 310, 311
Samuelson, L., 275(208), 309
Samuelson, R., 133(209), 144 (231), 176, 177
Sanders, A. P., 199(65), 302
Sandman, D. J., 197(50), 213 (89), 268(200), 275(208), 278(200), 301, 303, 309
Sandona, G., 296(251), 312
Sanghera, G. S., 101(35), 169
Santos, L. M., 157(271), 179
Satio, G., 182(8), 299
Sauerbrey, G. Z., 239(142), 306
Sauy, M., 293(236), 311
Sawyer, D. T., 109(65), 170
Schafer, D. E., 197(48), 301
Schectman, L. A., 266(196), 309
Scheller, F., 161(282), 162(282), 179
Scheutow, D., 197(40), 301
Schick, K. G., 116(115), 172
Schieffer, G. W., (211), 176
Schirber, J. E., 229(118), 233 (118), 305
Schneider, J., 66(145), 87
Schnurrbusch, M., 62(135), 63 (138), 87
Schrodel, R., 293(232), 311
Schumaker, R. R., 187(7), 229(119, 121), 233(119, 121), 299, 305

AUTHOR INDEX

Schubert, U., 235(171), 308
Schultz, A. J., 226(109), 304
Schutz, H. J., 182(5), 299
Schweitzer, D., 197(51), 229(122, 130), 230(125), 232(125, 163, 166), 235(130, 172), 302, 305, 306, 307, 308
Scott, B. A., 221(100), 250(184), 304, 308
Scott, J. C., 182(7), 229(119), 233(119), 299, 305
Scott, L. D., 157(267), 179
Scott, L. D. L., 284(211), 310
Seisshaar, D. E., 119(153), 174
Sekiguchi, T., 295(241, 242, 243, 244), 311
Selavka, C. M., 148(234), 177
Semersky, F. E., 162(283), 179
Senda, M., 11(14), 81
Sequaris, J. M., 41(103), 85
Shah, M., 4(17), 9(17), 11(17), 37(17), 82
Shalom, E. A., 197(52), 198(52), 302
Sharp, M., 270(201, 202, 203), 280(201, 202, 203), 281(201, 202, 203), 309
Shay, M., 238(140), 306
Shchegolev, I. F., 230(123, 127), 232(123, 127), 305
Sherwood, R. A., 112(73), 170
Shibaeva, R. P., 230(126), 232(126), 305
Shimon, J. W., 245(179), 245(259), 308
Shoup, R. E., 138(216, 217), 140(224), 176, 177
Sienko, M. J., 206(81), 303
Siggia, S., 119(133), 173
Silverman, B. D., 221(100), 250(184), 304, 308
Sim, K. W., 286(217), 287(217), 310
Simon, J., 182(2a, 2b), 293(235), 299, 311
Sioda, R. E., 2(1), 81
Sirasaka, Y., 162(284), 180

Skogerboe, R. K., 121(165), 174
Skoog, D. A., 256(186), 308
Skotheim, T. A., 182(10a), 183 (10a), 268(10a), 299
Skoulios, A., 182(2a, 2b), 299
Slattery, J. T., 130(199), 176
Sleszynski, N., 101(34), 114(34), 169
Small, C. E., 163(289), 167 (289), 180
Smart, J. C., 198(58, 63), 302
Smirnova, S. A., 198(62), 302
Smith, R. V., 130(201), 176
Smyth, F., 121(160), 174
Smyth, M. R., 42(106), 85
Snyder, L. R., 110(66), 170
Song, D., 35(73), 84
Soos, Z., 218(93), 303
Soos, Z. G., 223(101), 304
Soriaga, M. P., 13(23), 35(72, 73), 82, 84
Spencer, B., 198(58), 302
Staab, H. A., 198(56), 302
Stara, V., 39(26), 49(26), 52(125), 82, 86
Stark, J. C., 197(50), 301
Starkova, B., 119(131), 173
St. Claire, R. L., 143(228), 177
Stepnik, B., 114(101), 172
Stifel, D. N., 162(283), 179
Stokes, J. B., 231(158), 261(158), 307
Stulikova, M., 122(166), 174
Strafelda, F., 113(85), 171
Stanford, C., 286(230), 292(230), 311
Strasser, V. A., 119(146), 174
Strohl, A. N., 119(143), 174
Strohl, J. H., 113(87), 171
Stojek, S., 114(101), 172
Stulik, K., 63(139), 119(131, 132, 140), 87, 173
Sugawara, T., 201(73), 303
Sullivan, B. W., 207(85), 303
Sullivan, F. M., 164(293), 165(293), 180
Summan, A. M., 113(84), 171

Suzuki, S., 164(292), 180
Svanberg, L. R., 152(243), 178
Swatzfager, D. G. 108(62), 170
Szentirmay, M. N., 66(146), 71 (146), 87

Tachikawa, H., 266(194), 293(237), 309, 311
Takei, F., 295(241, 243, 244), 311
Takizawa, R., 62(136), 87
Tallman, D. E., 119(135, 136, 137), 129(135, 136, 137), 155(255), 173, 178
Tam, W., 182(11, 12), 299
Tanaka, C., 232(162), 307
Tanaka, J., 232(162), 307
Tanaka, M., 232(162), 295(241, 242, 243, 244), 307, 311
Tan, W. T., 11(16), 82
Tang, P. C., 116(110), 172
Tanaka, H., 231(149, 159), 242 (149), 244(159), 306, 307
Tanaka, S., 197(44, 47), 231(47), 301
Tapia, T., 36(79, 82), 41(104), 84, 85
Tasaka, Y., 197(47), 231(47, 149), 242(149, 152), 246(152), 301, 306, 307
Tay, B. T., 101(26, 27), 109(27), 115(102), 129(26), 130(26, 27, 210), 131(102), 132(102), 133 (102), 134(27, 210), 135(27), 136(27), 144(102, 233), 145(102), 147(233), 167(233), 168, 172, 176, 177
Tazhi, P., 26(48), 83
Tedoradze, G. A., 46(115), 86
Tenygl, J., 112(76), 114(76), 171
Thiak, P. C., 99(18), 126(18), 168
Thirsk, H. R., 243(176), 308
Thomaides, J. S., 197(45), 219(45), 301
Thomas, A., 62(135), 87
Thomas, J. D. R., 101(35), 169

Thompson, S. B., 112(80), 171
Thørgerson, N., 111(162), 121 (162), 143(162), 174
Thormann, W., 156(260), 178
Thorup, N., 231(154), 307
Tien, H. T., 293(239, 240), 294(240), 311
Tokuda, K., 103(44), 105(44), 295(248, 249, 250), 169, 311, 312
Tokumoto, M., 229(116), 232 (164), 233(170), 244(116), 305, 307, 308
Tomkiewicz, Y., 228(114), 231 (114), 305
Tonge, L. M., 235(148), 241 (148), 306
Toriumi, T., 182(8), 297
Torrance, J. B., 190(27, 29), 211(87), 214(91), 221(100), 250(184), 300, 303, 304, 308
Torrance, K., 25(36), 82
Toth, K., 113(89), 171
Tougas, T. P. 107(49), 119(144), 169, 174
Tournilhac, F., 182(26), 299
Trachtenberg, I., 17(27), 82
Trasatti, S., 7(5), 81
Trojanek, A., 144(230), 177
Tsang, C. H., 95(13), 168
Tse, D. C. S., 117(122), 162 (285), 172, 180
Tsuchida, E., 235(145, 146, 147), 240(145, 146, 147), 241(146), 306
Tucker, D. J., 112(80), 171
Turner, A. P. F., 157(267), 284(211), 179, 310
Tuzhi, P., 34(69), 36(82), 65(142), 68(149), 72(149), 73(149), 69 (152), 138(222), 140(222), 141 (222), 84, 87, 177

Underhill, A. E., 226(109), 304

AUTHOR INDEX

Valenta, P., 13(22), 18(22), 23 (32), 25(22), 31(58), 32(58), 41(103), 123(180), 82, 83, 85, 175
van Dalen, E., 123(179), 175
van den Berg, C. M. G., 28(53, 54, 55, 56, 57), 30(54, 55, 57), 32(54, 55, 56, 59), 33(53, 57, 62, 65, 67), 50(53, 56, 57), 83
van den Boasch, P. A., 112(83), 114(98), 171, 172
Van den Bosch, P., 156(260), 178
Van der Linden, 121(158), 174
van der Linden, W. E., 96(28), 101(28), 163(288), 168, 180
Van der Vijgh, 114(99), 172
Van Duyne, R. P., 182(16), 198 (60b), 300, 302
Van Grieken, R., 24(35), 82
Van Mil, L. J. W., 245(179), 259 (179), 308
vanNieuwkerk, H. J., 106(46), 108(46), 113(46), 169
van Oort, W. J., 143(227), 177
van Rooijen, H. W., 108(60), 170
Varadi, M., 100(20, 21), 168
Varineau, P. T., 239(143), 306
Varughese, K., 25(39), 26(48), 32(39), 33(48), 82, 83
Vazquez, J. E., 214(91), 303
Venturini, E. L., 232(169), 308
Villa, V., 36(81), 38(84, 86), 41(104, 105), 42(106a), 84, 85
Villarroel, D., 66(147), 67(147), 69(147), 87
Vitali, P., 130(200), 176
Voegell, R. H., 201(76), 303
Vojtiskova, M., 40(102)
Volicen, L., 140(225), 177
von Schultz, J. V., 231(153), 246 (153), 307
Voogt, W. H., 113(88), 114(93), 171
Vos, L., 24(35), 82

Wahdat, F., 25(38), 82
Walatka, V., 182(3a), 299

Wallace, G. C., 61(134), 112(81), 150(240), 151(241), 87, 171, 178
Wallace, W. L., 270(205), 309
Wang, H. H., 182(6), 229(118), 230(124, 128, 129), 232(124, 128, 129, 168, 169), 233(118), 299, 305, 307, 308
Wang, J., 2(1, 2), 10(13), 11(20), 13(21), 14(24), 16(25), 19(2), 25(39), 26(42, 43, 44, 45, 46, 47, 48, 49, 50), 27(49), 28 (44, 50, 52), 29(42), 31(2, 60, 61), 32(42, 43, 44, 45, 47, 63), 33(47), 34(69), 35 (71, 75), 36(78, 79, 80, 81, 82, 83, 84, 85, 86, 87), 37 (21, 71, 75, 78, 88, 89, 91, 93), 39(97), 41(24, 104, 105), 42(24, 106a), 43(108, 109), 44(109), 47(63, 87), 48(63, 75, 87), 49(21), 50(119, 121), 51(121, 122, 123), 52(123), 58(131), 65(142, 144), 68 (149), 69(152), 72(149), 73 (149), 75(131, 164a), 76(131), 77(169), 101(30, 31), 107(50, 51, 52), 114(30, 96), 119 (148, 151), 120(151), 121(164), 122(30, 31, 169, 170, 171), 123(177), 125(30, 31, 182, 183), 126(164), 127(190), 138(222), 140(222), 141(222), 144(232), 157(263), 81, 82, 83, 84, 85, 86, 87, 88, 169, 170, 171, 175, 177, 179
Wang, Y., 182(11, 12), 299
Ward, M. D., 182(9, 13), 190 (30, 31, 32), 193(30), 197 (41, 43, 46), 199(41, 43), 200(41), 201(30, 41, 77), 214(30, 31, 33), 216(30, 31), 217(33), 236(30, 150, 151), 239(141, 144), 242(30, 150, 151), 246(30, 151), 254(150), 257(151), 276(144), 295(150), 299, 300, 301, 303, 306, 307

Warren, L. F., 198(60a), 302
Washecheck, D. M., 226(108), 304
Watanabe, T., 201(73), 303
Watson, B., 162(283), 179
Webb, K. S., 230(128), 232(128), 305
Webber, A., 4(17), 9(17), 11(17), 37(17), 82
Weber, G., 20(29), 82
Weber, S. G., 96(22), 100(22), 102(39), 104(22), 107(54, 57, 58), 108(57), 112(54), 113(54), 127(54), 131(54), 133(54), 138(54, 220), 148(237, 238), 149(54, 237), 150(237), 153(54), 168, 169, 170, 177
Weberand, S. G., 117(123), 173
Webster, D. W., 199(70), 302
Wechter, C., 101(34), 114(34), 169
Wegner, G., 198(55), 235(173), 248(55, 173), 266(195), 302, 308, 309
Wehe, D., 229(130), 235(130), 306
Weissbuch, I., 245(179), 249(179), 309
Weisshar, D. E., 119(136, 137, 154), 120(154), 129(136, 137), 173
Welber, B., 221(100), 250(184), 304, 308
Wemer, H. P., 231(153), 246(153), 307
Werner, M., 229(130), 235(130), 306
West, D. M., 256(186), 308
Whangbo, M. H., 232(169), 308
Whangbo, M.-Y., 230(128), 232(128), 305
Wheeler, B. L., 266(196), 309
Wheland, R. C., 197(37), 198(37), 200(37), 208(86), 210(37, 86), 234(37), 237(37), 301, 303
White, H. S., 237(135, 136), 238(136), 276(135, 136), 306
White, J. G., 143(228, 229), 177
White, J. H., 35(73), 84
White, J. R., 73(163), 88
Whiten, K. K., 130(197), 156(197), 176

Wier, L. M., 56(128), 59(133), 60(133), 86, 87
Wightman, R. M., 103(41), 107(54), 112(54), 113(54), 117(121), 119(139), 127(54), 131(54), 133(54), 138(54), 143(226), 144(226), 149(54), 154(251, 252), 156(258, 279), 207(84), 169, 170, 177, 178, 303
Wilckens, R., 235(172), 308
Willard, H. H., 25(40), 83
Williams, D. F., 223(101), 224(101), 225(103), 235(101), 245(101, 103), 304
Williams, J. M., 182(6), 226(107, 108, 109), 229(117, 118, 120), 230(124, 128, 129), 232(124, 128, 129, 168, 169), 233(118, 120), 299, 304, 305, 307, 308
Willis, W. S., 120(155), 174
Wilson, J. M., 130(199), 176
Wilson, J. R., 162(287), 180
Wilson, P. H., 35(72), 84
Wilson, R. J., 198(60a), 302
Wingard, L. B., 284(209), 310
Winsaer, K., 116(116), 172
Wise, J. A., 38(38), 104(38), 114(38), 122(38), 169
Wohtmann, 164(293), 165(293), 180
Wojciechowski, M., 125(184), 175
Wood, P. J., 152(243), 178
Woods, R., 116(105), 172
Wrighton, G. C., 268(199), 309
Wrighton, M. S., 197(44), 301
Wudl, F., 197(48, 52), 198(52), 301, 302
Wuryanto, S., 232(161), 250(181, 182), 274(161, 181, 182), 307, 308

Yabubskii, E. B., 230(126), 232(126), 305
Yagubskii, E. B., 230(123, 127), 231(155), 232(123, 127), 305, 307

Yamada, J., 95(7, 10), 168
Yamahira, A., 197(49), 301
Yang, N. C., 198(54), 207(83), 302, 303
Yarnitzky, C., 114(96), 171
Yashida, Z., 122(168), 174
Yasuda, T., 164(292), 180
Yasuoka, N., 190(32), 217(32), 301
Ye, J., 150(239), 151(239), 177
Yoshimura, I., 295(249), 312
Yoshino, T., 117(119), 172
Yur'eva, L. P., 199(68), 302

Zadeii, J., 16(25), 26(44, 45, 46, 50), 28(44, 50), 31(61), 32(44, 45, 61, 63), 33(46, 50), 47(63), 48(63), 82, 83
Zaitseva, N. N., 199(68), 302
Zak, J., 119(141, 142), 173
Zapardiel, A., 49(118a), 86
Zatka, A., 113(91), 171
Zakurin, N. V., 198(61), 199(68), 302
Zhang, J. H., 182(14, 15), 218 (94, 95, 96), 236(15), 242 (15), 246(15), 257(15), 300, 303, 304
Zirino, A., 122(172), 175
Zoski, G. D., 275(208), 309
Zvarykina, A. V., 230(123, 127), 231(155), 232(123, 127), 305, 307

Subject Index

Aceteaminophen metabolites, 130
Adriamycin, 38
Adsorption isotherms, 6, 7
Adsorptive stripping tensammetry, 43–47
Adsorptive stripping voltammetry (ASV), 4–53, 126, 127
  adsorption of complexes for trace measurements of metals, 19–34
  adsorption at electrode/solution interface, 5–7
  adsorptive stripping tensammetry, 43–47
  combination of adsorptive voltammetry with catalytic effects, 47, 48
  method development—practical considerations, 15–19
  problems and solutions, 48–53
  trace measurements of organic compounds, 34–43
  voltammetric response of surface-confined analytes, 7–15
Analytical calibration, 12–14
Anodic stripping voltammetry (ASV), 99, 121–126
Anticancer drugs, trace measurements of, 38
Aromatic amines, 130

Benzodiazepines, 133
Biological macromolecules, adsorptive stripping measurements of, 39–43

Carbamate pesticides, 130
Carbon electrodes, 117–120
  conditioning, 119, 120
Carbon paste electrodes, 118
  ion-exchange modified, 75–80
  modified, 58, 59
Catalytic processes, combination of adsorptive voltammetry with, 47, 48
Catecholamines, 130
Catechol complexes, 28–31
Chemical derivatization, LC–EC and, 153, 154
Chemically modified electrodes (CMEs)
  in continuous-flow analysis, 156–160
  voltammetry following preconcentration at, 53–80
  preconcentration schemes, 59–80
  requirements, problems and solutions, 55–57
  ways to introduce preconcentrating agent, 58, 59
Chemical process-control systems, use of HDV techniques in, 162–164
Chlorambucil, 38
Composite carbon electrodes, 119
Computer control and automation of HDV techniques, 166, 167
Continuous-flow analysis, hydrodynamic voltammetry (HDV) in, 89–180
  analytical techniques and applications, 120–148

[Continuous-flow analysis, hydrodynamic voltammetry (HDV) in]
  flow injection-electrochemical detection, 120, 121
  liquid chromatography-electrochemical detection, 127–140
  pulse and potential-scan techniques, 140–148
  stripping voltammetry, 121–127
  background, 93–112
    practical considerations, 103–112
    theoretical considerations, 93–98
    wall-jet and thin-layer detectors, 98–103
  electrode materials, 112–120
    carbon electrodes, 117–120
    mercury electrodes, 112–115
    metal electrodes, 115, 116
  indirect methods, 148–154
    chemical derivatization, 153, 154
    indirect electrode reactions, 152, 153
    photoexcitation, 148, 149
    use of electroactive intermediary, 149–152
  new electrode systems, 154–162
    chemically modified electrodes, 159, 160
    enzyme electrodes, 160–162
    ultramicroelectrodes, 154–156
  on-line monitoring applications, 162–167
    computer control and automation, 166, 167
    patient monitoring, 164-166
    process monitoring, 162–164
Copper electrodes, 116
Covalent reactions, preconcentration at CMEs by, 77–80

Daunorubicin, 38
Differential-pulse voltammetry, 10
Dimethylglyoxime (DMG) complexes, 23–25
Di-O-hydoxyazo dyes, 25–28
Dropping mercury electrode (DME), 112–115
Dual-electrode LC–EC detection, 137–140

Electrochemical aspects of low-dimensional molecular solids, 181–312
  description of low-dimensional solids, 184–196
  electrochemical preparation of low-dimensional solids, 220–267
    basic principles, 220–223
    electrochemical doping of low-dimensional phthalocyanines, 263–267
    examples, 223–243
    mechanistic aspects, 243–254
    role of electrochemical parameters on crystallization, 254–263
  low-dimensional solids as electrode materials, 267–295
    characteristics of "synthetic metal" electrodes, 267–280
    potential applications of "synthetic metal" electrodes, 280–292
  redox properties of molecular solids, 196–220
    electrochemical properties, 196–202
    role of redox behavior in solid-state properties, 202–220

SUBJECT INDEX

Electrocrystallization
  complexes prepared by, 231–236
  of one-dimensional molecular solids, 223–243
  role of electrochemical parameters on, 254–263
Electrode materials
  for continuous-flow HDV, 112–120
    carbon electrodes, 117–120
    mercury electrodes, 112–115
    metal electrodes, 115, 116
  low-dimensional solids as 267–295
  "synthetic metal" electrodes, 267–292
Enzyme electrodes, 160–162
processes at, 284
Estrogens, 130

Flow injection-electrochemical detection (FI–EC), 90–92, 120, 121
  theory of, 98
  use of computer control and automation for, 166
5-Fluorouracil, 38

Glassy carbon electrodes, 117, 118
Gold electrodes, 116

Hanging mercury drop electrode (HMDE), 112
Hydrodynamic boundary layer, 97
Hydrodynamic voltammetry (HDV) in continuous-flow analysis, 89–180

[Hydrodynamic voltammetry (HDV) in continuous-flow analysis]
  analytical techniques and applications, 120–148
    flow injection-electrochemical detection, 120, 121
    liquid chromatography-electrochemical detection, 127–140
    pulse and potential-scan techniques, 140–148
    stripping voltammetry, 121–127
  background, 93–112
    practical considerations, 103–112
    theoretical considerations, 93–98
    wall-jet and thin-layer detectors, 98–103
  electrode materials, 112–120
    carbon electrodes, 117–120
    mercury electrodes, 112–115
    metal electrodes, 115, 116
  indirect methods, 148–154
    chemical derivatization, 153, 154
    indirect electrode reactions, 152, 153
    photoexcitation, 148, 149
    use of an electroactive intermediary, 149–152
  new electrode systems, 154–162
    chemically modified electrodes, 156–160
    enzyme electrodes, 160–162
    ultramicroelectrodes, 154–156
  on-line monitoring applications, 162–167
    computer control and automation, 166–167
    patient monitoring, 164–166
    process monitoring, 162–164

Indirect electrode reactions, 152, 153

Iodine, ET complexes with, 230
Ion-exchange modified carbon paste electrodes, 75–77
Ion-exchange voltammetry, 65–77
  electrostatic preconcentration at inorganic coatings, 73–75
  ion-exchange modified carbon paste electrodes, 75–77
  polyelectrolyte-coated electrodes as preconcentration surfaces, 68–73
Ion-sensing electrodes, potential applications of, 280–283

Ligand complexes for trace measurements of metals, 28–31
Ligand-containing complexing electrodes, preconcentration at, 59–65
Linear-scan voltammetry, 8, 9
Liquid chromatography-electrochemical detection (LC–EC), 90–92, 127–140
  chemical derivatization in, 153, 154
  dual electrode detection, 137–140
  normal-phase applications, 133–137
  oxidative detection, 129, 130
  photoexcitation in, 148, 149
  reductive detection, 113–133
  theory of, 98
  use of computer control and automation for, 166
Low-dimensional molecular solids, 184–196
  electrochemical preparation of, 220–267
    basic principles, 220–223
    electrochemical doping of low-dimensional phthalocyanines, 263–267

[Low-dimensional molecular solids]
    examples, 223–243
    mechanistic aspects, 243–254
    role of electrochemical parameters on crystallization, 254–263
  as electrode materials, 267–295
    characteristics of "synthetic metal" electrodes, 267–280
    potential applications of "synthetic metal" electrodes, 280–292
  redox properties of, 196–220
    electrochemical properties, 196–202
    role of redox behavior in solid-state properties, 202–220
Low-dimensional phthalocyanines, electrochemical doping of, 263–267

Mass transfer, analogy between momentum transfer and, 94–96
Mechanistic aspects of electrocrystallization of low-dimensional molecular solids, 243–254
Mercury electrodes, 112–115
Metal electrodes, 115, 116
Metals, trace measurements by adsorptive stripping voltammetry, 19–34
Methotrexate, 38
Mitomycin C, 38
Modified carbon paste electrodes, 58, 59
Momentum transfer, analogy between mass transfer and, 94–96
Morphine, 130
Morphine derivatives, 130

SUBJECT INDEX

Natural waters, metal speciation in, 31–34
Nickel electrodes, 116
Nitroaromatics, 133
Nitrosamines, 133
Noble metal electrodes, 115, 116
Nonelectrolytic preconcentration, voltammetry following, 1–88
  adsorptive stripping voltammetry, 4–53
    adsorption of complexes for trace measurements of metals, 19–34
    adsorption at electrode/solution interface, 5–7
    adsorptive stripping tensammetry, 43–47
    combination of adsorptive voltammetry with catalytic effects, 47, 48
    method development–practical considerations, 15–19
    problems and solutions, 48–53
    trace measurements of organic compounds, 34–43
    voltammetric response of surface-confined analytes, 7–15
  voltammetry following preconcentration at chemically modified electrodes, 53–80
    preconcentration schemes, 59–80
    requirements, problems and solutions, 55–57
    ways to introduce preconcentrating agents, 57–59
Normal phase LC–EC, 133–137

On-line monitoring applications, use of HDV techniques in, 162–167

Organic compounds, trace measurements by adsorptive stripping voltammetry, 34–43
Organic solid electrodes, enzymatic processes at, 286
Organometallics, 133
Oxidative LC–EC detection, 129, 130
Oxine complexes, 28–31

Patient monitoring, use of HDV techniques for, 164–166
Penicillamine, 130
Pesticides, 133
Phenols, 130
Photoexcitation, electrochemical detection and, 148, 149
Phthalocyanine polymers, electrochemical doping of, 263–267
cis-Platin, 38
Platinum electrodes, 116
Polymeric coatings for electrode modification, 57, 58
Polynuclear aromatic hydrocarbons, 130
Potential-scan techniques, 140–148
Pulse amperometric techniques, 140–148
Pyrolytic graphite electrodes, 117

Redox properties of molecular solids, 196–220
  electrochemical properties, 196–202
  role of redox behavior in solid-state properties, 202–220
Reductive LC–EC detection, 131–133
Rotating disk electrode (RDE), 94

Stripping voltammetry, 121–127
Sulfonamides, 130
Surface-confined analytes, voltammetric responses of, 7–15
Synthetic metal electrodes
 characteristics of, 267–280
 potential applications of, 280–292

Thin-layer electrode detector, 98–103

Ultramicroelectrodes in continuous-flow analysis, 154–156

Vinblastine, 38
Voltammetry following nonelectrolytic preconcentration, 1–88
 adsorptive stripping voltammetry, 4–53
  adsorption of complexes for trace measurements of metals, 19–34

[Voltammetry following nonelectrolytic preconcentration]
  adsorption at electrode/solution interface, 5–7
  adsorptive stripping tensammetry, 43–47
  combination of adsorptive voltammetry with catalytic effects, 47, 48
  method development—practical considerations, 15–19
  problems and solutions, 48–53
  trace measurements of organic compounds, 34–43
  voltammetric response of surface-confined analytes, 7–15
 voltammetry following preconcentration at chemically confined electrodes, 53–80
  preconcentration schemes, 59–80
  requirements, problems and solutions, 55–57
  ways to introduce preconcentrating agent, 57–59

Wall-jet electrode detector, 98–103